MW00780656

AN INTRODUCTION TO ANALYSIS OF FINANCIAL DATA WITH R

AN INTRODUCTION TO ANALYSIS OF FINANCIAL DATA WITH R

Ruey S. Tsay

University of Chicago

A JOHN WILEY & SONS, INC., PUBLICATION

Library of Congress Cataloging-in-Publication Data:

Tsay, Ruey S.,
  An introduction to analysis of financial data with R / Ruey S. Tsay.
    p. cm.
  Includes index.
  ISBN 978-0-470-89081-3 (cloth)
    1. Finance–Econometric models. 2. Time-series analysis. 3. Econometrics. 4. R (Computer program language) I. Title.
  HG106.T76 2012
  332.0285'133–dc23

                                                                          2012002912

Printed in the United States of America.

10 9 8 7 6 5 4 3 2 1

To Teresa

CONTENTS

PREFACE

I am often asked by readers of *Analysis of Financial Time Series*: Can you make the analysis of financial data easier? I am also asked by my students: How to simplify the empirical work and what is the relevance of statistics to finance? These are important questions. They motivate me to write this introductory text.

To simplify empirical data analysis, I decided to use R for all analyses. My decision is based on several reasons. First, R is free and available for most operating systems. Second, many researchers have developed nice packages for analyzing financial data, especially RMetrics has many useful packages. Third, the capability of R packages improves dramatically and rapidly. This trend is expected to continue. Fourth, I wrote some simple R scripts to perform specific analyses in several places. These scripts serve two purposes. They attain to the special need I have in presenting the concepts and methods. More importantly, they demonstrate that once a reader has some experience with R, he/she can easily tailor R commands and scripts into his/her need to simplify analysis of financial data.

To simplify the concept of econometric and statistical theory, I tried to present it in a concise manner and used extensively real examples in demonstration. The book has seven chapters; two of them are case studies. These two chapters demonstrate the relevance of statistics in finance. The other chapters are organized to help readers understand the concepts of and gain experience in analyzing financial data. Chapter 1 introduces financial data and discusses their summary statistics and visualization. It also introduces R so that readers can start to explore financial data. Chapter 2 provides basic knowledge of linear time series analysis. It covers simple econometric models that are useful in business, finance and economics. I tried to make the chapter as comprehensive as possible while keeping it concise. It includes exponential smoothing for forecasting and methods for model comparison. Chapter 3 considers three case studies. The models used are not simple, but they are designed to help readers understand the value and limitations of linear time series models in applications. Chapter 4 studies different approaches to calculate asset volatility and various volatility models. The approaches discussed include methods that use daily open, high, low and close stock prices. Again, I tried to make the chapter as comprehensive as possible while avoiding much of the heavy theory. Chapter 5 considers some applications of volatility models in finance. It is intended to help readers gain better insight into the term structure of volatility and use of volatility in financial applications. Chapter 6 deals with high-frequency financial data, including simple models for price changes and trading intensity and realized volatility. Finally, Chapter 7 studies quantitative methods for risk management, including value at risk and conditional value at risk. The chapter covers important econometric and statistical methods to assess risk, including those based on extreme value theory and quantile regression.

The book contains many plots and demonstrations. The goal is to simplify the analysis of financial data and to make the results easily understandable. Like many authors, I struggle to obtain a balance between the length of the book and new developments in financial econometrics. Omission of some important topics is unavoidable. There is some overlap with *Analysis of Financial Time Series* in coverage, but all examples are new.

I like to express my sincere thanks to my wife. Without her love and support, this book could not be written. I also like to thank my children; they are my inspiration and help me editing some chapters. Many readers and students constantly give me feedback and suggestions. Their input is invaluable. Finally, I like to thank Steve Quigley, Jacqueline Palmieri and their Wiley team for their support and encouragement.

The web page of the book is

http://faculty.chicagobooth.edu/ruey.tsay/teaching/introTS.

R. S. T.

Chicago, Illinois
October 2012

1

FINANCIAL DATA AND THEIR PROPERTIES

The importance of quantitative methods in business and finance has increased substantially in recent years because we are in a data-rich environment and the economies and financial markets are more integrated than ever before. Data are collected systematically for thousands of variables in many countries and at a finer timescale. Computing facilities and statistical packages for analyzing complicated and high dimensional financial data are now widely available. As a matter of fact, with an internet connection, one can easily download financial data from open sources within a software package such as R. All of these good features and capabilities are free and widely accessible.

The objective of this book is to provide basic knowledge of financial time series, introduce statistical tools useful for analyzing financial data, and gain experience in financial applications of various econometric methods. We begin with the basic concepts of financial data to be analyzed throughout the book. The software R is introduced via examples. We also discuss different ways to visualize financial data in R. Chapter 2 reviews basic concepts of linear time series analysis such as stationarity and autocorrelation function, introduces simple linear models for handling serial dependence of the data, and discusses regression models with time series errors, seasonality, unit-root nonstationarity, and long-memory processes. The chapter also considers

An Introduction to Analysis of Financial Data with R, First Edition. Ruey S. Tsay.
© 2013 John Wiley & Sons, Inc. Published 2013 by John Wiley & Sons, Inc.

exponential smoothing for forecasting and methods for model comparison. Chapter 3 considers some applications of the models introduced in Chapter 2 in the form of case studies. The goal is to help readers understand better data analysis, empirical modeling, and making inference. It also points out the limitations of linear time series models in long-term prediction. Chapter 4 focuses on modeling conditional heteroscedasticity (i.e., the conditional variance of an asset return). It introduces various econometric models for describing the evolution of asset volatility over time. The chapter also discusses alternative methods to volatility modeling, including use of daily high and low prices of an asset. In Chapter 5, we demonstrate some applications of volatility models using, again, some case studies. All steps for building volatility models are given, and the merits and weaknesses of various volatility models are discussed, including the connection to diffusion limit of continuous time models. Chapter 6 is concerned with analysis of high frequency financial data. It starts with special characteristics of high frequency data and gives models and methods that can be used to analyze such data. It shows that nonsynchronous trading and bid-ask bounce can introduce serial correlations in a stock return. It also studies the dynamic of time duration between trades and some econometric models for analyzing transaction data. In particular, we discuss the use of logistic linear regression and probit models to study the stock price movements in consecutive trades. Finally, the chapter studies the realized volatility using intraday log returns. Chapter 7 discusses risk measures of a financial position and their use in risk management. It introduces value at risk and conditional value at risk to quantify the risk of a financial position within a holding period. It also provides various methods for calculating risk measures for a financial position, including RiskMetrics, econometric modeling, extreme value theory, quantile regression, and peaks over thresholds.

The book places great emphasis on application and empirical data analysis. Every chapter contains real examples, and, in many occasions, empirical characteristics of financial time series are used to motivate the development of econometric models. In some cases, simple R scripts are given on the web page for specific analysis. Many real data sets are also used in the exercises of each chapter.

1.1 ASSET RETURNS

Most financial studies involve returns, instead of prices, of assets. Campbell et al. (1997) give two main reasons for using returns. First, for average investors, return of an asset is a complete and scale-free summary of the investment opportunity. Second, return series are easier to handle than price series because the former have more attractive statistical properties. There are, however, several definitions of an asset return.

Let P_t be the price of an asset at time index t. We discuss some definitions of returns that are used throughout the book. Assume for the moment that the asset pays no dividends.

One-Period Simple Return. Holding the asset for one period from date $t-1$ to date t would result in a *simple gross return*

$$1 + R_t = \frac{P_t}{P_{t-1}} \quad \text{or} \quad P_t = P_{t-1}(1 + R_t). \tag{1.1}$$

The corresponding one-period *simple net return* or *simple return* is

$$R_t = \frac{P_t}{P_{t-1}} - 1 = \frac{P_t - P_{t-1}}{P_{t-1}}. \tag{1.2}$$

For demonstration, Table 1.1 gives five daily closing prices of Apple stock in December 2011. From the table, the 1-day gross return of holding the stock from December 8 to December 9 is $1 + R_t = 393.62/390.66 \approx 1.0076$ so that the corresponding daily simple return is 0.76%, which is (393.62-390.66)/390.66.

Multiperiod Simple Return. Holding the asset for k periods between dates $t-k$ and t gives a k-period simple gross return

$$
\begin{aligned}
1 + R_t[k] &= \frac{P_t}{P_{t-k}} = \frac{P_t}{P_{t-1}} \times \frac{P_{t-1}}{P_{t-2}} \times \cdots \times \frac{P_{t-k+1}}{P_{t-k}} \\
&= (1 + R_t)(1 + R_{t-1}) \cdots (1 + R_{t-k+1}) \\
&= \prod_{j=0}^{k-1}(1 + R_{t-j}).
\end{aligned}
$$

Thus, the k-period simple gross return is just the product of the k one-period simple gross returns involved. This is called a *compound return*. The k-period simple net return is $R_t[k] = (P_t - P_{t-k})/P_{t-k}$.

To illustrate, consider again the daily closing prices of Apple stock of Table 1.1. Since December 2 and 9 are Fridays, the weekly simple gross return of the stock is $1 + R_t[5] = 393.62/389.70 \approx 1.0101$ so that the weekly simple return is 1.01%.

In practice, the actual time interval is important in discussing and comparing returns (e.g., monthly return or annual return). If the time interval is not given, then it is implicitly assumed to be one year. If the asset was held for k years, then the annualized (average) return is defined as

$$\text{Annualized}\{R_t[k]\} = \left[\prod_{j=0}^{k-1}(1 + R_{t-j}) \right]^{1/k} - 1.$$

TABLE 1.1. Daily Closing Prices of Apple Stock from December 2 to 9, 2011

Date	12/02	12/05	12/06	12/07	12/08	12/09
Price($)	389.70	393.01	390.95	389.09	390.66	393.62

This is a geometric mean of the k one-period simple gross returns involved and can be computed by

$$\text{Annualized}\{R_t[k]\} = \exp\left[\frac{1}{k}\sum_{j=0}^{k-1}\ln(1 + R_{t-j})\right] - 1,$$

where $\exp(x)$ denotes the exponential function and $\ln(x)$ is the natural logarithm of the positive number x. Because it is easier to compute arithmetic average than geometric mean and the one-period returns tend to be small, one can use a first-order Taylor expansion to approximate the annualized return and obtain

$$\text{Annualized}\{R_t[k]\} \approx \frac{1}{k}\sum_{j=0}^{k-1}R_{t-j}. \qquad (1.3)$$

Accuracy of the approximation in Equation (1.3) may not be sufficient in some applications, however.

Continuous Compounding. Before introducing continuously compounded return, we discuss the effect of compounding. Assume that the interest rate of a bank deposit is 10% per annum and the initial deposit is $1.00. If the bank pays interest once a year, then the net value of the deposit becomes $1(1+0.1) = \$1.1$, 1 year later. If the bank pays interest semiannually, the 6-month interest rate is 10%/2 = 5% and the net value is $ 1(1 + 0.1/2)^2 = \$1.1025$ after the first year. In general, if the bank pays interest m times a year, then the interest rate for each payment is 10%/m and the net value of the deposit becomes $\$1(1 + 0.1/m)^m$, 1 year later. Table 1.2 gives the results for some commonly used time intervals on a deposit of $1.00 with interest rate of 10% per annum. In particular, the net value approaches

TABLE 1.2. Illustration of the Effects of Compounding: the Time Interval is 1 Year and the Interest Rate is 10% Per Annum

Type	Number of Payments	Interest Rate per Period	Net Value
Annual	1	0.1	$1.10000
Semiannual	2	0.05	$1.10250
Quarterly	4	0.025	$1.10381
Monthly	12	0.0083	$1.10471
Weekly	52	$\frac{0.1}{52}$	$1.10506
Daily	365	$\frac{0.1}{365}$	$1.10516
Continuously	∞		$1.10517

$1.1052, which is obtained by exp(0.1) and referred to as the *result of continuous compounding*. The effect of compounding is clearly seen.

In general, the net asset value A of continuous compounding is

$$A = C \exp(r \times n), \tag{1.4}$$

where r is the interest rate per annum, C is the initial capital, and n is the number of years. From Equation (1.4), we have

$$C = A \exp(-r \times n), \tag{1.5}$$

which is referred to as the *present value* of an asset that is worth A dollars n years from now, assuming that the continuously compounded interest rate is r per annum.

Continuously Compounded Return. The natural logarithm of the simple gross return of an asset is called the *continuously compounded return* or *log return*:

$$r_t = \ln(1 + R_t) = \ln \frac{P_t}{P_{t-1}} = p_t - p_{t-1}, \tag{1.6}$$

where $p_t = \ln(P_t)$. Continuously compounded returns r_t enjoy some advantages over the simple net returns R_t. First, consider multiperiod returns. We have

$$r_t[k] = \ln(1 + R_t[k]) = \ln[(1 + R_t)(1 + R_{t-1}) \cdots (1 + R_{t-k+1})]$$

$$= \ln(1 + R_t) + \ln(1 + R_{t-1}) + \cdots + \ln(1 + R_{t-k+1})$$

$$= r_t + r_{t-1} + \cdots + r_{t-k+1}.$$

Thus, the continuously compounded multiperiod return is simply the sum of continuously compounded one-period returns involved. Second, statistical properties of log returns are more tractable.

To demonstrate, we again consider the daily closing prices of Apple stock of Table 1.1. The daily log return from December 8 to December 9 is $r_t = \log(393.62) - \log(390.66) \approx 0.75\%$ and the weekly log return from December 2 to December 9 is $r_t[5] = \log(393.62) - \log(389.70) \approx 1.00\%$. One can easily verify that the weekly log return is the sum of the five daily log returns involved.

Portfolio Return. The simple net return of a portfolio consisting of N assets is a weighted average of the simple net returns of the assets involved, where the weight on each asset is the percentage of the portfolio's value invested in that asset. Let p be a portfolio that places weight w_i on asset i. Then, the simple return of p at time t is $R_{p,t} = \sum_{i=1}^{N} w_i R_{it}$, where R_{it} is the simple return of asset i.

The continuously compounded returns of a portfolio, however, do not have the above convenient property. If the simple returns R_{it} are all small in magnitude, then we have $r_{p,t} \approx \sum_{i=1}^{N} w_i r_{it}$, where $r_{p,t}$ is the continuously compounded return of the portfolio at time t. This approximation is often used to study portfolio returns.

Dividend Payment. If an asset pays dividends periodically, we must modify the definitions of asset returns. Let D_t be the dividend payment of an asset between dates $t-1$ and t, and P_t be the price of the asset at the end of period t. Thus, dividend is not included in P_t. Then, the simple net return and continuously compounded return at time t become

$$R_t = \frac{P_t + D_t}{P_{t-1}} - 1, \quad r_t = \ln(P_t + D_t) - \ln(P_{t-1}).$$

Excess Return. Excess return of an asset at time t is the difference between the asset's return and the return on some reference asset. The reference asset is often taken to be riskless such as a short-term U.S. Treasury bill return. The simple excess return and log excess return of an asset are then defined as

$$Z_t = R_t - R_{0t}, \quad z_t = r_t - r_{0t}, \tag{1.7}$$

where R_{0t} and r_{0t} are the simple and log returns of the reference asset, respectively. In the finance literature, the excess return is thought of as the payoff on an arbitrage portfolio that goes long in an asset and short in the reference asset with no net initial investment.

Remark. A long financial position means owning the asset. A short position involves selling an asset one does not own. This is accomplished by borrowing the asset from an investor who has purchased it. At some subsequent date, the short seller is obligated to buy exactly the same number of shares borrowed to pay back the lender. Because the repayment requires equal shares rather than equal dollars, the short seller benefits from a decline in the price of the asset. If cash dividends are paid on the asset while a short position is maintained, these are paid to the buyer of the short sale. The short seller must also compensate the lender by matching the cash dividends from his own resources. In other words, the short seller is also obligated to pay cash dividends on the borrowed asset to the lender. ☐

Summary of Relationship. The relationships between simple return R_t and continuously compounded (or log) return r_t are

$$r_t = \ln(1 + R_t), \quad R_t = e^{r_t} - 1.$$

If the returns R_t and r_t are in percentages, then

$$r_t = 100 \ln\left(1 + \frac{R_t}{100}\right), \quad R_t = 100\left(e^{r_t/100} - 1\right).$$

Temporal aggregation of the returns produces

$$1 + R_t[k] = (1 + R_t)(1 + R_{t-1}) \cdots (1 + R_{t-k+1}),$$

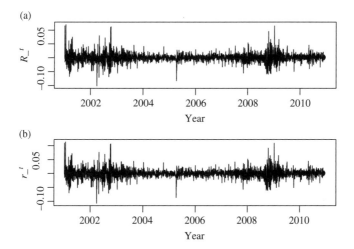

Figure 1.1. Time plots of daily returns of IBM stock from January 2, 2001 to December 31, 2010: Panels (a) and (b) show simple and log returns, respectively.

$$r_t[k] = r_t + r_{t-1} + \cdots + r_{t-k+1}.$$

If the continuously compounded interest rate is r per annum, then the relationship between present and future values of an asset is

$$A = C \exp(r \times n), \quad C = A \exp(-r \times n).$$

Example 1.1. If the monthly log return of an asset is 4.46%, then the corresponding monthly simple return is $100[\exp(4.46/100) - 1] = 4.56\%$. Also, if the monthly log returns of the asset within a quarter are 4.46%, -7.34%, and 10.77%, respectively, then the quarterly log return of the asset is $(4.46 - 7.34 + 10.77)\% = 7.89\%$. $\quad\square$

Figure 1.1 shows the time plots of daily simple and log returns of IBM stock from January 2, 2001 to December 31, 2010. There are 2515 observations. From the plots, the behavior of log returns is similar to that of the simple returns. As a matter of fact, the correlation coefficient between the simple and log returns is 0.9997. This is understandable because, when x is close to zero, $\log(1 + x) \approx x$ and daily simple returns of IBM stock are small in the sampling period.

1.2 BOND YIELDS AND PRICES

Bonds are a financial instrument that will pay the face value (or par value) to its holder at the time of maturity. Some bonds also pay interest periodically referring to as *coupon payment*. Zero-coupon bonds do not pay periodic interest. Bond yield is the return an investor will receive by holding a bond to maturity. In finance, several types of bond yield are used. The common ones are the current yield and yield to maturity (YTM).

Current Yield. The current yield denotes the percentage return that the annual coupon payment provides the investor. Mathematically, we have

$$\text{Current yield} = \frac{\text{Annual interest paid in dollars}}{\text{Market price of the bond}} \times 100\%.$$

For example, if an investor paid \$90 for a bond with face value of \$100, also known as *par value*, and the bond paid a coupon rate of 5% per annum, then the current yield of the bond is $c_t = (0.05 \times 100)/90 \times 100\% = 5.56\%$. We use the subscript t to signify that the yield is typically time dependent. From the definition, current yield does not include any capital gains or losses of the investment. For zero-coupon bonds, the yield is calculated as follows:

$$\text{Current yield} = \left(\frac{\text{Face value}}{\text{Purchase price}} \right)^{1/k} - 1,$$

where k denotes time to maturity in years. For instance, if an investor purchased a zero-coupon bond with face value \$100 for \$90 and the bond will mature in 2 years, then the yield is $c_t = (100/90)^{1/2} - 1 = 5.41\%$.

Yield to Maturity. The current yield does not consider the time value of money, because it does not consider the present value of the coupon payments the investor will receive in the future. Therefore, a more commonly used measurement of bond investment is the *YTM*. The calculation of YTM, however, is more complex. Simply put, YTM is the yield obtained by equating the bond price to the present value of all future payments. Suppose that the bond holder will receive k payments between purchase and maturity. Let y and P be the YTM and price of the bond, respectively. Then,

$$P = \frac{C_1}{1+y} + \frac{C_2}{(1+y)^2} + \cdots + \frac{C_k + F}{(1+y)^k},$$

where F denotes the face value and C_i is the ith cash flow of coupon payment. Suppose that the coupon rate is α per annum, the number of payments is m per year, and the time to maturity is n years. In this case, cash flow of coupon payment is $F\alpha/m$, and the number of payments is $k = mn$. The bond price and YTM can be formulated as

$$P = \frac{\alpha F}{m} \left[\frac{1}{(1+y)} + \frac{1}{(1+y)^2} + \cdots + \frac{1}{(1+y)^k} \right] + \frac{F}{(1+y)^k}$$

$$= \frac{\alpha F}{my} \left[1 - \frac{1}{(1+y)^k} \right] + \frac{F}{(1+y)^k}.$$

The table below gives some results between bond price and YTM assuming that $F = \$100$, coupon rate is 5% per annum payable semiannually, and time to maturity is 3 years.

Yield to Maturity (%)	Semiannual Rate (%)	Bond Price ($)
6	3.0	97.29
7	3.5	94.67
8	4.0	92.14
9	4.5	89.68
10	5.0	87.31

From the table, we see that as the YTM increases the bond price decreases. In other words, YTM is inversely proportional to the bond price. In practice, we observed bond price so that YTM must be calculated. The solution is not easy to find in general, but calibration can be used to obtain an accurate approximation. As an example, suppose that one paid $94 to purchase the bond shown in the prior table. From which, we see that the YTM must be in the interval [7,8]%. With trial and error, we have

Yield to Maturity (%)	Semiannual Rate (%)	Bond Price ($)
7.1	3.55	94.41
7.2	3.6	94.16
7.3	3.65	93.90
7.25	3.625	94.03
7.26	3.63	94.00

Therefore, the YTM is approximately 7.26% per annum for the investor. Many financial institutions provide online programs that calculate bond YTM and price, for example, Fidelity Investments.

U.S. Government Bonds. The U.S. Government issues various bonds to finance its debts. These bonds include Treasury bills, Treasury notes, and Treasury bonds. A simple description of these bonds is given below.

- Treasury bills (T-Bills) mature in one year or less. They do not pay interest prior to maturity and are sold at a discount of the face value (or par value) to create a positive YTM. The commonly used maturities are 28 days (1 month), 91 days (3 months), 182 days (6 months), and 364 days (1 year). The minimum purchase is $100. The discount yield of T-Bills is calculated via

$$\text{Discount yield } (\%) = \frac{F - P}{F} \times \frac{360}{\text{Days till maturity}} \times 100(\%),$$

where F and P denote the face value and purchase price, respectively. The U.S. Treasury Department announces the amounts of offering for 13- and 26-week bills each Thursday for auction on the following Monday and settlement on Thursday. Offering amount for 4-week bills are announced on Monday for auction the next day and settlement on Thursday. Offering amounts for 52-week bills are announced every fourth Thursday for auction the next Tuesday and settlement on Thursday.

- Treasury notes (T-Notes) mature in 1–10 years. They have a coupon payment every 6 months and face value of $1000. These notes are quoted on the secondary market at percentage of face value in thirty-seconds of a point. For example, a quote 95 : 08 on a note indicates that it is trading at a discount $(95 + 8/32) \times 1000 = \$952.5$. The 10-year Treasury note has become the security most frequently quoted when discussing the U.S. government bond market; see the Chicago Board Options Exchange (CBOE) 10-year Notes of the next section. Figures 1.5 and 1.7 show, respectively, the time plots of the daily yield and its return of the 10-year T-Notes.

- Treasury bonds (T-Bonds) have longer maturities, ranging from 20 to 30 years. They have a coupon payment every 6 months and are commonly issued with maturities 30 years. The 30-year bonds were suspended for a 4-year and 6-month period starting October 31, 2001, but they were reintroduced in February 2006 and are now issued quarterly.

1.3   IMPLIED VOLATILITY

Stock options are financial contracts. A call option on Stock A gives its holder the right, but not obligation, to buy certain shares of Stock A at a prespecified price within a given period of time. A put option, on the other hand, gives its holder the right, but not obligation, to sell certain shares of the stock at a prespecified price within a given period of time. The prespecified price is called the *strike price* and the time period is referred to as *time to maturity*. In the United States, a stock option typically involves 100 shares of the underlying stock. The options are traded at the options markets such as CBOE. There are many types of options. The well-known ones are the European options, which can only be exercised at the time of maturity, and the American options, which can be exercised any time before maturity. See Hull (2011) for further details. If an option would result in a positive cash flow to its holder if it were exercised immediately, we say that the option is *in-the-money*. If an option would result in a negative cash flow to its holder if it were exercised immediately, we say that the option is *out-of-the-money*. Finally, if an option would result in zero cash flow to its holder if it were exercised immediately, we say that the option is *at-the-money*.

The price of an option depends on many factors such as strike price, risk-free interest rate, and the current price and volatility of the stock. See, for instance, the famous Black–Scholes formula. This closed-form solution was derived under the assumption that the stock price follows a geometric Brownian motion. For the purpose of this chapter, it suffices to say that the only factor in the Black–Scholes formula that is not directly observable is the volatility of the stock. By volatility, we mean the conditional standard deviation of the stock price. In practice, we can use the observed price of an option and the Black–Scholes formula to back out the value of the stock volatility. This volatility is referred to as the *implied volatility*. Similar to the YTM of bonds, calibration is often used to obtain the implied volatility or an approximation of it.

The most well-known implied volatility is the volatility index (VIX) of CBOE. The index was originally designed in 1993 to measure the market's expectation of 30-day volatility implied by at-the-money S&P 100 index option prices. However, the index was updated by CBOE and Goldman Sachs in 2003 to reflect a new measure of expected volatility. It is now based on the S&P 500 index (SPX) and estimates expected volatility by averaging the weighted prices of SPX puts and calls over a wide range of strike prices. See CBOE VIX white paper for further information. This new VIX is often regarded as the market fear factor and has played an important role in the financial markets. As a matter of fact, VIX futures and options are now traded on CBOE.

Figure 1.2 shows the time plot of the updated VIX index from January 2, 2004 to November 21, 2011 for 1988 observations. From the plot, the financial market was very volatile in late 2008 and in the beginning of 2009. The volatility was also high in 2011. We shall analyze the VIX index in later chapters. Also, see Chapter 4 for more information on asset volatility.

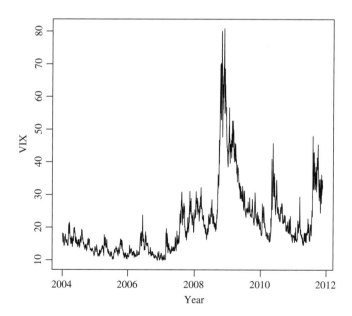

Figure 1.2. Time plot of the VIX index of Chicago Board Options Exchange from January 2, 2004 to November 21, 2011.

1.4   R PACKAGES AND DEMONSTRATIONS

Before studying some real examples of financial data, we briefly introduce the R program to be used extensively in the book. The package and commands used to perform the analysis will be given when needed. Our goal is to make the empirical analysis as easy as possible so that readers can reproduce the results shown in the book.

R is a free software available from http://www.r-project.org. It runs on many operating systems, including Linux, MacOS X, and Windows. One can click *CRAN* on its web page to select a nearby *CRAN Mirror* to download and install the software and selected packages. The simplest way to install the program is to follow the online instructions and to use the default options. Because R is an open-source software, it contains hundreds of packages developed by researchers around the world for various statistical analyses. For financial time series analysis, the Rmetrics of Dr. Diethelm Wuertz and his associates has many useful packages, including fBasics, fGarch, and fPortfolio. We use many functions of these packages in the book. We also use some other packages that are powerful and easy to use in R, for example, the `evir` package for extreme value analysis in R. Further information concerning installing R and the commands used can be found either on the web page of the book or on the author's teaching web page. There exist several introductory books for R; see, for instance, Adler (2010) and Crawley (2007). The R commands are case sensitive and must be followed exactly.

1.4.1   Installation of R Packages

Using default options in R installation creates an icon on the desktop of a computer. One can start the R program simply by double clicking the R icon. For Windows, a `RGui` window will appear with command menu and the R `Console`. To install packages, one can click on the command `Packages` to select `Install packages`. A pop-up window appears asking users to select an R mirror (similar to R installation mentioned before). With a selected mirror, another pop-up window appears that contains all available packages. One can click on the desired packages for installation.

With packages installed, one can load them into R by clicking on the command `Packages` followed by clicking `Load packages`. A pop-up window appears that contains all installed packages for users to choose. An alternative approach to load a package is to use the command `library`. See the demonstration in the following discussion.

1.4.2   The Quantmod Package

To begin with, we consider a useful R package for downloading financial data directly from some open sources, including Yahoo Finance, Google Finance, and the Federal Reserve Economic Data (FRED) of Federal Reserve Bank of St. Louis. The package is `quantmod` by Jeffry A. Ryan. It is highly recommended that one installs it. The package requires three additional packages that need to be installed as well. They are `TTR`, `xts`, and `zoo`.

Once installed, the quantmod package allows users, with internet connection, to use tick symbols to access daily stock data from Yahoo and Google Finance and to use series name to access over 1000 economic and financial time series from FRED. The command is getSymbols. The package also has some nice functions, for example, obtaining time series plots of closing price and trading volume. The command is chartSeries. The default option of these two commands is sufficient for basic analysis of financial time series. One can use subcommands to further enhance the capabilities of the package such as specifying the time span of interest in get-Symbols. Interested readers may consult the document associated with the package for description of the commands available. Here, we provide a simple demonstration. Figure 1.3 shows the time plots of daily closing price and trading volume of Apple stock from January 3, 2007 to December 2, 2011. The plot also shows the price and volume of the last observation. The subcommand theme = "white" of chartSeries is used to set the background of the time plot. The default is black. Figure 1.4 shows the time plot of monthly U.S. unemployment rates from January 1948 to November 2011. Figure 1.5 shows the time plot of daily interest rate of 10-year treasures notes from January 3, 2007 to December 2, 2011. These are the interest rates from the CBOE obtained from Yahoo Finance. As there is no volume, the subcommand TA = NULL is used to omit the time plot of volume in chartSeries. The commands head and tail show, respectively, the first and the last six rows of the data.

R Demonstration with quantmod package Output edited. > denotes R prompt and explanation starts with %.

```
> library(quantmod)    % Load the package
> getSymbols("AAPL")   % Download daily prices of Apple stock from Yahoo
[1] "AAPL" % I ran R on 2011-12-03 so that the last day was 12-02.
> dim(AAPL)   % (dimension): See the size of the downloaded data.
[1] 1241     6
> head(AAPL)   % See the first 6 rows of the data
              Open       High      Low       Close      Volume    Adjusted
2007-01-03    86.29      86.58     81.90     83.80      44225700     83.80
2007-01-04    84.05      85.95     83.82     85.66      30259300     85.66
....
2007-01-10    94.75      97.80     93.45     97.00      105460000    97.00
> tail(AAPL)    % See the last 6 rows of the data
              Open       High      Low       Close      Volume    Adjusted
2011-11-25    368.42     371.15    363.32    363.57     9098600      363.57
.....
2011-12-01    382.54     389.00    380.75    387.93     13709400     387.93
2011-12-02    389.83     393.63    388.58    389.70     13537700     389.70

> chartSeries(AAPL,theme="white")   % Plot the daily price and volume
% The subcommand theme is used to obtain white background of the plot.
> chartSeries(AAPL)%Not shown giving the same plot with black background.
% The next command specifies the data span of interest
> getSymbols("AAPL",from="2005-01-02", to="2010-12-31")
```

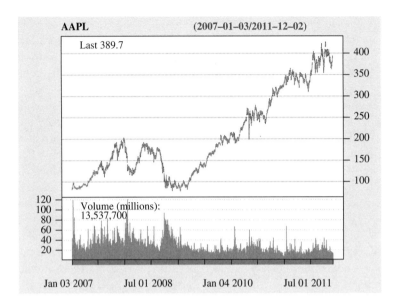

Figure 1.3. Time plots of daily closing price and trading volume of Apple stock from January 3, 2007 to December 2, 2011.

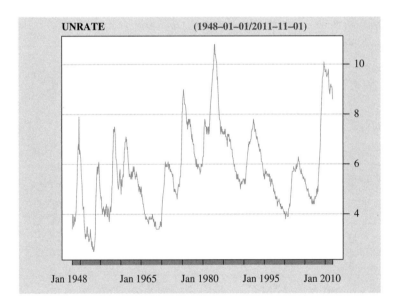

Figure 1.4. Time plot of U.S. monthly unemployment rates from January 1948 to November 2011.

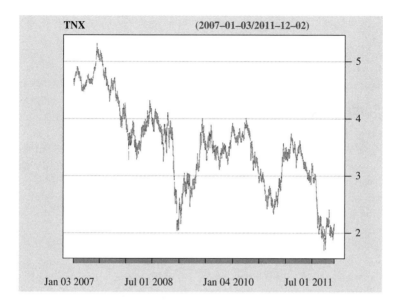

Figure 1.5. Time plot of Chicago Board Options Exchange interest rates of 10-year Treasury notes from January 3, 2007 to December 2, 2011.

```
[1] "AAPL"
> head(AAPL)
           AAPL.Open AAPL.High AAPL.Low AAPL.Close AAPL.Volume Adjusted
2005-01-03    64.78     65.11     62.60      63.29     24714000    31.65
2005-01-04    63.79     65.47     62.97      63.94     39171800    31.97
 ......
> getSymbols("UNRATE",src="FRED")%Download unemployment rates from FRED.
[1] "UNRATE"
> head(UNRATE)
            UNRATE
1948-01-01    3.4
1948-02-01    3.8
 ......
1948-06-01    3.6
> chartSeries(UNRATE,theme="white")  % Plot monthly unemployment rates
% The subcommand "src" is used to specify the data source.
% The default is Yahoo.
> getSymbols("INTC",src="google")  % Download data from Google.
[1] "INTC"
> head(INTC)
           INTC.Open INTC.High INTC.Low INTC.Close INTC.Volume
2007-01-03    20.45     20.88     20.14      20.35     68665100
2007-01-04    20.63     21.33     20.56      21.17     87795400
 .....
2007-01-10    21.09     21.62     21.03 .    21.52     75522200

> getSymbols("^TNX") % Download CBOE 10-year Treasures Notes
[1] "TNX"
> head(TNX)
           TNX.Open TNX.High TNX.Low TNX.Close Volume Adjusted
```

```
2007-01-03    4.66    4.69    4.64    4.66    0    4.66
2007-01-04    4.66    4.66    4.60    4.62    0    4.62
.....
2007-01-10    4.67    4.70    4.66    4.68    0    4.68
> chartSeries(TNX,theme="white",TA=NULL) % Obtain plot without volume.
```

1.4.3   Some Basic R Commands

After starting R, the first thing to do is to set the working directory. By working directory, we mean the computer directory where data sets reside and output will be stored. This can be done in two ways. The first method is to click on the command File. A pop-up window appears that allows one to select the desired directory. The second method is to type in the desired directory in the R Console using the command setwd, which stands for set working directory. See the demonstration in the following discussion.

R is an object-oriented program. It handles many types of object. For the purposes of the book, we do not need to study details of an object in R. Explanations will be given when needed. It suffices now to say that R allows one to assign values to variables and refer to them by name. The assignment operator is $<-$, but $=$ can also be used. For instance, $x<-10$ assigns the value 10 to the variable "x." Here, R treats "x" as a sequence of real numbers with the first element being 10. There are several ways to load data into the R working space, depending on the data format. For simple text data, the command is read.table. For .csv files, the command is read.csv. The data file is specified in either a single or double quotes; see the R demonstration. R treats the data as an object and refer to them by the assigned name. For both loading commands, R stores the data in a matrix framework. As such, one can use the command dim (i.e., dimension) to see the size of the data. Finally, the basic operations in R are similar to those we commonly use and the command to exit R is q().

R Demonstration

```
> setwd("C:/Users/rst/book/introTS/data") % Set my working directory
> library(fBasics) % Load package
> x <- 10   % Assign value, here "x" is a variable.
> x   % See the value of x.
[1] 10 % Here [1] signifies the first element.
> 1 + 2 % Basic operation: addition
[1] 3
> 10/2 % Basic operation: division
[1] 5
% Use * and ^ for multiplication and power, respectively.
% Use log for the natural logarithm.
> da=read.table('d-ibm-0110.txt',header=T) % Load text data with names.
> head(da) % See the first 6 rows
      date      return
1 20010102 -0.002206
2 20010103  0.115696
  ....
```

```
6 20010109 -0.010688
> dim(da) % Dimension of the data object "da".
[1] 2515     2
> da <- read.csv("d-vix0411.csv",header=T) % Load csv data with names.
> head(da)  % See the first 6 rows
        Date VIX.Open VIX.High VIX.Low VIX.Close
1 1/2/2004    17.96    18.68   17.54     18.22
2 1/5/2004    18.45    18.49   17.44     17.49
 ....
6 1/9/2004    16.15    16.88   15.57     16.75
```

1.5 EXAMPLES OF FINANCIAL DATA

In this section, we examine some of the return series in finance. Figure 1.6 shows the time plot of daily log returns of Apple stock from January 4, 2007 to December 2, 2011. As defined before, daily log returns are simply the change series of log prices. In R, a change series can easily be obtained by taking the *difference* of the log prices. Specifically, $r_t = \ln(P_1) - \ln(P_{t-1})$, where P_t is the stock price at time t. Note that in the demonstration, I used *adjusted* daily price to compute log returns because adjusted price takes into consideration the stock splits, if any, during the sample period. From the plot, we see that (i) there exist some large outlying observations and (ii) the returns were volatile in certain periods but stable in others. The latter characteristic is referred to as *volatility clustering* in asset returns. The former, on the other hand, are indicative that the returns have heavy tails.

Figure 1.7 shows the time plot of daily changes in YTM of the 10-year Treasury notes also from January 4, 2007 to December 2, 2011. The changes in YTM exhibit similar characteristics as those of daily returns of Apple stock. Figure 1.8 provides the time plot of daily log returns of the Dollar–Euro exchange rate. Again, the log returns of exchange rates have the same features as those of the daily log returns of stock. The daily Dollar–Euro exchange rate is given in Figure 1.9. The exchange rates are downloaded from the database FRED.

R Demonstration

```
> library(quantmod)
> getSymbols("AAPL",from="2007-01-03",to="2011-12-02") %Specify period
[1] "AAPL"
> AAPL.rtn=diff(log(AAPL$AAPL.Adjusted)) % Compute log returns
> chartSeries(AAPL.rtn,theme="white")
> getSymbols("^TNX",from="2007-01-03",to="2011-12-02")
[1] "TNX"

> TNX.rtn=diff(TNX$TNX.Adjusted) % Compute changes
> chartSeries(TNX.rtn,theme="white")
> getSymbols("DEXUSEU",src="FRED") % Obtain exchange rates from FRED
[1] "DEXUSEU"
> head(DEXUSEU)
          DEXUSEU
1999-01-04  1.1812
```

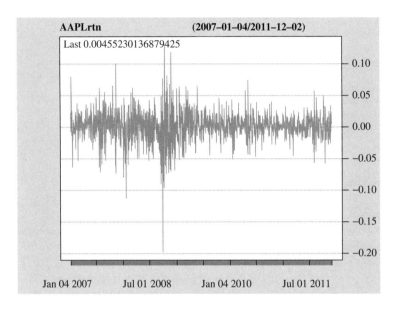

Figure 1.6. Time plot of daily log returns of Apple stock from January 3, 2007 to December 2, 2011.

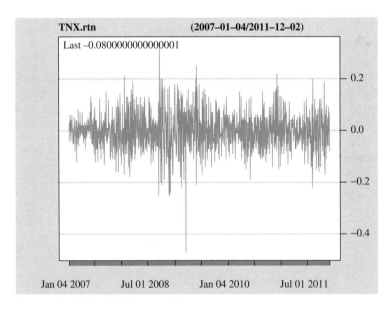

Figure 1.7. Time plot of daily changes in the YTM for the U.S. 10-year Treasury notes from January 3, 2007 to December 2, 2011.

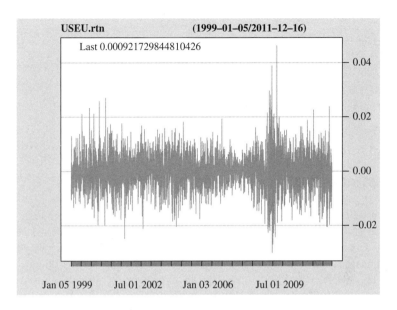

Figure 1.8. Time plot of daily log returns of the dollar–euro exchange rates from January 5, 1999 to December 16, 2011. The rate is dollars per Euro.

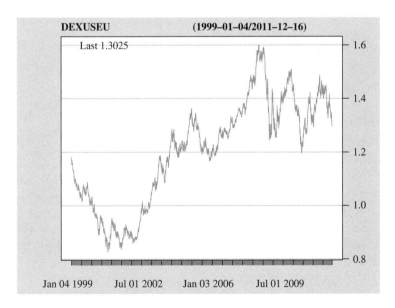

Figure 1.9. Time plot of daily dollar–euro exchange rates from January 4, 1999 to December 16, 2011. The rate is dollars per Euro.

```
1999-01-05  1.1760
    ....
1999-01-11  1.1534
> tail(DEXUSEU)
           DEXUSEU
2011-12-09  1.3368
    ....
2011-12-16  1.3025
> USEU.rtn=diff(log(DEXUSEU$DEXUSEU))
> chartSeries(DEXUSEU,theme="white")
> chartSeries(USEU.rtn,theme="white")
```

1.6 DISTRIBUTIONAL PROPERTIES OF RETURNS

To gain a better understanding on asset returns, we begin with their distributional properties. The objective here is to study the behavior of the returns across assets and over time. Consider a collection of N assets held for T time periods, say, $t = 1, \ldots, T$. For each asset i, let r_{it} be its log return at time t. The log returns under study are $\{r_{it}; i = 1, \ldots, N; t = 1, \ldots, T\}$. One can also consider the simple returns $\{R_{it}; i = 1, \ldots, N; t = 1, \ldots, T\}$ and the log excess returns $\{z_{it}; i = 1, \ldots, N; t = 1, \ldots, T\}$.

1.6.1 Review of Statistical Distributions and Their Moments

We briefly review some basic properties of statistical distributions and the moment equations of a random variable. Let R^k be the k-dimensional Euclidean space. A point in R^k is denoted by $x \in R^k$. Consider two random vectors $X = (X_1, \ldots, X_k)'$ and $Y = (Y_1, \ldots, Y_q)'$. Let $P(X \in A, Y \in B)$ be the probability that X is in the subspace $A \subset R^k$ and Y is in the subspace $B \subset R^q$. For most of the cases considered in this book, both random vectors are assumed to be continuous.

Joint Distribution. The function

$$F_{X,Y}(x,y;\theta) = P(X \leq x, Y \leq y;\theta),$$

where $x \in R^p, y \in R^q$, and the inequality "$\leq$" is a component-by-component operation and is a joint distribution function of X and Y with parameter θ. Behavior of X and Y is characterized by $F_{X,Y}(x,y;\theta)$. For instance, the linear dependence between X and Y is shown by the covariance of the joint distribution. If the joint probability density function $f_{x,y}(x,y;\theta)$ of X and Y exists, then

$$F_{X,Y}(x,y;\theta) = \int_{-\infty}^{x} \int_{-\infty}^{y} f_{x,y}(w,z;\theta)dz\,dw.$$

In this case, X and Y are continuous random vectors.

Marginal Distribution. The marginal distribution of X is given by

$$F_X(x; \theta) = F_{X,Y}(x, \infty, \cdots, \infty; \theta).$$

Thus, the marginal distribution of X is obtained by integrating out Y. A similar definition applies to the marginal distribution of Y.

If $k = 1$, X is a scalar random variable and the distribution function becomes

$$F_X(x) = P(X \leq x; \theta),$$

which is known as the *cumulative distribution function* (CDF) of X. The CDF of a random variable is nondecreasing (i.e., $F_X(x_1) \leq F_X(x_2)$ if $x_1 \leq x_2$) and satisfies $F_X(-\infty) = 0$ and $F_X(\infty) = 1$. For a given probability p, the smallest real number x_p such that $p \leq F_X(x_p)$ is called the *pth quantile* of the random variable X. More specifically,

$$x_p = \inf_x \{x \,|\, p \leq F_X(x)\}.$$

We use the CDF to compute the *p*-value of a test statistic in the book.

Conditional Distribution. The conditional distribution of X given $Y \leq y$ is given by

$$F_{X|Y \leq y}(x; \theta) = \frac{P(X \leq x, Y \leq y; \theta)}{P(Y \leq y; \theta)}.$$

If the probability density functions involved exist, then the conditional density of X given $Y = y$ is

$$f_{x|y}(x; \theta) = \frac{f_{x,y}(x, y; \theta)}{f_y(y; \theta)}, \tag{1.8}$$

where the marginal density function $f_y(y; \theta)$ is obtained by

$$f_y(y; \theta) = \int_{-\infty}^{\infty} f_{x,y}(x, y; \theta) dx.$$

From Equation (1.8), the relation among joint, marginal, and conditional distributions is

$$f_{x,y}(x, y; \theta) = f_{x|y}(x; \theta) \times f_y(y; \theta). \tag{1.9}$$

This identity is used extensively in time series analysis (e.g., in maximum likelihood estimation). Finally, X and Y are independent random vectors if and only if $f_{x|y}(x; \theta) = f_x(x; \theta)$. In this case, $f_{x,y}(x, y; \theta) = f_x(x; \theta) f_y(y; \theta)$.

Moments of a Random Variable. The ℓth moment of a continuous random variable X is defined as

$$m'_\ell = E(X^\ell) = \int_{-\infty}^{\infty} x^\ell f(x)\mathrm{d}x,$$

where E stands for expectation and $f(x)$ is the probability density function of X. The first moment is called the *mean* or *expectation* of X. It measures the central location of the distribution. We denote the mean of X by μ_x. For an asset, an interesting question is whether the mean of its return is zero. In other words, we often consider the hypothesis testing $H_0 : \mu_x = 0$ versus $H_a : \mu \neq 0$ or $H_0 : \mu_x \leq 0$ versus $H_a : \mu_x > 0$.

The ℓth central moment of X is defined as

$$m_\ell = E[(X - \mu_x)^\ell] = \int_{-\infty}^{\infty} (x - \mu_x)^\ell f(x)\mathrm{d}x$$

provided that the integral exists. The second central moment, denoted by σ_x^2, measures the variability of X and is called the *variance* of X. The positive square root, σ_x, of variance is the *standard deviation* of X. For asset returns, variance (or standard deviation) is a measure of uncertainty and, hence, is often used as a risk measure. The first two moments of a random variable uniquely determine a normal distribution. For other distributions, higher order moments are also of interest.

The third central moment measures the symmetry of X with respect to its mean, whereas the fourth central moment measures the tail behavior of X. In statistics, *skewness* and *kurtosis*, which are normalized third and fourth central moments of X, are often used to summarize the extent of asymmetry and tail thickness. Specifically, the skewness and kurtosis of X are defined as

$$S(x) = E\left[\frac{(X - \mu_x)^3}{\sigma_x^3}\right], \quad K(x) = E\left[\frac{(X - \mu_x)^4}{\sigma_x^4}\right].$$

The quantity $K(x) - 3$ is called the *excess kurtosis* because $K(x) = 3$ for a normal distribution. Thus, the excess kurtosis of a normal random variable is zero. A distribution with positive excess kurtosis is said to have heavy tails, implying that the distribution puts more mass on the tails of its support than a normal distribution does. In practice, this means that a random sample from such a distribution tends to contain more extreme values. Such a distribution is said to be *leptokurtic*. On the other hand, a distribution with negative excess kurtosis has short tails (e.g., a uniform distribution over a finite interval). Such a distribution is said to be *platykurtic*. In finance, the first fourth moments of a random variable are used to describe the behavior of asset returns. This does not imply that higher order moments are not important; they are much harder to study.

In application, moments of a random variable can be estimated by their sample counterparts. Let $\{x_1, \ldots, x_T\}$ be a random sample of X with T observations. The sample mean is

$$\hat{\mu}_x = \frac{1}{T} \sum_{t=1}^{T} x_t, \tag{1.10}$$

the sample variance is

$$\hat{\sigma}_x^2 = \frac{1}{T-1} \sum_{t=1}^{T} (x_t - \hat{\mu}_x)^2, \tag{1.11}$$

the sample skewness is

$$\hat{S}(x) = \frac{1}{(T-1)\hat{\sigma}_x^3} \sum_{t=1}^{T} (x_t - \hat{\mu}_x)^3, \tag{1.12}$$

and the sample kurtosis is

$$\hat{K}(x) = \frac{1}{(T-1)\hat{\sigma}_x^4} \sum_{t=1}^{T} (x_t - \hat{\mu}_x)^4. \tag{1.13}$$

Under rather weak conditions, the sample mean $\hat{\mu}_x$ is a consistent estimate of μ_x, meaning that $\hat{\mu}_x$ converges to μ_x as $T \to \infty$. More specifically, we have $\hat{\mu}_x \sim N(\mu_x, \sigma_x^2/T)$ for a sufficiently large T. This result is often used to test any hypothesis about μ_x. For instance, consider $H_0 : \mu_x = 0$ versus $H_a : \mu_x \neq 0$. The test statistic is

$$t = \frac{\sqrt{T}\hat{\mu}_x}{\hat{\sigma}_x},$$

which follows a Student's-t distribution with $T - 1$ degrees of freedom. For a sufficiently large T, the test statistic approaches a standard normal distribution. The decision rule is then to reject H_0 at the $100\alpha\%$ level if $|t| > Z_{1-\alpha/2}$, where $Z_{1-\alpha/2}$ is the $(1 - \alpha/2)$th quantile of the standard normal distribution. Most statistical packages now provide p-value for each test statistic. The decision rule is then to reject H_0 at the $100\alpha\%$ level if the p-value is less than α.

If X is a normal random variable, then $\hat{S}(x)$ and $\hat{K}(x) - 3$ are distributed asymptotically as normal with zero mean and variances $6/T$ and $24/T$, respectively; see Snedecor and Cochran (1980, p. 78). These asymptotic properties can be used to test the normality of asset returns. Given an asset return series $\{r_1, \ldots, r_T\}$, to test the skewness of the returns, we consider the null hypothesis $H_o : S(r) = 0$ versus the alternative hypothesis $H_a : S(r) \neq 0$. The t-ratio statistic of the sample skewness in Equation (1.12) is

$$t = \frac{\hat{S}(r)}{\sqrt{6/T}}.$$

The decision rule is to reject the null hypothesis at the $100\alpha\%$ significance level, if $|t| > Z_{1-\alpha/2}$.

Similarly, one can test the excess kurtosis of the return series using the hypotheses $H_o : K(r) - 3 = 0$ versus $H_a : K(r) - 3 \neq 0$. The test statistic is

$$t = \frac{\hat{K}(r) - 3}{\sqrt{24/T}},$$

which is asymptotically a standard normal random variable. The decision rule is to reject H_o if and only if the p-value of the test statistic is less than the significance level α. Jarque and Bera (1987) combine the two prior tests and use the test statistic

$$\mathrm{JB} = \frac{\hat{S}^2(r)}{6/T} + \frac{(\hat{K}(r) - 3)^2}{24/T},$$

which is asymptotically distributed as a chi-squared random variable with 2 degrees of freedom, to test for the normality of r_t. One rejects H_o of normality if the p-value of the JB statistic is less than the significance level.

Example 1.2. Consider the daily simple returns of the 3M stock from January 2, 2001 to September 30, 2011. The data are obtained from the Center for Research of Security Prices (CRSP), University of Chicago. Figure 1.10 shows the time plot of the data. Here, we use the command `basicStats` of `fBasics` in Rmetrics to obtain summary statistics of the returns and to perform some basic hypothesis testing. From

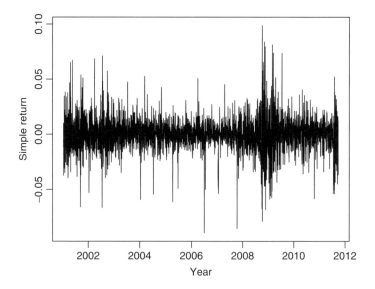

Figure 1.10. Time plot of daily simple returns of 3M stock from January 2, 2001 to September 30, 2011.

the output, we see that there are 2704 data points, the sample mean of the simple return is 0.0278%, and the sample standard error is 0.0155. The sample skewness and excess kurtosis are 0.0279 and 4.631, respectively. Next, consider the hypothesis $H_0 : \mu = 0$ versus $H_a : \mu \neq 0$, where μ denotes the mean of the daily 3M simple return. The test statistic is

$$t = \frac{0.000278}{0.0155/\sqrt{2704}} = 0.933,$$

with p-value 0.35, which is greater than 0.05. Thus, the null hypothesis of zero mean cannot be rejected at the 5% level. For the skewness, the hypothesis is $H_0 : S = 0$ versus $H_a : S \neq 0$. The test statistic is

$$t = \frac{0.0279}{\sqrt{6/2704}} = 0.59,$$

with p-value 0.55. Again, one cannot reject zero skewness at the 5% level. For the excess kurtosis, the hypothesis is $H_0 : K - 3 = 0$ versus $H_a : k - 3 \neq 0$. For the 3M simple returns, the test statistic is

$$t = \frac{4.631}{\sqrt{24/2704}} = 49.15,$$

which is large compared with a standard normal random variable. Thus, the p-value is close to zero and one reject the null hypothesis of $K = 3$. In other words, the daily simple returns of 3M stock have heavy tails. Finally, the Jarque–Bera test statistic is 2422, which is very large compared with a chi-square distribution with 2 degrees of freedom. Therefore, the normality assumption for the daily 3M simple returns is rejected. This is not surprising as the returns have heavy tails. □

R Demonstration Output edited.

```
> library(fBasics) % Load package
> da=read.table("d-mmm-0111.txt",header=T) % Load data


> head(da) % Show the first 6 rows of data
        date        rtn
1 20010102 -0.010892
 . . . .
6 20010109 -0.015727
> mmm=da[,2]   % Obtain 3m simple returns
> basicStats(mmm) %Compute summary statistics
                 mmm
nobs         2704.000000 % Sample size
NAs             0.000000 % No of missing values
```

```
Minimum          -0.089569 % Minimum
Maximum           0.098784 % Maximum
1. Quartile      -0.007161 % 25th percentile
3. Quartile       0.007987 % 75th percentile
Mean              0.000278 % Sample mean
Median            0.000350 % Sample median
Sum               0.751082 % Sample total
SE Mean           0.000298 % Standard error of Sample mean
                            % = sqrt(sample variance/sample size)
LCL Mean         -0.000306 % Lower bound of 95% C.I.
UCL Mean          0.000862 % Upper bound of 95% C.I.
Variance          0.000240 % Sample variance
Stdev             0.015488 % Sample standard error
Skewness          0.027949 % Sample skewness
Kurtosis          4.630925 & % Sample excess kurtosis

% Commands for individual moments
> mean(mmm)
[1] 0.000277767
> var(mmm)
[1] 0.0002398835
> stdev(mmm) % standard deviation
[1] 0.01548817
% Simple tests
> t.test(mmm)  % Testing mean return = 0
        One Sample t-test
data:   mmm
t = 0.9326, df = 2703, p-value = 0.3511
alternative hypothesis: true mean is not equal to 0
95 percent confidence interval:
 -0.0003062688  0.0008618028 % See prior summary statistics.
% p-value > 0.05; one cannot reject the null hypothesis.

> s3=skewness(mmm)
> T=length(mmm) % Sample size
> T
[1] 2704
> t3=s3/sqrt(6/T) % Skewness test
> t3
[1] 0.593333
> pp=2*(1-pnorm(t3)) % Compute p-value
> pp
[1] 0.5529583 % Cannot reject the null of symmetry.
> s4=kurtosis(mmm)
> t4=s4/sqrt(24/T) % Kurtosis test
> t4
[1] 49.15475 % Value is huge; reject the null. Has heavy tails.

> normalTest(mmm,method='jb') % JB-test
```

```
Title:  Jarque - Bera Normalality Test
Test Results:
  STATISTIC:   X-squared: 2422.4384
  P VALUE:  Asymptotic p Value: < 2.2e-16 % Reject normality
```

1.7 VISUALIZATION OF FINANCIAL DATA

Graphs are useful tools in analyzing financial data. Besides the time series plot shown before, we discuss some additional plots to display financial data in this section. To gain a better visualization of the distribution of asset returns, we can exam either the histogram or empirical density function of the data. Consider, for instance, the daily simple returns of 3M stock from January 2, 2001 to September 30, 2011 for 2704 observations. The summary statistics of the data are given before. Figure 1.11 shows a histogram of the data. This is obtained by dividing the data range into 30 bins. The plot confirms that the returns appear to be symmetric with respect to its mean zero. The solid line of Figure 1.12 shows the empirical density function of the 3M returns. This is obtained by a nonparametric smoothing method. The empirical density function can be regarded as a refined version of the histogram. The dashed line of Figure 1.12 shows the density function of a normal distribution that has the same mean and standard deviation as those of the 3M data. The plot provides a visual inspection of the normality assumption for the daily 3M simple returns. The empirical density function has a higher peak and longer tails than the normal density. This phenomenon is common for daily stock returns. In general, the deviation between the solid and dashed line indicates that the daily simple returns of 3M stock are not normally distributed. This, again, is consistent with the result of normality test shown before.

To study the price variability of a stock, we consider the daily open, high, low, and close prices of the stock. Figure 1.13 shows a time plot of these statistics for Apple stock from January 3 to June 30, 2011. This plot is referred to as a *Bar Chart* in the literature. We use a R script `ohlc.R` to obtain the plot. This script is a modified version of that given in Klemelä (2009). In the plot, the vertical bar shows the daily range of the stock price, the horizontal line points to the left gives the opening price, and the horizontal line points to the right denotes the closing price. For this graph to be informative, one cannot show too many days in the plot. Figure 1.14 shows the daily closing price of Apple stock along with a moving-average price of the past 21 trading days from January 2, 2010 to December 8, 2011. This is referred to as a *moving-average chart*. The use of 21 days is arbitrary; it is roughly the number of trading days in a month. The moving-average chart provides information about stock price relative to its recent history. In statistics, averaging is a simple way to reduce the random variability.

Turn to multiple asset returns. Figure 1.15 shows the time plots of monthly log returns of IBM stock and the S&P composite index from January 1926 to September 2011. These returns are obtained from CRSP. Except for the *Great Depression* period, returns of individual stock are in general more volatile than the market index. The time plots exhibit certain simultaneous drops or jumps between IBM stock and the

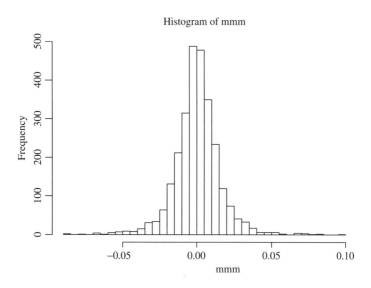

Figure 1.11. Histogram of daily simple returns of 3M stock from January 2, 2001 to September 30, 2011.

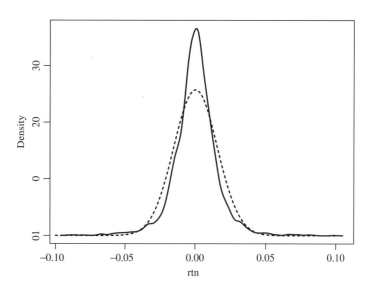

Figure 1.12. Empirical density function of daily simple returns of 3M stock from January 2, 2001 to September 30, 2011. The dashed line denotes the density function of a normal distribution with the same mean and variance.

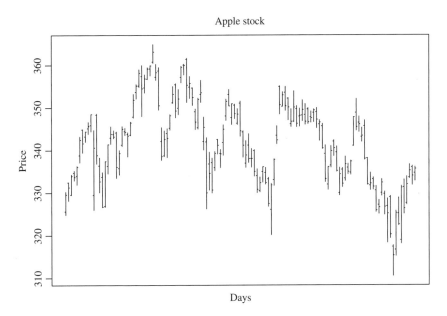

Figure 1.13. Bar chart for daily Apple stock prices from January 3 to June 30, 2011. The vertical bar shows the daily price range, the left horizontal line gives the opening price, and the right horizontal line denotes the closing price.

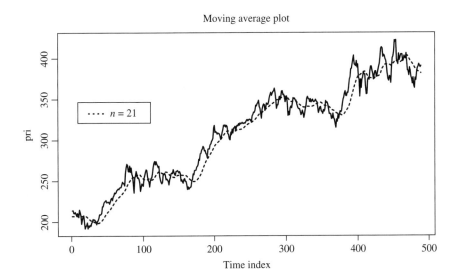

Figure 1.14. Moving-average plot of daily Apple stock from January 2, 2010 to December 8, 2011. The moving average denotes the average price of the most recent 21 trading days (inclusive).

market returns. Figure 1.16 shows the *scatter* plot of the two log returns. The plot also shows a least squares linear regression line between the two returns. From the plot, it is clear that, as expected, IBM and market returns have a positive relationship. This linear relationship can be measured by the correlation between the two returns. In this particular instance, the correlation is 0.64. Alternatively, one can consider the *Market Model*

$$r_t = \alpha + \beta m_t + \epsilon_t,$$

where r_t and m_t are the individual and market return, respectively, and ϵ_t is the error term. The parameters α and β denote the excess return, with respect to the market, and β coefficient, respectively. For monthly log returns of IBM stock, we have $r_t = 0.008 + 0.807m_t + \epsilon_t$. These two parameters are significantly different from zero at the usual 5% level. For more information on Market model, see the capital asset pricing model (CAPM) of Sharpe (1964).

R Demonstration

```
> library(fBasics)
> da=read.table("d-mmm-0111.txt",header=T) % Load data
> mmm=da[,2] % Locate 3M simple returns
> hist(mmm,nclass=30) % Histogram
> d1=density(mmm) % Obtain density estimate
> range(mmm)   % Range of 3M returns
[1] -0.089569  0.098784
> x=seq(-.1,.1,.001) % Create a sequence of x with increment 0.001.
      % The next command creates normal density
> y1=dnorm(x,mean(mmm),stdev(mmm))
> plot(d1$x,d1$y,xlab='rtn',ylab='density',type='l')
> lines(x,y1,lty=2)
 % ohlc plot
> library(quantmod)
> getSymbols("AAPL",from="2011-01-03",to="2011-06-30")
> X=AAPL[,1:4] % Locate open, high, low, and close prices
> xx=cbind(as.numeric(X[,1]),as.numeric(X[,2]),as.numeric(X[,3]),
          as.numeric(X[,4]))
> source("ohlc.R") % Compile the R script
> ohlc(xx,xl="days",yl="price",title="Apple Stock")
 % Moving average plot
> source("ma.R")   % Compile R script
> getSymbols("AAPL",from="2010-01-02",to="2011-12-08")
> x1=as.numeric(AAPL$AAPL.Close) % Locate close price
> ma(x1,21)
 % Bivariate and Scatter plots
> da=read.table("m-ibmsp-2611.txt",header=T)
> head(da)
       data        ibm         sp
1 19260130 -0.010381   0.022472
 . . . . .
6 19260630  0.068493   0.043184
> ibm=log(da$ibm+1) % Transform to log returns
```

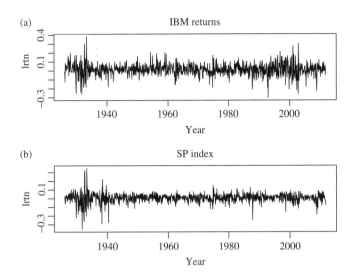

Figure 1.15. Time pots of monthly log returns of IBM stock and the S&P composite index from January 1926 to September 2011. (a) The IBM returns.

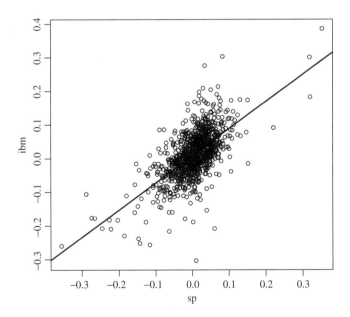

Figure 1.16. Scatter plot of monthly log returns between IBM stock (Y-axis) and S&P composite index from January 1926 to September 2011. The solid line denotes the least squares fit.

```
> sp=log(da$sp+1)
> tdx=c(1:nrow(da))/12+1926 % Create time index
> par(mfcol=c(2,1))
> plot(tdx,ibm,xlab='year',ylab='lrtn',type='l')
> title(main='(a) IBM returns')
> plot(tdx,sp,xlab='year',ylab='lrtn',type='l') % X-axis first.
> title(main='(b) SP index')
> cor(ibm,sp)   % Obtain sample correlation
[1] 0.6409642
> m1=lm(ibm~ sp)   % Fit the Market Model (linear model)
> summary(m1)
Call: lm(formula = ibm ~ sp)
Coefficients:
            Estimate Std. Error t value Pr(>|t|)
(Intercept) 0.007768   0.001672   4.645 3.84e-06 ***
sp          0.806685   0.030144  26.761 < 2e-16  ***
---
Residual standard error: 0.05348 on 1027 degrees of freedom
Multiple R-squared: 0.4108,     Adjusted R-squared: 0.4103
> plot(sp,ibm,cex=0.8)   % Obtain scatter plot
> abline(0.008,.807) % Add the linear regression line
```

1.8 SOME STATISTICAL DISTRIBUTIONS

Several statistical distributions have been proposed in the literature for the marginal distributions of asset returns, including normal distribution, lognormal distribution, stable distribution, and scale mixture of normal distributions. We briefly discuss these distributions.

1.8.1 Normal Distribution

A traditional assumption made in financial study is that the simple returns $\{R_{it}|t = 1, \cdots, T\}$ are independently and identically distributed (iid) as normal with fixed mean and variance. This assumption makes statistical properties of asset returns tractable. But it encounters several difficulties. First, the lower bound of a simple return is -1. Yet the normal distribution may assume any value in the real line and, hence, has no lower bound. Second, if R_{it} is normally distributed, then the multiperiod simple return $R_{it}[k]$ is not normally distributed because it is a product of one-period returns. Third, the normality assumption is not supported by many empirical asset returns, which tend to have a positive excess kurtosis.

1.8.2 Lognormal Distribution

Another commonly used assumption is that the log returns r_t of an asset are iid as normal with mean μ and variance σ^2. The simple returns are then iid lognormal random variables with mean and variance given by

$$E(R_t) = \exp\left(\mu + \frac{\sigma^2}{2}\right) - 1, \quad \text{Var}(R_t) = \exp\left(2\mu + \sigma^2\right)\left[\exp\left(\sigma^2\right) - 1\right]. \quad (1.14)$$

These two equations are useful in studying asset returns (e.g., in forecasting using models built for log returns). Alternatively, let m_1 and m_2 be the mean and variance, respectively, of the simple return R_t, which is distributed as lognormal. Then, the mean and variance of the corresponding log return r_t are

$$E(r_t) = \ln\left(\frac{m_1 + 1}{\sqrt{1 + \dfrac{m_2}{(1 + m_1)^2}}}\right), \quad \operatorname{Var}(r_t) = \ln\left(1 + \frac{m_2}{(1 + m_1)^2}\right).$$

Because the sum of a finite number of iid normal random variables is normal, $r_t[k]$ is also normally distributed under the normal assumption for $\{r_t\}$. In addition, there is no lower bound for r_t, and the lower bound for R_t is satisfied using $1 + R_t = \exp(r_t)$. However, the lognormal assumption is not consistent with all the properties of historical stock returns. In particular, many stock returns exhibit a positive excess kurtosis.

1.8.3 Stable Distribution

The stable distributions are a natural generalization of normal in that they are stable under addition, which meets the need of continuously compounded returns r_t. Furthermore, stable distributions are capable of capturing excess kurtosis shown by historical stock returns. However, nonnormal stable distributions do not have a finite variance, which is in conflict with most finance theories. In addition, statistical modeling using nonnormal stable distributions is difficult. An example of nonnormal stable distributions is the Cauchy distribution, which is symmetric with respect to its median but has infinite variance.

1.8.4 Scale Mixture of Normal Distributions

Recent studies of stock returns tend to use scale mixture or finite mixture of normal distributions. Under the assumption of scale mixture of normal distributions, the log return r_t is normally distributed with mean μ and variance σ^2 [i.e., $r_t \sim N(\mu, \sigma^2)$]. However, σ^2 is a random variable that follows a positive distribution (e.g., σ^{-2} follows a gamma distribution). An example of finite mixture of normal distributions is

$$r_t \sim (1 - X)N(\mu, \sigma_1^2) + XN(\mu, \sigma_2^2),$$

where X is a Bernoulli random variable such that $P(X = 1) = \alpha$ and $P(X = 0) = 1 - \alpha$ with $0 < \alpha < 1$, σ_1^2 is small, and σ_2^2 is relatively large. For instance, with $\alpha = 0.05$, the finite mixture says that 95% of the returns follow $N(\mu, \sigma_1^2)$ and 5% follow $N(\mu, \sigma_2^2)$. The large value of σ_2^2 enables the mixture to put more mass at the tails of its distribution. The low percentage of returns that are from $N(\mu, \sigma_2^2)$ says that the majority of the returns follow a simple normal distribution. Advantages of mixtures of normal include that they maintain the tractability of normal, have finite

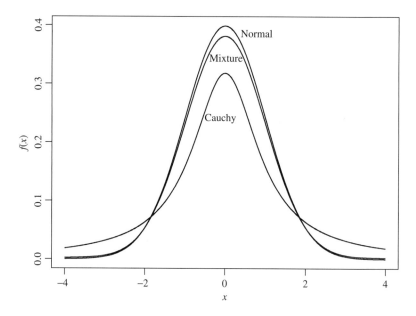

Figure 1.17. Comparison of finite mixture, stable, and standard normal density functions.

higher order moments, and can capture the excess kurtosis. Yet it is hard to estimate the mixture parameters (e.g., the α in the finite mixture case).

Figure 1.17 shows the probability density functions of a finite mixture of normal, Cauchy, and standard normal random variable. The finite mixture of normal is $(1 - X)N(0, 1) + X \times N(0, 16)$ with X being Bernoulli such that $P(X = 1) = 0.05$, and the density function of Cauchy is

$$f(x) = \frac{1}{\pi(1 + x^2)}, \qquad -\infty < x < \infty.$$

It is seen that the Cauchy distribution has fatter tails than the finite mixture of normal, which, in turn, has fatter tails than the standard normal.

1.8.5 Multivariate Returns

Let $\mathbf{r}_t = (r_{1t}, \ldots, r_{Nt})'$ be the log returns of N assets at time t. The multivariate analyses are concerned with the joint distribution of $\{\mathbf{r}_t\}_{t=1}^T$. In the presence of serial dependence, statistical analysis is then focused on the specification of the conditional distribution function $F(\mathbf{r}_t | \mathbf{r}_{t-1}, \ldots, \mathbf{r}_1, \boldsymbol{\theta})$. In particular, how the conditional expectation and conditional covariance matrix of $\mathbf{r}_t$ evolve over time are of special interest in portfolio selection and risk management.

The mean vector and covariance matrix of a random vector $X = (X_1, \ldots, X_p)$ are defined as

$$E(x) = \mu_x = [E(X_1), \ldots, E(X_p)]',$$

$$\mathrm{Cov}(x) = \Sigma_x = E[(x - \mu_x)(x - \mu_x)'],$$

provided that the expectations involved exist. When the data $\{x_1, \ldots, x_T\}$ of X are available, the sample mean and covariance matrix are defined as

$$\widehat{\mu}_x = \frac{1}{T}\sum_{t=1}^{T} x_t, \quad \widehat{\Sigma}_x = \frac{1}{T-1}\sum_{t=1}^{T}(x_t - \widehat{\mu}_x)(x_t - \widehat{\mu}_x)'.$$

These sample statistics are consistent estimates of their theoretical counterparts provided that the covariance matrix of X exists. In the finance literature, multivariate normal distribution is often used for the log return r_t.

To demonstrate, consider again the monthly log returns of IBM stock and S&P 500 composite index from January 1926 to September 2011 shown in Figure 1.16. Let $r_t = (r_{1t}, r_{2t})'$ with r_{1t} and r_{2t} being the monthly log return of IBM stock and S&P index, respectively. Then, we have 1029 observations for r_t. The sample mean and covariance matrix of r_t are

$$\widehat{\mu} = \begin{bmatrix} 0.0113 \\ 0.0044 \end{bmatrix}, \quad \widehat{\Sigma} = \begin{bmatrix} 4849 & 2470 \\ 2470 & 3062 \end{bmatrix} \times 10^{-6}.$$

To check the validity of the bivariate normality assumption, we can use statistical simulation. Specifically, we can generate 1029 data points from a bivariate normal distribution with mean $\widehat{\mu}$ and covariance matrix $\widehat{\Sigma}$. In R, this can be done using the command rmnorm of the package mnormt. Figure 1.18 shows the scatter plot of such a simulation. By comparing this scatter plot with Figure 1.16, we see that significant differences exist between the two plots, indicating that the normality assumption is questionable.

R Demonstration

```
> da=read.table("m-ibmsp-2611.txt",header=T) % Load   data
> dim(da)
[1] 1029    3
> ibm=log(da$ibm+1) % Compute log returns
> sp=log(da$sp+1)
> rt=cbind(ibm,sp) % Obtain bivariate returns
> m1=apply(rt,2,mean) % Obtain sample means
> v1=cov(rt) % Obtain sample covariance matrix
> m1
         ibm              sp
0.011303024 0.004381644
```

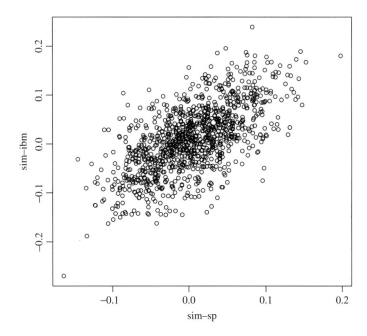

Figure 1.18. Scatterplot of 1029 data points simulated from a bivariate normal distribution based on the sample mean and covariance of monthly log returns of IBM stock and S&P index.

```
> v1
            ibm             sp
ibm 0.004849390 0.002469738
sp  0.002469738 0.003061590
> library(mnormt) % Load package
> x=rmnorm(1029,mean=m1,varcov=v1) % Simulation
> dim(x)
[1] 1029    2
> plot(x[,2],x[,1],xlab='sim-sp',ylab='sim-ibm',cex=0.8)
```

EXERCISES

1. Consider the daily simple returns of American Express (AXP), CRSP value-weighted index (VW), CRSP equal-weighted index (EW), and the S&P composite index (SP) from September 01, 2001 to September 30, 2011. Returns of indices include dividends. The data are in the file d-axp3dx-0111.txt (date, axp, vw, ew, sp).

   (a) Compute the sample mean, standard deviation, skewness, excess kurtosis, minimum, and maximum of each simple return series.

   (b) Transform the simple returns to log returns. Compute the sample mean, standard deviation, skewness, excess kurtosis, minimum, and maximum of each log return series.

   (c) Test the null hypothesis that the mean of the log returns of AXP stock is zero. Use 5% significance level to draw your conclusion.

2. Answer the same questions as Problem 1 but using monthly returns for General Electric (GE), CRSP value-weighted index (VW), CRSP equal-weighted index (EW), and S&P composite index from January 1940 to September 2011. The returns include dividend distributions. Data file is `m-ge3dx-4011.txt` (date, ge, vw, ew, sp).

3. Consider the monthly stock returns of S&P composite index from January 1940 to September 2011 in Problem 2. Perform the following tests and draw conclusions using the 5% significance level.

   (a) Test $H_0 : \mu = 0$ versus $H_a : \mu \neq 0$, where μ denotes the mean return.

   (b) Test $H_0 : m_3 = 0$ versus $H_a : m_3 \neq 0$, where m_3 denotes the skewness.

   (c) Test $H_0 : K = 3$ versus $H_a : K \neq 3$, where K denotes the kurtosis.

4. Consider the daily log returns of American Express stock from September 1, 2001 to September 30, 2011 as in Problem 1. Use the 5% significance level to perform the following tests: (i) Test the null hypothesis that the skewness measure of the returns is zero and (ii) test the null hypothesis that the excess kurtosis of the returns is zero.

5. Daily foreign exchange rates (spot rates) can be obtained from the Federal Reserve Bank in Chicago. The data are the noon buying rates in New York City certified by the Federal Reserve Bank of New York. Consider the exchange rates between the U.S. dollar and the British pound and Japanese yen from January 2, 2007 to November 30, 2011. The data are also available on the web. (i) Compute the daily log return of each exchange rate. (ii) Compute the sample mean, standard deviation, skewness, excess kurtosis, minimum, and maximum of the log returns of each exchange rate. (iii) Obtain a density plot of the daily log returns of Dollar–Yen exchange rate. (iv) Test $H_0 : \mu = 0$ versus $H_a : \mu \neq 0$, where μ denotes the mean of the daily log return of Dollar–Yen exchange rate. Use the 5% significance level to draw the conclusion.

REFERENCES

Adler J. R in a Nutshell. Sebastopol (CA): O'Reilly Media; 2010.

Campbell JY, Lo AW, MacKinlay AC. The Econometrics of Financial Markets. Princeton (NJ): Princeton University Press; 1997.

Crawley MJ. The R Book. Hoboken (NJ): John Wiley & Sons; 2007.

Hull JC. Options, Futures, and Other Derivatives. 8th ed. Upper Saddle River (NJ): Prentice Hall; 2011.

Jarque CM, Bera AK. A test of normality of observations and regression residuals. Int Stat Rev 1987; 55:163172.

Klemelä J. Smoothing of Multivariate Data: Density Estimation and Visualization. Hoboken (NJ): John Wiley & Sons; 2009.

Sharpe W. Capital asset prices: a theory of market equilibrium under conditions of risk. J Finance 1964; 19:425–442.

Snedecor GW, Cochran WG. chapStatistical Methods. 7th ed. Ames (IA): Iowa State University Press; 1980.

2

LINEAR MODELS FOR FINANCIAL TIME SERIES

In this chapter, we discuss the methods and linear models useful in modeling and forecasting financial time series. We use real examples to introduce important statistical concepts, illustrate step-by-step data analysis, and discuss financial applications. For general concepts of linear time series analysis, see Tsay (2010, Chapter 2), Box et al. (1994, Chapters 2 and 3), Brockwell and Davis (2002, Chapters 1–3), Shumway and Stoffer (2000), and Woodward et al. (2012).

The models introduced include (i) simple autoregressive (AR) models, (ii) simple moving average (MA) models, (iii) mixed autoregressive moving average (ARMA) models, (iv) unit-root models including unit-root tests, (v) exponential smoothing, (vi) seasonal models, (vii) regression models with time series errors, and (viii) fractionally differenced models for long-range dependence. For each class of models, we study their fundamental properties, introduce methods for model selection, consider ways to produce prediction, and discuss their applications. The chapter also discusses methods for comparing different models, for example, backtesting and model averaging in prediction.

Let $\{x_t\}$ be a collection of certain financial measurements over time. Figure 2.1 shows the daily closing price of Apple stock from January 3, 2003 to April 5, 2010. The daily prices exhibit certain degrees of variability and show an upward

An Introduction to Analysis of Financial Data with R, First Edition. Ruey S. Tsay.
© 2013 John Wiley & Sons, Inc. Published 2013 by John Wiley & Sons, Inc.

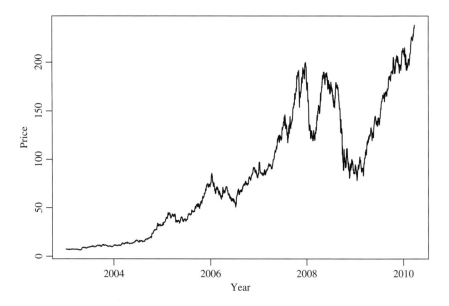

Figure 2.1. Daily closing prices of Apple stock from January 3, 2003 to April 5, 2010.

movement during the sample period. Figure 2.2 shows the quarterly earnings per share of Coca-Cola Company from 1983 to 2009. The quarters are marked in the plot. Besides an upward trend, the earnings also exhibit a clear annual pattern, referred to as *seasonality* in the time series analysis. It will be seen later that many economic and financial time series exhibit a clear seasonal pattern. Figure 2.3 gives the monthly log returns of the S&P 500 index from January 1926 to December 2009. From the plot, it is seen that the returns fluctuate around 0 and, except for a few extreme values, are within a fixed range. Figure 2.4 shows two time series. They are the weekly US 3-month and 6-month treasury bill rates from January 2, 1959 to April 16, 2010. The rates are from the secondary market. The upper plot is the 6-month rate and the lower plot is the 3-month rate. The two series move closely, and also exhibit certain differences. As expected, the 6-month rate was higher in general, but the 3-month rate appeared to be higher in some periods, for example, the early 1980s. This phenomenon is referred to as an *inverted yield curve* in the term structure of interest rates. In these four examples, the series x_t is observed at (roughly) equally spaced time intervals. They are the examples of financial time series that we analyze in this chapter. Our goal is to study the dynamic dependence of the series so that proper inference of the series can be made.

2.1 STATIONARITY

The foundation of statistical inference in time series analysis is the concept of weak stationarity. As shown in Figure 2.3, the monthly log returns of the S&P 500 index vary around 0 over time. In fact, one can divide the time span into several subperiods,

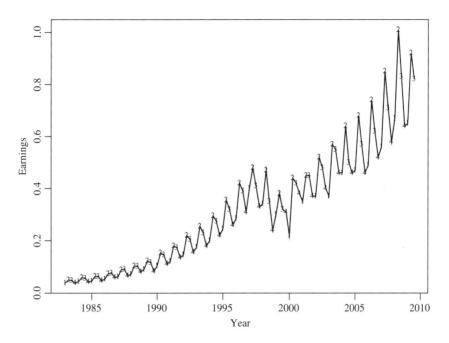

Figure 2.2. Quarterly earnings per share of Coca-Cola Company from the first quarter of 1983 to the third quarter of 2009.

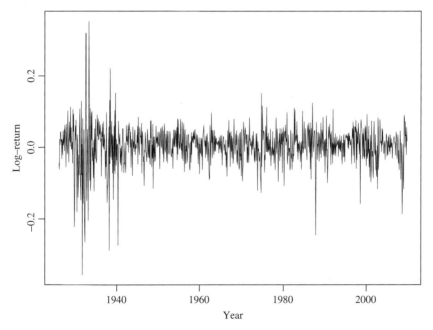

Figure 2.3. Monthly log returns of S&P 500 index from January 1926 to December 2009.

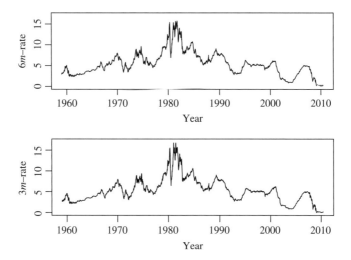

Figure 2.4. Weekly rates, from the secondary market, of the US 3-month and 6-month treasury bills from January 2, 1959 to April 16, 2010.

and the resulting sample means of the subperiods would all be close to 0. In statistics, this phenomenon suggests that the mean of the returns is constant over time or simply the expected return is time invariant. Furthermore, Figure 2.3 also shows that, except for the Great Depression era, the range of the monthly log returns is approximately $[-0.2, 0.2]$ throughout the sample span. In statistics, this characteristic indicates that the variance of the log returns is constant over time. Putting these two time-invariant properties together, we say that the log returns $\{x_t\}$ is weakly stationary. Formally, a time series x_t is weakly stationary if its first two moments (mean and variance) are time invariant. The weak stationarity is important because they provide the basic framework for prediction. For the monthly log returns of the S&P 500 index, we can predict with reasonable confidence that the future monthly returns will be around 0 and vary between -0.2 and 0.2.

On the other hand, consider the quarterly earnings per share of Coca-Cola Company shown in Figure 2.2. If one divides the time span into few subperiods, the resulting sample means differ substantially from one subperiod to the other. Therefore, the earnings are not weakly stationary. This is not surprising because one would expect that the quarterly earnings of a good company increase over time. The time plot in Figure 2.2 also shows that the variability of the earnings increased over time. Therefore, the variance of quarterly earnings is also time varying. Consequently, the series of quarterly earnings is not stationary. We shall discuss models and methods for modeling such nonstationary series later.

In the discussion of weak stationarity, we implicitly assume that the first two moments of x_t are finite. For ease in presentation, we denote the mean and variance of x_t by $E(x_t) = \mu$ and $E(x_t - \mu)^2 = \gamma_0$, respectively. Since these two statistics are time invariant, they are constant and can be represented by simple notation μ and γ_0.

A stationary time series x_t has many other important properties. For a given integer k, define the lag-k autocovariance of x_t as $\gamma_k = \text{Cov}(x_t, x_{t-k})$. One can easily show, using the Cauchy–Schwarz inequality, that γ_k exists and is also time invariant. In other words, for a weakly stationary time series x_t, γ_k depends on k only, and it measures the linear dependence between x_t and x_{t-k}. That is, it measures the dynamic dependence of x_t on its past x_{t-k}. Linear time series analysis focuses on studying the dynamic dependence of the series x_t. The autocovariance γ_k has two important properties: (i) $\gamma_0 = \text{Var}(x_t)$ and (ii) $\gamma_{-k} = \gamma_k$. The second property holds because $\gamma_{-k} = \text{Cov}(x_t, x_{t-(-k)}) = \text{Cov}(x_{t-(-k)}, x_t) = \text{Cov}(x_{t+k}, x_t) = \text{Cov}(x_{t_1}, x_{t_1-k}) = \gamma_k$, where $t_1 = t + k$.

2.2 CORRELATION AND AUTOCORRELATION FUNCTION

Figure 2.5 shows the scatter plot of the monthly simple returns of IBM stock and the S&P 500 index. The two returns appear to be positively related. The degree of this linear dependence is often measured by the Pearson's correlation coefficient or simply correlation coefficient. In statistics, the correlation coefficient between two random variables X and Y is defined as

$$\rho_{x,y} = \frac{\text{Cov}(X, Y)}{\sqrt{\text{Var}(X)\text{Var}(Y)}} = \frac{E[(X - \mu_x)(Y - \mu_y)]}{\sqrt{E(X - \mu_x)^2 E(Y - \mu_y)^2}},$$

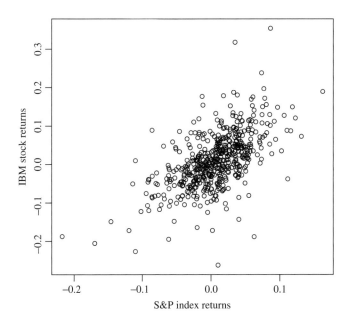

Figure 2.5. Scatter plot of the monthly simple returns: IBM stock versus S&P 500 index. The time span is from January 1967 to December 2009.

where μ_x and μ_y are the mean of X and Y, respectively, and the variances are assumed to be finite. This coefficient measures the strength of linear dependence between X and Y, and it can be shown that $-1 \leq \rho_{x,y} \leq 1$ and $\rho_{x,y} = \rho_{y,x}$. The two random variables are uncorrelated if $\rho_{x,y} = 0$. In addition, if both X and Y are normal random variables, then $\rho_{x,y} = 0$ if and only if X and Y are independent.

When the sample $\{(x_t, y_t)|t = 1, \ldots, T\}$ is available, the correlation can be consistently estimated by its sample counterpart,

$$\hat{\rho}_{x,y} = \frac{\sum_{t=1}^{T}(x_t - \bar{x})(y_t - \bar{y})}{\sqrt{\sum_{t=1}^{T}(x_t - \bar{x})^2 \sum_{t=1}^{T}(y_t - \bar{y})^2}},$$

where $\bar{x} = \sum_{t=1}^{T} x_t/T$ and $\bar{y} = \sum_{t=1}^{T} y_t/T$ are the sample mean of X and Y, respectively.

Remark. In theory, the Pearson correlation coefficient is between -1 and 1. However, for some random variables, the actual range of the coefficient can be shorter. Some alternative dependence measures have been proposed in the literature. The two most popular alternatives are Spearman's rho and Kendall's tau. Spearman's rho is known as *Spearman's rank correlation* because it is the correlation coefficient based on the ranks of the marginal variables. Kendall's tau denotes the difference between concordance and discordance. Suppose that (X_1, Y_1) and (X_2, Y_2) are two independent and identically distributed (iid) continuous bivariate random variables. Kendall's tau is defined as

$$\tau = P[(X_1 - X_2)(Y_1 - Y_2) > 0] - P[(X_1 - X_2)(Y_1 - Y_2) < 0].$$

For illustration, consider the scatter plot of Figure 2.5. The Pearson correlation between the two monthly simple returns is 0.5857. The Spearman's rho of the data is 0.5861 and the Kendall's tau is 0.4196. □

```
> da=read.table("m-ibmsp6709.txt", header=T)
> head(da)
      date         ibm          sp
1 19670131   0.075370   0.078178
 ....
6 19670630   0.067024   0.017512
> ibm=da$ibm
> sp5=da$sp
> cor(sp5,ibm)
[1] 0.5856544
> cor(sp5,ibm,method='spearman')
[1] 0.5860817
> cor(sp5,ibm,method='kendall')
[1] 0.4196587
```

Autocorrelation Function (ACF). Consider a weakly stationary time series x_t. The correlation coefficient between x_t and x_{t-k} is called the *lag-k autocorrelation* of x_t and is commonly denoted by ρ_k. Specifically, we define

$$\rho_k = \frac{\text{Cov}(x_t, x_{t-k})}{\sqrt{\text{Var}(x_t)\text{Var}(x_{t-k})}} = \frac{\text{Cov}(x_t, x_{t-k})}{\text{Var}(x_t)} = \frac{\gamma_k}{\gamma_0}, \tag{2.1}$$

where $\text{Var}(x_{t-k}) = \text{Var}(x_t)$, because x_t is weakly stationary. From the definition, we have $\rho_0 = 1, \rho_\ell = \rho_{-\ell}$, and $-1 \leq \rho_\ell \leq 1$. The collection of autocorrelations, $\{\rho_k\}$, is called the *autocorrelation function (ACF)* of x_t. A weakly stationary time series x_t is not serially correlated if and only if $\rho_k = 0$ for all $k > 0$.

For a given sample $\{x_t | t = 1, \ldots, T\}$, let $\bar{x}$ be the sample mean (i.e., $\bar{x} = \sum_{t=1}^{T} x_t/T$). Then the lag-1 sample autocorrelation of x_t is

$$\hat{\rho}_1 = \frac{\sum_{t=2}^{T}(x_t - \bar{x})(x_{t-1} - \bar{x})}{\sum_{t=1}^{T}(x_t - \bar{x})^2}.$$

Under some general conditions, $\hat{\rho}_1$ is a consistent estimate of ρ_1. For example, if $\{x_t\}$ is a sequence of iid random variables and $E(x_t^2) < \infty$, then $\hat{\rho}_1$ is asymptotically normal with mean 0 and variance $1/T$ (Brockwell and Davis, 2009, Theorem 7.2.2). In general, the lag-k sample autocorrelation of x_t is defined as

$$\hat{\rho}_k = \frac{\sum_{t=k+1}^{T}(x_t - \bar{x})(x_{t-k} - \bar{x})}{\sum_{t=1}^{T}(x_t - \bar{x})^2}, \quad 0 \leq k < T - 1. \tag{2.2}$$

If $\{x_t\}$ is a sequence of iid random variables satisfying $E(x_t^2) < \infty$, then $\hat{\rho}_k$ is asymptotically normal with mean 0 and variance $1/T$ for any fixed positive integer k. More generally, if x_t is a weakly stationary time series satisfying $x_t = \mu + \sum_{i=0}^{q} \psi_i a_{t-i}$, where $\psi_0 = 1$ and $\{a_j\}$ is a sequence of iid random variables with mean 0, then $\hat{\rho}_k$ is asymptotically normal with mean 0 and variance $(1 + 2\sum_{i=1}^{q} \rho_i^2)/T$ for $k > q$. This is referred to as the *Bartlett's formula* in the time series literature (Box et al., 1994). For more information on the asymptotic distribution of sample autocorrelations $\hat{\rho}_k$, see Fuller (1995, Chapter 6) and Brockwell and Davis (2009).

Example 2.1. Consider the monthly simple returns of the Decile 10 portfolio of CRSP from January 1967 to December 2009. There are 516 observations, that is, $T = 516$. The portfolio consists of the smallest 10% of the stocks, in market capitalization, on NYSE/AMEX/NASDAQ and is rebalanced annually. Figure 2.6a shows the time plot of the return series, whereas Figure 2.6b gives the sample ACF of the series. The two horizontal dashed lines of the ACF plot are the two standard error limits, that is, $\pm 2/\sqrt{T}$. The ACF plot starts with $\hat{\rho}_0 = 1$. The plot clearly shows that the lag-1 ACF is significantly different from 0 at the 5% level. The R commands used in the analysis are given as follows:

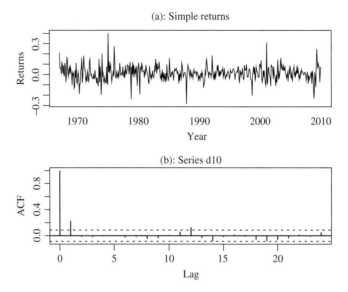

Figure 2.6. Monthly simple returns of CRSP Decile 10 portfolio from January 1967 to December 2009: Panel (a) shows the time plot of the returns and (b) gives the sample ACF of the returns.

```
> da=read.table("m-dec12910.txt",header=T)
> head(da)
       date       dec1       dec2       dec9      dec10
1 19670131   0.068568   0.080373   0.180843 0.211806
2 19670228   0.008735   0.011044   0.048767 0.064911
> d10=da$dec10   # select the Decile 10 returns
> dec10=ts(d10,frequency=12,start=c(1967,1))
> par(mfcol=c(2,1))
> plot(dec10,xlab='year',ylab='returns')
> title(main='(a): Simple returns')
> acf(d10,lag=24) # command to obtain sample ACF of the data
```

□

Testing Individual ACF. For a given positive integer k, the previous result can be used to test $H_0 : \rho_k = 0$ versus $H_a : \rho_k \neq 0$. The test statistic is

$$t\text{-ratio} = \frac{\hat{\rho}_k}{\sqrt{(1 + 2 \sum_{i=1}^{k-1} \hat{\rho}_i^2)/T}}.$$

If $\{x_t\}$ is a stationary Gaussian series satisfying $\rho_j = 0$ for $j > k$, the t-ratio is asymptotically distributed as a standard normal random variable. Hence, the decision rule of the test is to reject H_0 if $|t\text{-ratio}| > Z_{\alpha/2}$, where $Z_{\alpha/2}$ is the $100(1 - \alpha/2)$th percentile of the standard normal distribution. Alternatively, one can use the p-value of the t-ratio to draw a conclusion. If the p-value is less than the type I error, say 0.05, then the null hypothesis is rejected. If the p-value is greater than or equal to the type I error, then one cannot reject H_0.

For simplicity, many software packages use $1/T$ as the asymptotic variance of $\hat{\rho}_k$ for all $k \neq 0$. The t-ratio then becomes $\sqrt{T}\hat{\rho}_k$. This simplification essentially assumes that the underlying time series is a sequence of iid random variables.

Example 2.1 (continued). For various reasons, for example, tax consideration or year-end portfolio adjustment, small stocks in the United States tend to show a positive return in January. This is referred to as the *January effect* of small stocks. A simple approach to verify the existence of the January effect in small stocks is to test the null hypothesis $H_0 : \rho_{12} = 0$ versus the alternative hypothesis $H_a : \rho_{12} \neq 0$, using the monthly simple returns of CRSP Decile 10 portfolio of Example 2.1. From the data, we have $\hat{\rho}_{12} = 0.13$. Using $1/\sqrt{T}$ as the asymptotic standard error of $\hat{\rho}_{12}$, we obtain a t-ratio $t = \sqrt{T}\hat{\rho}_{12} = 2.96$, which is greater than the 5% critical value 1.96. Therefore, we reject the null hypothesis, that is, $\rho_{12} = 0$ at the 5% significance level. In other words, the data confirm the existence of January effect in small stock returns.

```
> f1=acf(d10,lag=24)
> f1$acf
 [1,]  1.000000000   # lag-0
 [2,]  0.227386585   # lag-1
  .    . . . . . . .
[13,]  0.130411045   # lag-12
[14,] -0.036881195   # lag-13
> tt=f1$acf[13]*sqrt(516)
> tt
[1] 2.962369
```

□

In finite samples, $\hat{\rho}_k$ is a biased estimator of ρ_k. The bias is in the order of $1/T$, which can be substantial when the sample size T is small. In most financial applications, T is relatively large so that the bias is not serious.

Portmanteau Test. The statistics $\hat{\rho}_1, \hat{\rho}_2, \ldots$ defined in Equation (2.2) is called the *sample ACF* of x_t. It plays an important role in linear time series analysis. As a matter of fact, a linear time series model can be characterized by its ACF, and linear time series modeling makes use of the sample ACF to specify a model that can capture the dynamic dependence of the data. In many financial applications, we are interested in testing jointly that several autocorrelations of x_t are 0. Box and Pierce (1970) propose the Portmanteau statistic

$$Q_*(m) = T \sum_{\ell=1}^{m} \hat{\rho}_\ell^2$$

as a test statistic for the null hypothesis $H_0 : \rho_1 = \cdots = \rho_m = 0$ against the alternative hypothesis $H_a : \rho_i \neq 0$ for some $i \in \{1, \ldots, m\}$. Under the assumption that

$\{x_t\}$ is a sequence of iid random variables with certain moment conditions, $Q_*(m)$ is asymptotically a chi-squared random variable with m degrees of freedom.

Ljung and Box (1978) modify the $Q_*(m)$ statistic as below to increase the power of the test in finite samples,

$$Q(m) = T(T+2) \sum_{\ell=1}^{m} \frac{\hat{\rho}_\ell^2}{T - \ell}. \tag{2.3}$$

The decision rule is to reject H_0 if $Q(m) > \chi_\alpha^2$, where χ_α^2 denotes the $100(1 - \alpha)$th percentile of a chi-squared distribution with m degrees of freedom. Most software packages will provide the p-value of $Q(m)$. The decision rule then is to reject H_0 if the p-value is less than α, the type I error or significance level.

Example 2.2. Consider the monthly simple and log returns of IBM stock from January 1967 to December 2009. The sample size is 516. Figure 2.7 shows the sample ACFs of the monthly simple and log returns of IBM stock. The two sample ACFs are close to each other, and they are all within the two standard error limits, suggesting that the

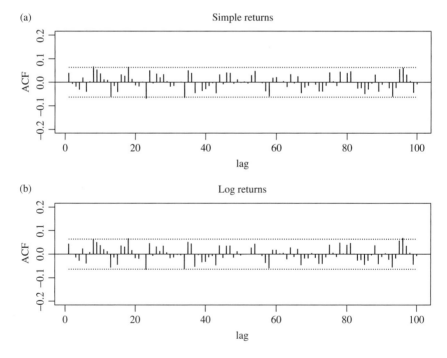

Figure 2.7. Sample autocorrelation functions of the monthly simple and log returns of IBM stock from January 1967 to December 2009. Panel (a) is the ACF of the simple returns and (b) is the ACF of the log returns.

serial correlations of monthly IBM stock returns are very small, if any. To verify that the returns have no serial correlations, we test $H_0 : \rho_1 = \rho_2 = \cdots = \rho_m = 0$ versus $H_a : \rho_i \neq 0$ for some $i \in \{1, \ldots, m\}$ with $m = 12$ and 24. For the simple returns, we have $Q(12) = 7.57$ with p-value 0.82 and $Q(24) = 25.49$ with p-value 0.38. For the log returns, we have $Q(12) = 7.40$ with p-value 0.83 and $Q(24) = 25.39$ with p-value 0.38. Therefore, the Ljung–Box statistics cannot reject the null hypothesis of no serial correlations in the IBM stock returns.

```
> da=read.table("m-ibmsp6709.txt",header=T)
> ibm=da$ibm
> lnibm=log(ibm+1) # Transfer to log returns
> Box.test(ibm,lag=12,type='Ljung')
        Box-Ljung test
data:   ibm
X-squared = 7.5666, df = 12, p-value = 0.818

> Box.test(lnibm,lag=12,type='Ljung')
        Box-Ljung test
data:   lnibm
X-squared = 7.4042, df = 12, p-value = 0.8298
```
□

In practice, the choice of m may affect the performance of the $Q(m)$ statistic. Several values of m are often used. Simulation studies suggest that the choice of $m \approx \ln(T)$ provides better power performance. This general rule needs modification in analyzing seasonal time series for which autocorrelations with lags at the multiples of the seasonality are more important. For instance, lags 12 and 24 are important for monthly time series.

Example 2.1 (continued). For the monthly simple returns of the Decile 10 portfolio of Example 2.1, the Ljung–Box statistics show that $Q(12) = 41.06$ with p-value 4.79×10^{-5} and $Q(24) = 56.25$ with p-value 2.12×10^{-4}. These p-values are small so that the null hypothesis of no serial correlations is rejected at the 5% significance level. Consequently, there exist serial correlations in the monthly simple returns of the Decile 10 portfolio. □

Let x_t be a series of asset returns. In the finance literature, a version of the capital asset pricing model (CAPM) theory is that $\{x_t\}$ is not predictable and should have no autocorrelations. Testing for zero autocorrelations has been used as a tool to check the efficient market hypothesis. However, the way by which stock prices are determined and index returns are calculated might introduce autocorrelations in the observed return series. This is particularly so in analysis of high frequency financial data. We discuss some of these issues, such as bid–ask bounce and nonsynchronous trading, in Chapter 6.

2.3   WHITE NOISE AND LINEAR TIME SERIES

White Noise. A time series x_t is called a *white noise* if $\{x_t\}$ is a sequence of iid random variables with finite mean and variance. In particular, if x_t is normally distributed with mean 0 and variance σ^2, the series is called a *Gaussian white noise*. For a white noise series, all the ACFs are 0. In practice, if all sample ACFs are close to 0, then the series is a white noise series. On the basis of Figures 2.7 and 2.6b, the monthly returns of IBM stock are close to white noise, whereas those of the Decile 10 portfolio are not.

In the following text, we discuss some simple statistical models that are useful in modeling the dynamic structure of a time series. The concepts presented are also useful later in modeling volatility of asset returns.

Linear Time Series. A time series x_t is said to be linear if it can be written as

$$x_t = \mu + \sum_{i=0}^{\infty} \psi_i a_{t-i}, \tag{2.4}$$

where μ is the mean of x_t, $\psi_0 = 1$, and $\{a_t\}$ is a sequence of iid random variables with mean 0 and a well-defined distribution (i.e., $\{a_t\}$ is a white noise series). It will be seen later that a_t denotes the new information at time t of the time series and is often referred to as the *innovation* or *shock* at time t. Thus, a time series is linear if it can be written as a linear combination of past innovations. In this book, we are mainly concerned with the case where the innovation a_t is a continuous random variable. Not all financial time series are linear, but linear models can often provide accurate approximations in real applications.

For a linear time series in Equation (2.4), the dynamic structure of x_t is governed by the coefficients ψ_i, which are called the ψ-*weights* of x_t in the time series literature. If x_t is weakly stationary, we can obtain its mean and variance easily by using properties of $\{a_t\}$ as

$$E(x_t) = \mu, \qquad \mathrm{Var}(x_t) = \sigma_a^2 \sum_{i=0}^{\infty} \psi_i^2, \tag{2.5}$$

where σ_a^2 is the variance of a_t. Because $\mathrm{Var}(x_t) < \infty$, $\{\psi_i^2\}$ must be a convergent sequence, implying that $\psi_i^2 \to 0$ as $i \to \infty$. Consequently, for a stationary series, impact of the remote shock a_{t-i} on the return x_t vanishes as i increases.

The lag-ℓ autocovariance of x_t is

$$\gamma_\ell = \mathrm{Cov}(x_t, x_{t-\ell}) = E\left[\left(\sum_{i=0}^{\infty} \psi_i a_{t-i} \right) \left(\sum_{j=0}^{\infty} \psi_j a_{t-\ell-j} \right) \right]$$

$$= E\left(\sum_{i,j=0}^{\infty} \psi_i \psi_j a_{t-i} a_{t-\ell-j}\right) = \sum_{j=0}^{\infty} \psi_{j+\ell} \psi_j E(a_{t-\ell-j}^2)$$

$$= \sigma_a^2 \sum_{j=0}^{\infty} \psi_j \psi_{j+\ell}. \tag{2.6}$$

Consequently, the ψ-weights are related to the autocorrelations of x_t as follows:

$$\rho_\ell = \frac{\gamma_\ell}{\gamma_0} = \frac{\sum_{i=0}^{\infty} \psi_i \psi_{i+\ell}}{1 + \sum_{i=1}^{\infty} \psi_i^2}, \qquad \ell \geq 0, \tag{2.7}$$

where $\psi_0 = 1$. Linear time series models are econometric and statistical models employed to describe the pattern of the ψ-weights of x_t. For a weakly stationary time series, $\psi_i \to 0$ as $i \to \infty$ and, hence, ρ_ℓ converges to 0 as ℓ increases. For asset returns, this means that, as expected, the linear dependence of the current return x_t on the remote past return $x_{t-\ell}$ diminishes for large ℓ.

2.4 SIMPLE AUTOREGRESSIVE MODELS

When x_t has a statistically significant lag-1 autocorrelation, the lagged value x_{t-1} might be useful in predicting x_t. A simple model that makes use of such predictive power is

$$x_t = \phi_0 + \phi_1 x_{t-1} + a_t, \tag{2.8}$$

where $\{a_t\}$ is assumed to be a white noise series with mean 0 and variance σ_a^2. This model is in the same form as the well-known simple linear regression model, in which x_t is the dependent variable and x_{t-1} is the explanatory variable. In the time series literature, model (Eq. 2.8) is referred to as an *AR model* of order 1 or simply an AR(1) model. This simple model is also widely used in stochastic volatility modeling when x_t is replaced by its log volatility (Chapter 4).

The AR(1) model in Equation (2.8) has several properties similar to those of the simple linear regression model. However, there are some significant differences between the two models, which we discuss later. Here, it suffices to note that an AR(1) model implies that, conditional on the past return x_{t-1}, we have

$$E(x_t|x_{t-1}) = \phi_0 + \phi_1 x_{t-1}, \qquad \text{Var}(x_t|x_{t-1}) = \text{Var}(a_t) = \sigma_a^2.$$

For asset returns, the above results imply that given the past return x_{t-1}, the current return is centered around $\phi_0 + \phi_1 x_{t-1}$ with standard deviation σ_a. This is a Markov property such that conditional on x_{t-1}, the return x_t is not correlated with x_{t-i} for $i > 1$. Obviously, there are situations in which x_{t-1} alone cannot determine the conditional expectation of x_t and a more flexible model must be sought. A straight-forward generalization of the AR(1) model is the AR(p) model

$$x_t = \phi_0 + \phi_1 x_{t-1} + \cdots + \phi_p x_{t-p} + a_t, \qquad (2.9)$$

where p is a nonnegative integer and $\{a_t\}$ is defined in Equation (2.8). This model says that, given the past data, the first p lagged variables $x_{t-i} (i = 1, \ldots, p)$ jointly determine the conditional expectation of x_t. The AR(p) model is in the same form as a multiple linear regression model with lagged values serving as the explanatory variables.

2.4.1   Properties of AR Models

For effective use of AR models, it pays to study their basic properties. We discuss properties of AR(1) and AR(2) models in detail and give the results for the general AR(p) model.

AR(1) Model. We begin with the sufficient and necessary condition for weak stationarity of the AR(1) model in Equation (2.8). Assuming that the series is weakly stationary, we have $E(x_t) = \mu$, $\mathrm{Var}(x_t) = \gamma_0$, and $\mathrm{Cov}(x_t, x_{t-j}) = \gamma_j$, where μ and γ_0 are constants and γ_j is a function of j, not t. We can easily obtain the mean, variance, and autocorrelations of the series as follows. Taking the expectation of Equation (2.8) and using $E(a_t) = 0$, we obtain

$$E(x_t) = \phi_0 + \phi_1 E(x_{t-1}).$$

Under the stationarity condition, $E(x_t) = E(x_{t-1}) = \mu$ and hence

$$\mu = \phi_0 + \phi_1 \mu \quad \text{or} \quad E(x_t) = \mu = \frac{\phi_0}{1 - \phi_1}.$$

This result has two implications for x_t. First, the mean of x_t exists if $\phi_1 \neq 1$. Second, the mean of x_t is 0 if and only if $\phi_0 = 0$. Thus, for a stationary AR(1) process, the constant term ϕ_0 is related to the mean of x_t via $\phi_0 = (1 - \phi_1)\mu$, and $\phi_0 = 0$ implies that $E(x_t) = 0$.

Next, using $\phi_0 = (1 - \phi_1)\mu$, the AR(1) model can be rewritten as

$$x_t - \mu = \phi_1 (x_{t-1} - \mu) + a_t. \qquad (2.10)$$

By repeated substitutions, the prior equation implies that

$$x_t - \mu = a_t + \phi_1 a_{t-1} + \phi_1^2 a_{t-2} + \cdots$$

$$= \sum_{i=0}^{\infty} \phi_1^i a_{t-i}. \qquad (2.11)$$

This equation expresses an AR(1) model in the form of Equation (2.4) with $\psi_i = \phi_1^i$. Thus, $x_t - \mu$ is a linear function of a_{t-i} for $i \geq 0$. Using this property and the independence of the series $\{a_t\}$, we obtain $E[(x_t - \mu)a_{t+1}] = 0$. By the stationarity assumption, we have $\text{Cov}(x_{t-1}, a_t) = E[(x_{t-1} - \mu)a_t] = 0$. This latter result can also be seen from the fact that x_{t-1} occurred before time t and a_t, being a shock at time t, does not depend on any past information. Taking the square and the expectation of Equation (2.10), we obtain

$$\text{Var}(x_t) = \phi_1^2 \text{Var}(x_{t-1}) + \sigma_a^2,$$

where σ_a^2 is the variance of a_t, and we make use of the fact that the covariance between x_{t-1} and a_t is 0. Under the stationarity assumption, $\text{Var}(x_t) = \text{Var}(x_{t-1})$, so that

$$\text{Var}(x_t) = \frac{\sigma_a^2}{1 - \phi_1^2}$$

provided that $\phi_1^2 < 1$. The requirement of $\phi_1^2 < 1$ results from the fact that the variance of a random variable is nonnegative and x_t is weakly stationary. Consequently, the weak stationarity of an AR(1) model implies that $-1 < \phi_1 < 1$, that is, $|\phi_1| < 1$. Yet if $|\phi_1| < 1$, then by Equation (2.11) and the independence of the $\{a_t\}$ series, we can show that the mean and variance of x_t are finite and time invariant; see Equation (2.5). In addition, by Equation (2.6), all the autocovariances of x_t are finite. Therefore, the AR(1) model is weakly stationary. In summary, the necessary and sufficient condition for the AR(1) model in Equation (2.8) to be weakly stationary is $|\phi_1| < 1$.

Using $\phi_0 = (1 - \phi_1)\mu$, one can rewrite a stationary AR(1) model as

$$x_t = (1 - \phi_1)\mu + \phi_1 x_{t-1} + a_t.$$

This model is often used in the finance literature with ϕ_1 measuring the persistence of the dynamic dependence of an AR(1) time series.

Autocorrelation Function of an AR(1) Model. Multiplying Equation (2.10) by a_t, using the independence between a_t and x_{t-1}, and taking expectation, we obtain

$$E[a_t(x_t - \mu)] = \phi_1 E[a_t(x_{t-1} - \mu)] + E(a_t^2) = E(a_t^2) = \sigma_a^2,$$

where σ_a^2 is the variance of a_t. Multiplying Equation (2.10) by $(x_{t-\ell} - \mu)$, taking expectation, and using the prior result, we have

$$\gamma_\ell = \begin{cases} \phi_1 \gamma_1 + \sigma_a^2 & \text{if } \ell = 0 \\ \phi_1 \gamma_{\ell-1} & \text{if } \ell > 0, \end{cases}$$

where we use $\gamma_\ell = \gamma_{-\ell}$. Consequently, for a weakly stationary AR(1) model in Equation (2.8), we have

$$\text{Var}(x_t) = \gamma_0 = \frac{\sigma^2}{1 - \phi_1^2} \quad \text{and} \quad \gamma_\ell = \phi_1 \gamma_{\ell-1}, \quad \text{for} \quad \ell > 0.$$

From the latter equation, the ACF of x_t satisfies

$$\rho_\ell = \phi_1 \rho_{\ell-1}, \quad \text{for} \quad \ell > 0.$$

Because $\rho_0 = 1$, we have $\rho_\ell = \phi_1^\ell$. This result says that the ACF of a weakly station-ary AR(1) series decays exponentially with rate ϕ_1 and starting value $\rho_0 = 1$. For a positive ϕ_1, the plot of ACF of an AR(1) model shows a nice exponential decay. For a negative ϕ_1, the plot consists of two alternating exponential decays with rate ϕ_1^2. Figure 2.8 shows the ACF of two AR(1) models with $\phi_1 = 0.8$ and $\phi_1 = -0.8$.

AR(2) Model. An AR(2) model assumes the form

$$x_t = \phi_0 + \phi_1 x_{t-1} + \phi_2 x_{t-2} + a_t. \tag{2.12}$$

Using the same technique as that of the AR(1) case, we obtain

$$E(x_t) = \mu = \frac{\phi_0}{1 - \phi_1 - \phi_2}$$

provided that $\phi_1 + \phi_2 \neq 1$. Using $\phi_0 = (1 - \phi_1 - \phi_2)\mu$, we can rewrite the AR(2) model as

$$(x_t - \mu) = \phi_1(x_{t-1} - \mu) + \phi_2(x_{t-2} - \mu) + a_t.$$

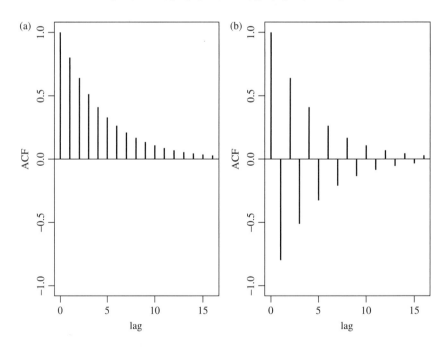

Figure 2.8. The autocorrelation function of an AR(1) model: (a) for $\phi_1 = 0.8$ and (b) for $\phi_1 = -0.8$.

Multiplying the prior equation by $(x_{t-\ell} - \mu)$, we have

$$(x_{t-\ell} - \mu)(x_t - \mu) = \phi_1(x_{t-\ell} - \mu)(x_{t-1} - \mu)$$
$$+ \phi_2(x_{t-\ell} - \mu)(x_{t-2} - \mu) + (x_{t-\ell} - \mu)a_t.$$

Taking expectation and using $E[(x_{t-\ell} - \mu)a_t] = 0$ for $\ell > 0$, we obtain

$$\gamma_\ell = \phi_1\gamma_{\ell-1} + \phi_2\gamma_{\ell-2}, \quad \text{for} \quad \ell > 0.$$

This result is referred to as the *moment equation* of a stationary AR(2) model. Dividing the above equation by γ_0, we have the property

$$\rho_\ell = \phi_1\rho_{\ell-1} + \phi_2\rho_{\ell-2}, \quad \text{for} \quad \ell > 0, \tag{2.13}$$

for the ACF of x_t. In particular, the lag-1 ACF satisfies

$$\rho_1 = \phi_1\rho_0 + \phi_2\rho_{-1} = \phi_1 + \phi_2\rho_1.$$

Therefore, for a stationary AR(2) series x_t, we have $\rho_0 = 1$,

$$\rho_1 = \frac{\phi_1}{1 - \phi_2}$$
$$\rho_\ell = \phi_1\rho_{\ell-1} + \phi_2\rho_{\ell-2}, \quad \ell \geq 2.$$

The result of Equation (2.13) says that the ACF of a stationary AR(2) series satisfies the second-order difference equation

$$(1 - \phi_1 B - \phi_2 B^2)\rho_\ell = 0,$$

where B is called the *backshift* operator such that $B\rho_\ell = \rho_{\ell-1}$. This difference equation determines the properties of the ACF of a stationary AR(2) time series. It also determines the behavior of the forecasts of x_t. In the time series literature, some people use the notation L instead of B for the backshift operator. Here, L stands for *lag* operator. For instance, $Lx_t = x_{t-1}$ and $L\psi_k = \psi_{k-1}$.

Corresponding to the prior difference equation, there is a second-order polynomial equation

$$1 - \phi_1 z - \phi_2 z^2 = 0. \tag{2.14}$$

Solutions of this equation are

$$z = \frac{\phi_1 \pm \sqrt{\phi_1^2 + 4\phi_2}}{-2\phi_2}.$$

In the time series literature, inverses of the two solutions are referred to as the *characteristic roots* of the AR(2) model. Denote the two characteristic roots by ω_1 and ω_2. If both ω_i are real valued, then the second-order difference equation of the model can be factored as $(1 - \omega_1 B)(1 - \omega_2 B)$, and the AR(2) model can be regarded as an AR(1) model operates on top of another AR(1) model. The ACF of x_t is then a mixture of two exponential decays. If $\phi_1^2 + 4\phi_2 < 0$, then ω_1 and ω_2 are complex numbers (called a *complex conjugate pair*), and the plot of ACF of x_t would show a picture of damping sine and cosine waves. In business and economic applications, complex characteristic roots are important. They give rise to the behavior of business cycles. It is then common for economic time series models to have complex-valued characteristic roots. For an AR(2) model in Equation (2.12) with a pair of complex characteristic roots, the *average* length of the stochastic cycles is

$$k = \frac{2\pi}{\cos^{-1}[\phi_1/(2\sqrt{-\phi_2})]},$$

where the cosine inverse is stated in radian. If one writes the complex solutions as $a \pm bi$, where $i = \sqrt{-1}$, then we have $\phi_1 = 2a, \phi_2 = -(a^2 + b^2)$, and

$$k = \frac{2\pi}{\cos^{-1}(a/\sqrt{a^2 + b^2})},$$

where $\sqrt{a^2 + b^2}$ is the absolute value of $a \pm bi$. See Example 2.3 for an illustration.

Figure 2.9 shows the ACF of four stationary AR(2) models. Part (b) is the ACF of the AR(2) model $(1 - 0.6B + 0.4B^2)x_t = a_t$. Because $\phi_1^2 + 4\phi_2 = 0.36 + 4 \times (-0.4) = -1.24 < 0$, this particular AR(2) model contains two complex characteristic roots, and hence its ACF exhibits damping sine and cosine waves. The other three AR(2) models have real-valued characteristic roots. Their ACFs decay exponentially.

Example 2.3. As an illustration, consider the quarterly growth rate of US gross national product (GNP), seasonally adjusted, from the second quarter of 1947 to the first quarter of 2010 for 252 observations. The log series of GNP, in billions of dollars, and its growth rate are shown in Figure 2.10. A horizontal line of zero is added to the time plot of the growth rate. The plot clearly shows that most of the growth rates are positive and the largest drop in GNP occurred in the 2008 recession.

On the basis of the model building procedure of the next section, we employ an AR(3) model for the data. The fitted model is

$$(1 - 0.438B - 0.206B^2 + 0.156B^3)(x_t - 0.016) = a_t, \quad \hat{\sigma}_a = 9.55 \times 10^{-5}. \quad (2.15)$$

The standard errors of the estimates are 0.062, 0.067, 0.063, and 0.001, respectively. See the attached R output for further information. Model (2.15) gives rise to a third-order polynomial equation

$$1 - 0.438z - 0.206z^2 + 0.156z^3 = 0,$$

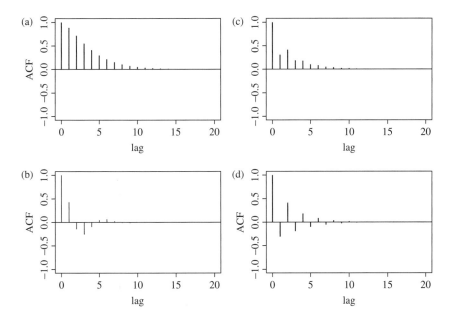

Figure 2.9. The autocorrelation function of an AR(2) model: (a) $\phi_1 = 1.2$ and $\phi_2 = -0.35$, (b) $\phi_1 = 0.6$ and $\phi_2 = -0.4$, (c) $\phi_1 = 0.2$ and $\phi_2 = 0.35$, and (d) $\phi_1 = -0.2$ and $\phi_2 = 0.35$.

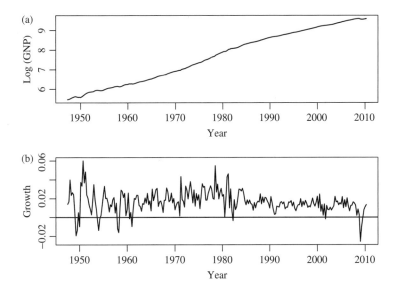

Figure 2.10. Time plots of US quarterly gross national product from 1947.I to 2010.I: (a) Log GNP series and (b) growth rate. The data are seasonally adjusted and in billions of dollars.

which has three solutions, namely, $1.616 + 0.864i, 1.616 - 0.864i$, and -1.909. The real solution corresponds to a factor $[1 - (1/ - 1.909)z] = (1 + 0.524z)$ that shows an exponentially decaying feature of the GNP growth rate. Focusing on the complex conjugate pair $1.616 \pm 0.864i$, we obtain the absolute value $\sqrt{1.616^2 + 0.864^2} = 1.833$ and

$$k = \frac{2\pi}{\cos^{-1}(1.616/1.833)} \approx 12.80.$$

Therefore, the fitted AR(3) model confirms the existence of business cycles in the US economy, and the average length of the cycles is 12.8 quarters, which is about 3 years. This result is reasonable as the US economy went through expansion and contraction and the length of expansion is generally believed to be around 3 years. If one uses a nonlinear model to separate US economy into "expansion" and "contraction" periods, the data show that the average duration of contraction periods is about three quarters and that of expansion periods is about 3 years; see, for instance, the analysis in Tsay (2010, Chapter 4). The average duration of 12.8 quarters is a compromise between the two separate durations. The periodic feature obtained here is common among growth rates of national economies. For example, similar features can be found for many economies in the Organization for Economic Cooperation and Development (OECD) countries.

```
> da=read.table("q-gnp4710.txt",header=T)
> head(da)
  Year Mon Day VALUE
1 1947   1   1 238.1
  . . .
6 1948   4   1 268.7
> G=da$VALUE
> LG=log(G)
> gnp=diff(LG)
> dim(da)
[1] 253    4
> tdx=c(1:253)/4+1947 # create the time index
> par(mfcol=c(2,1))
> plot(tdx,LG,xlab='year',ylab='GNP',type='l')
> plot(tdx[2:253],gnp,type='l',xlab='year',ylab='growth')
> acf(gnp,lag=12)
> pacf(gnp,lag=12) # compute PACF
> m1=arima(gnp,order=c(3,0,0))
> m1
Call:
arima(x = gnp, order = c(3, 0, 0))

Coefficients:
          ar1      ar2      ar3   intercept
       0.4386   0.2063  -0.1559      0.0163
s.e.   0.0620   0.0666   0.0626      0.0012
```

```
sigma^2 estimated as 9.549e-05:log likelihood=808.6,aic=-1607.1
> tsdiag(m1,gof=12)  # model checking discussed later
> p1=c(1,-m1$coef[1:3]) # set-up the polynomial
> r1=polyroot(p1) # solve the polynomial equation
> r1
[1]  1.616116+0.864212i -1.909216-0.000000i  1.616116-0.864212i
> Mod(r1)
[1] 1.832674 1.909216 1.832674  # compute absolute values
> k=2*pi/acos(1.616116/1.832674) # compute length of the period
> k
[1] 12.79523
```

□

Stationarity. The stationarity condition of an AR(2) time series is that the absolute values of its two characteristic roots are less than 1, that is, its two characteristic roots are less than 1 in modulus. Equivalently, the two solutions of the characteristic equation are greater than 1 in modulus. Under such a condition, the recursive equation in Equation (2.13) ensures that the ACF of the model converges to 0 as the lag ℓ increases. This convergence property is a necessary condition for a stationary time series. In fact, the condition also applies to the AR(1) model, where the polynomial equation is $1 - \phi_1 z = 0$. The characteristic root is $w = 1/z = \phi_1$, which must be less than 1 in modulus for x_t to be stationary. As shown before, $\rho_\ell = \phi_1^\ell$ for a stationary AR(1) model. The condition implies that $\rho_\ell \to 0$ as $\ell \to \infty$.

AR(p) Model. The results of AR(1) and AR(2) models can readily be generalized to the general AR(p) model in Equation (2.9). The mean of a stationary series is

$$E(x_t) = \frac{\phi_0}{1 - \phi_1 - \cdots - \phi_p}$$

provided that the denominator is not 0. The associated characteristic equation of the model is

$$1 - \phi_1 z - \phi_2 z^2 - \cdots - \phi_p z^p = 0.$$

If all the solutions of this equation are greater than 1 in modulus, then the series x_t is stationary. Again, inverses of the solutions are the *characteristic roots* of the model. Thus, stationarity requires that all characteristic roots are less than 1 in modulus. For a stationary AR(p) series, the ACF satisfies the difference equation

$$(1 - \phi_1 B - \phi_2 B^2 - \cdots - \phi_p B^p)\rho_\ell = 0, \quad \text{for} \quad \ell > 0.$$

The plot of ACF of a stationary AR(p) model would then show a mixture of damping sine and cosine patterns and exponential decays depending on the nature of its characteristic roots.

2.4.2   Identifying AR Models in Practice

In application, the order p of an AR time series is unknown. It must be specified empirically. This is referred to as the *order determination* (or order specification) of AR models, and it has been extensively studied in the time series literature. Two general approaches are available for determining the value of p. The first approach is to use the partial autocorrelation function (PACF), and the second approach uses some information criteria.

Partial Autocorrelation Function (PACF). The PACF of a stationary time series is a function of its ACF and is a useful tool for determining the order p of an AR model. A simple, yet effective way to introduce PACF is to consider the following AR models in consecutive orders:

$$x_t = \phi_{0,1} + \phi_{1,1}x_{t-1} + e_{1t},$$
$$x_t = \phi_{0,2} + \phi_{1,2}x_{t-1} + \phi_{2,2}x_{t-2} + e_{2t},$$
$$x_t = \phi_{0,3} + \phi_{1,3}x_{t-1} + \phi_{2,3}x_{t-2} + \phi_{3,3}x_{t-3} + e_{3t},$$
$$x_t = \phi_{0,4} + \phi_{1,4}x_{t-1} + \phi_{2,4}x_{t-2} + \phi_{3,4}x_{t-3} + \phi_{4,4}x_{t-4} + e_{4t},$$
$$\vdots \quad \vdots$$

where $\phi_{0,j}, \phi_{i,j}$, and $\{e_{jt}\}$ are, respectively, the constant term, the coefficient of x_{t-i}, and the error term of an AR(j) model. These models are in the form of a multiple linear regression and can be estimated by the least squares (LS) method. As a matter of fact, they are arranged in a sequential order that enables us to apply the idea of partial F test in multiple linear regression analysis. The estimate $\hat{\phi}_{1,1}$ of the first equation is called the *lag-1 sample PACF* of x_t. The estimate $\hat{\phi}_{2,2}$ of the second equation is the lag-2 sample PACF of x_t. The estimate $\hat{\phi}_{3,3}$ of the third equation is the lag-3 sample PACF of x_t, and so on.

From the definition, the lag-2 PACF $\hat{\phi}_{2,2}$ shows the added contribution of x_{t-2} to x_t over the AR(1) model $x_t = \phi_0 + \phi_1 x_{t-1} + e_{1t}$. The lag-3 PACF shows the added contribution of x_{t-3} to x_t over an AR(2) model, and so on. Therefore, for an AR(p) model, the lag-p sample PACF should not be 0, but $\hat{\phi}_{j,j}$ should be close to 0 for all $j > p$. We make use of this property to determine the order p. For a stationary Gaussian AR(p) model, it can be shown that the sample PACF has the following properties:

- $\hat{\phi}_{p,p}$ converges to ϕ_p as the sample size T goes to infinity.
- $\hat{\phi}_{\ell,\ell}$ converges to 0 for all $\ell > p$.
- The asymptotic variance of $\hat{\phi}_{\ell,\ell}$ is $1/T$ for $\ell > p$.

These results say that, for an AR(p) series, the sample PACF cuts off at lag p.

TABLE 2.1. Sample Partial Autocorrelation Function and Some Information Criteria for the Monthly Simple Returns of CRSP Value-Weighted Index From January 1926 to December 2008

p	1	2	3	4	5	6
PACF	0.115	−0.030	−0.102	0.033	0.062	−0.050
AIC	−5.838	−5.837	−5.846	−5.845	−5.847	−5.847
BIC	−5.833	−5.827	−5.831	−5.825	−5.822	−5.818
p	7	8	9	10	11	12
PACF	0.031	0.052	0.063	0.005	−0.005	0.011
AIC	−5.846	−5.847	−5.849	−5.847	−5.845	−5.843
BIC	−5.812	−5.807	−5.805	−5.798	−5.791	−5.784

As an example, consider the monthly simple returns of CRSP value-weighted index from January 1926 to December 2008. Table 2.1 gives the first 12 lags of sample PACF of the series. With $T = 996$, the asymptotic standard error of the sample PACF is approximately 0.032. Therefore, using the 5% significant level, we identify an AR(3) or AR(9) model for the data (i.e., $p = 3$ or 9). If the 1% significant level is used, we specify an AR(3) model.

As another example, Figure 2.11 shows the ACF and PACF of the GNP growth rate series of Example 2.3. The two dotted lines of the plots denote the approximate two standard error limits $\pm 2/\sqrt{252}$. The PACF plot suggests an AR(3) model for the data because the first three lags of sample PACF appear to be significant at the 5% level. There is a marginally significant PACF at lag 9. We do not consider high order models here for simplicity.

Information Criteria. There are several information criteria available to determine the order p of an AR process. All of them are likelihood based. For example, the well-known *Akaike Information Criterion (AIC)* (Akaike, 1973) is defined as

$$\text{AIC} = \frac{-2}{T} \ln(\text{likelihood}) + \frac{2}{T} \times (\text{number of parameters}), \qquad (2.16)$$

where the likelihood function is evaluated at the maximum likelihood estimates and T is the sample size. For a Gaussian AR(ℓ) model, AIC reduces to

$$\text{AIC}(\ell) = \ln(\tilde{\sigma}_\ell^2) + \frac{2\ell}{T},$$

where $\tilde{\sigma}_\ell^2$ is the maximum likelihood estimate of σ_a^2, which is the variance of a_t, and T is the sample size. The first term of the AIC in Equation (2.16) measures the goodness of fit of the AR(ℓ) model to the data, whereas the second term is called the *penalty function* of the criterion because it penalizes a candidate model by the number of parameters used. Different penalty functions result in different information criteria.

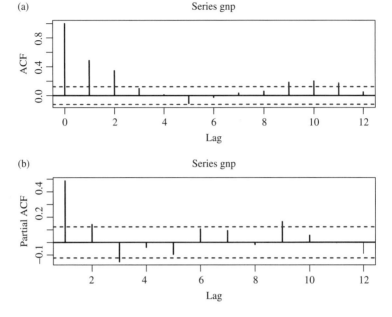

Figure 2.11. Sample autocorrelation and partial autocorrelation functions of the US quarterly real GNP growth rate from 1947.II to 2010.I. The dotted lines give approximate pointwise 95% confidence interval.

Another commonly used criterion function is the Schwarz–Bayesian criterion (BIC, Bayesian information criterion). For a Gaussian AR(ℓ) model, the criterion is

$$\mathrm{BIC}(\ell) = \ln(\tilde{\sigma}_\ell^2) + \frac{\ell \ln(T)}{T}.$$

The penalty for each parameter used is 2 for AIC and $\ln(T)$ for BIC. Thus, compared with AIC, BIC tends to select a lower AR model when the sample size is moderate or large.

Selection Rule. To use AIC to select an AR model in practice, one computes AIC(ℓ) for $\ell = 0, \ldots, P$, where P is a prespecified positive integer and selects the order k that has the minimum AIC value. The same rule applies to BIC.

Table 2.1 also gives the AIC and BIC for $p = 1, \ldots, 12$. The AIC values are close to each other with minimum -5.849 occurring at $p = 9$, suggesting that an AR(9) model is preferred by the criterion. The BIC, on the other hand, attains its minimum value -5.833 at $p = 1$ with -5.831 as a close second at $p = 3$. Thus, the BIC selects an AR(1) model for the value-weighted return series. This example shows that different approaches or criteria to order determination may result in different choices of p. There is no evidence to suggest that one approach outperforms the other in a real application. Substantive information of the problem under study and simplicity are two factors that also play an important role in choosing an AR model for a given time series.

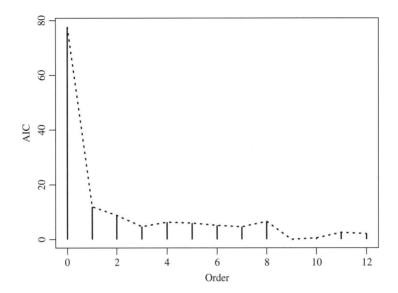

Figure 2.12. Akaike information criterion for the quarterly growth rates of US real GNP from 1947.II to 2010.I. The maximum order is $P = 12$.

Again, consider the growth rate series of US quarterly real GNP of Example 2.3. The AIC values with $P = 12$ are shown in Figure 2.12. The criterion identifies an AR(9) model for the series, but it also gives some justification for an AR(3) model. The plot indicates that AIC would specify an AR(3) model if one focuses on lower-order models. Note that the AIC value of the `ar` command in R has been adjusted so that the minimum AIC is 0.

```
> mm1=ar(gnp,method='mle')
> mm1$order % Find the identified order
[1] 9
> names(mm1)
 [1] "order"    "ar"         "var.pred"   "x.mean"    "aic"
 [6] "n.used"   "order.max"  "partialacf" "resid"     "method"
[11] "series"   "frequency"  "call"       "asy.var.coef"
> print(mm1$aic,digits=3)
    0      1      2       3       4       5       6       7       8       9
77.8  11.92   8.79   4.669   6.265   5.950   5.101   4.596   6.541   0.000
     10      11      12
  0.509   2.504   2.057

> aic=mm1$aic   % For plotting below.
> length(aic)
[1] 13
> plot(c(0:12),aic,type='h',xlab='order',ylab='aic')
> lines(0:12,aic,lty=2)
```

Parameter Estimation. For a specified AR(p) model in Equation (2.9), the conditional LS method, which starts with the $(p + 1)$th observation, is often used to estimate the parameters. Specifically, conditioning on the first p observations, we have

$$x_t = \phi_0 + \phi_1 x_{t-1} + \cdots + \phi_p x_{t-p} + a_t, \quad t = p + 1, \ldots, T,$$

which is in the form of a multiple linear regression and can be estimated by the LS method. Denote the estimate of ϕ_i by $\hat{\phi}_i$. The *fitted model* is

$$\hat{x}_t = \hat{\phi}_0 + \hat{\phi}_1 x_{t-1} + \cdots + \hat{\phi}_p x_{t-p}$$

and the associated residual is

$$\hat{a}_t = x_t - \hat{x}_t.$$

The series $\{\hat{a}_t\}$ is called the *residual series*, from which we obtain

$$\hat{\sigma}_a^2 = \frac{\sum_{t=p+1}^{T} \hat{a}_t^2}{T - 2p - 1}.$$

If the conditional Gaussian likelihood method is used, the estimates of ϕ_i remain unchanged, but the estimate of σ_a^2 becomes $\tilde{\sigma}_a^2 = \hat{\sigma}_a^2 \times (T - 2p - 1)/(T - p)$. In some packages, $\tilde{\sigma}_a^2$ is defined as $\hat{\sigma}_a^2 \times (T - 2p - 1)/T$. For illustration, consider an AR(3) model for the monthly simple returns of the value-weighted index in Table 2.1. The fitted model is

$$x_t = 0.0091 + 0.116 x_{t-1} - 0.019 x_{t-2} - 0.104 x_{t-3} + \hat{a}_t, \quad \hat{\sigma}_a = 0.054.$$

The standard errors of the coefficients are 0.002, 0.032, 0.032, and 0.032, respectively. Except for the lag-2 coefficient, all parameters are statistically significant at the 1% level.

For this example, the AR coefficients of the fitted model are small, indicating that the serial dependence of the series is weak, even though it is statistically significant at the 1% level. The significance of $\hat{\phi}_0$ of the entertained model implies that the expected mean return of the series is positive. In fact, $\hat{\mu} = 0.0091/(1 - 0.116 + 0.019 + 0.104) = 0.009$, which is small, but has an important long-term implication. It implies that the long-term return of the index can be substantial. Using the multi-period simple return defined in Chapter 1, the average annual simple gross return is $[\prod_{t=1}^{996}(1 + x_t)]^{12/996} - 1 \approx 0.093$. In other words, the monthly simple returns of the CRSP value-weighted index grew about 9.3% per annum from 1926 to 2008, supporting the common belief that equity market performs well in the long term. A one-dollar investment at the beginning of 1926 would be worth about \$1593 at the end of 2008.

```
> vw=read.table('m-ibm3dx.txt',header=T)[,3]
> t1=prod(vw+1)
```

```
> t1
[1] 1592.953
> t1^(12/996)-1
[1] 0.0929
```

Model Checking. A fitted model must be examined carefully to check for possible model inadequacy. If the model is adequate, then the residual series should behave as a white noise. The ACF and the Ljung–Box statistics in Equation (2.3) of the residuals can be used to check the closeness of $\hat{a}_t$ to a white noise. For an AR(p) model, the Ljung–Box statistic $Q(m)$ follows asymptotically a chi-squared distribution with $m - g$ degrees of freedom, where g denotes the number of AR coefficients used in the model. The adjustment in the degrees of freedom is made based on the number of constraints added to the residuals $\hat{a}_t$ from fitting the AR(p) to an AR(0) model. If a fitted model is found to be inadequate, it must be refined. For instance, if some of the estimated AR coefficients are not significantly different from 0, then the model should be simplified by removing those insignificant parameters. If residual ACF shows additional serial correlations, then the model should be extended to take care of the those correlations.

Remark. Most time series packages do not adjust the degrees of freedom when applying the Ljung–Box statistics $Q(m)$ to a residual series. This does not follow the theory but is understandable when $m \leq g$. In R, the command `tsdiag` can be used to perform diagnostic checking after fitting a model via the command `arima`. More details are discussed later.   □

Consider the residual series of the fitted AR(3) model for the monthly value-weighted simple returns. We have $Q(12) = 16.35$ with p-value 0.060 based on its asymptotic chi-squared distribution with 9 degrees of freedom. Thus, the null hypothesis of no residual serial correlation in the first 12 lags is barely not rejected at the 5% level. However, since the lag-2 AR coefficient is not significant at the 5% level, one can refine the model as

$$x_t = 0.0088 + 0.114x_{t-1} - 0.106x_{t-3} + a_t, \quad \hat{\sigma}_a = 0.0536,$$

where all the estimates are now significant at the 1% level. The residual series gives $Q(12) = 16.83$ with p-value 0.078 (based on χ^2_{10}). The model is adequate in modeling the dynamic linear dependence of the data.

```
> vw=read.table('m-ibm3dx2608.txt',header=T)[,3]
> m3=arima(vw,order=c(3,0,0))
> m3
Call:
arima(x = vw, order = c(3, 0, 0))

Coefficients:
         ar1      ar2      ar3  intercept
```

```
          0.1158   -0.0187   -0.1042     0.0089
s.e.      0.0315    0.0317    0.0317     0.0017

sigma^2 estimated as 0.002875:log likelihood=1500.86, aic=-2991.7

> (1-.1158+.0187+.1042)*mean(vw)  % Compute the intercept phi(0).
[1] 0.00896761
> sqrt(m3$sigma2) % Compute standard error of residuals
[1] 0.0536189

> Box.test(m3$residuals,lag=12,type='Ljung')
        Box-Ljung test

data:  m3$residuals       % R uses 12 degrees of freedom
X-squared = 16.3525, df = 12, p-value = 0.1756

> pv=1-pchisq(16.35,9) % Compute p value using 9 degrees of freedom
> pv
[1] 0.05992276
 % To fix the AR(2) coefficient to zero:
> m3=arima(vw,order=c(3,0,0),fixed=c(NA,0,NA,NA))
 % The subcommand "fixed" enables users to fix parameter values,
 % where NA means estimation and 0 means fixing the parameter to 0.
 % The ordering of the parameters can be found using m3$coef.
> m3
Call:
arima(x = vw, order = c(3, 0, 0), fixed = c(NA, 0, NA, NA))

Coefficients:
          ar1   ar2      ar3   intercept
       0.1136     0  -0.1063      0.0089
s.e.   0.0313     0   0.0315      0.0017

sigma^2 estimated as 0.002876: log likelihood=1500.69, aic=-2993.38
> (1-.1136+.1063)*.0089  % compute phi(0)
[1] 0.00883503
> sqrt(m3$sigma2)  % compute residual standard error
[1] 0.05362832

> Box.test(m3$residuals,lag=12,type='Ljung')
        Box-Ljung test

data:  m3$residuals
X-squared = 16.8276, df = 12, p-value = 0.1562

> pv=1-pchisq(16.83,10)
> pv
[1] 0.0782113
```

2.4.3 Goodness of Fit

A commonly used statistic to measure *goodness of fit* of a stationary model is the R-square (R^2) defined as

$$R^2 = 1 - \frac{\text{Residual sum of squares}}{\text{Total sum of squares}}.$$

For a stationary AR(p) time series model with T observations $\{x_t | t = 1, \ldots, T\}$, the measure becomes

$$R^2 = 1 - \frac{\sum_{t=p+1}^{T} \hat{a}_t^2}{\sum_{t=p+1}^{T} (x_t - \bar{x})^2},$$

where $\bar{x} = \sum_{t=p+1}^{T} x_t / (T - p)$. It is easy to show that $0 \le R^2 \le 1$. Typically, a larger R^2 indicates that the model provides a closer fit to the data. However, this is only true for a stationary time series. For the unit-root nonstationary series discussed later in this chapter, R^2 of an AR(1) fit converges to 1 when the sample size increases to infinity, regardless of the true underlying model of x_t.

For a given data set, it is well known that R^2 is a nondecreasing function of the number of parameters used. To overcome this weakness, an *adjusted* R^2 is proposed, which is defined as

$$\text{Adj}(R^2) = 1 - \frac{\text{Variance of residuals}}{\text{Variance of } x_t}$$

$$= 1 - \frac{\hat{\sigma}_a^2}{\hat{\sigma}_x^2},$$

where $\hat{\sigma}_x^2$ is the sample variance of x_t. This new measure takes into account the number of parameters used in the fitted model. However, it is no longer between 0 and 1.

2.4.4 Forecasting

Forecasting is an important application of time series analysis. For the AR(p) model in Equation (2.9), suppose that we are at the time index h and are interested in forecasting $x_{h+\ell}$, where $\ell \ge 1$. The time index h is called the *forecast origin* and the positive integer ℓ is the *forecast horizon*. Let $\hat{x}_h(\ell)$ be the forecast of $x_{h+\ell}$ using the minimum squared error loss function. In other words, the forecast $\hat{x}_k(\ell)$ is chosen such that

$$E\{[x_{h+\ell} - \hat{x}_h(\ell)]^2 | F_h\} \le \min_g E[(x_{h+\ell} - g)^2 | F_h],$$

where g is a function of the information available at time h (inclusive), that is, a function of F_h. We referred to $\hat{x}_h(\ell)$ as the ℓ-step ahead forecast of x_t at the forecast origin h. In the prior equation, F_h denotes the collection of information available at the forecast origin h.

1-Step Ahead Forecast. From the AR(p) model, we have

$$x_{h+1} = \phi_0 + \phi_1 x_h + \cdots + \phi_p x_{h+1-p} + a_{h+1}.$$

Under the minimum squared error loss function, the point forecast of x_{h+1} given F_h is the conditional expectation

$$\hat{x}_h(1) = E(x_{h+1}|F_h) = \phi_0 + \sum_{i=1}^{p} \phi_i x_{h+1-i}$$

and the associated forecast error is

$$e_h(1) = x_{h+1} - \hat{x}_h(1) = a_{h+1}.$$

Consequently, the variance of the 1-step ahead forecast error is $\mathrm{Var}[e_h(1)] = \mathrm{Var}(a_{h+1})$ $= \sigma_a^2$. If a_t is normally distributed, then a 95% 1-step ahead interval forecast of x_{h+1} is $\hat{x}_h(1) \pm 1.96 \times \sigma_a$. For the linear model in Equation (2.4), a_{t+1} is also the 1-step ahead forecast error at the forecast origin t. In the econometric literature, a_{t+1} is referred to as the *shock* to the series at time $t+1$.

In practice, estimated parameters are often used to compute point and interval forecasts. This results in a *conditional forecast* because such a forecast does not take into consideration the uncertainty in the parameter estimates. In theory, one can consider parameter uncertainty in forecasting, but it is much more involved. A natural way to consider parameter and model uncertainty in forecasting is Bayesian forecasting with Markov chain Monte Carlo (MCMC) methods. See Chapter 12 of Tsay (2010) for further discussion. For simplicity, we assume that the model is given in this chapter. When the sample size used in estimation is sufficiently large, then the conditional forecast is close to the unconditional one.

2-Step Ahead Forecast. Next, consider the forecast of x_{h+2} at the forecast origin h. From the AR(p) model, we have

$$x_{h+2} = \phi_0 + \phi_1 x_{h+1} + \cdots + \phi_p x_{h+2-p} + a_{h+2}.$$

Taking conditional expectation, we have

$$\hat{x}_h(2) = E(x_{h+2}|F_h) = \phi_0 + \phi_1 \hat{x}_h(1) + \phi_2 x_h + \cdots + \phi_p x_{h+2-p}$$

and the associated forecast error

$$e_h(2) = x_{h+2} - \hat{x}_h(2) = \phi_1[x_{h+1} - \hat{x}_h(1)] + a_{h+2} = a_{h+2} + \phi_1 a_{h+1}.$$

The variance of the forecast error is $\mathrm{Var}[e_h(2)] = (1 + \phi_1^2)\sigma_a^2$. Interval forecasts of x_{h+2} can be computed in the same way as those for x_{h+1}. It is interesting to see that $\mathrm{Var}[e_h(2)] \geq \mathrm{Var}[e_h(1)]$, meaning that as the forecast horizon increases the uncertainty in forecast also increases. This is in agreement with common sense that we are more uncertain about x_{h+2} than x_{h+1} at the time index h for a linear time series.

Multistep Ahead Forecast. In general, we have

$$x_{h+\ell} = \phi_0 + \phi_1 x_{h+\ell-1} + \cdots + \phi_p x_{h+\ell-p} + a_{h+\ell}.$$

The ℓ-step ahead forecast based on the minimum squared error loss function is the conditional expectation of $x_{h+\ell}$ given F_h, which can be obtained as

$$\hat{x}_h(\ell) = \phi_0 + \sum_{i=1}^{p} \phi_i \hat{x}_h(\ell - i),$$

where it is understood that $\hat{x}_h(i) = x_{h+i}$ if $i \leq 0$. This forecast can be computed recursively using forecasts $\hat{x}_h(i)$ for $i = 1, \ldots, \ell - 1$. The ℓ-step ahead forecast error is $e_h(\ell) = x_{h+\ell} - \hat{x}_h(\ell)$. It can be shown that for a stationary AR(p) model, $\hat{x}_h(\ell)$ converges to $E(x_t)$ as $\ell \to \infty$, meaning that for such a series long-term point forecast approaches its unconditional mean. This property is referred to as the *mean reversion* in the finance literature. For an AR(1) model, the speed of mean reversion is measured by the *half-life* defined as $\ell = \ln(0.5)/\ln(|\phi_1|)$. The variance of the forecast error then approaches the unconditional variance of x_t. Note that for an AR(1) model in Equation (2.8), let $\tilde{x}_t = x_t - E(x_t)$ be the mean-adjusted series. It is easy to see that the ℓ-step ahead forecast of $\tilde{x}_{h+\ell}$ at the forecast origin h is $\hat{\tilde{x}}_h(\ell) = \phi_1^\ell \tilde{x}_h$. The half-life is the forecast horizon such that $\hat{\tilde{x}}_h(\ell) = \frac{1}{2}\tilde{x}_h$. That is, $\phi_1^\ell = \frac{1}{2}$. Thus, $\ell = \ln(0.5)/\ln(|\phi_1|)$.

Table 2.2 contains the 1-step to 12-step ahead forecasts and the standard errors of the associated forecast errors at the forecast origin 984 for the monthly simple return of the value-weighted index using an AR(3) model that was reestimated using the first 984 observations. The fitted model is

$$x_t = 0.0098 + 0.1024 x_{t-1} - 0.0201 x_{t-2} - 0.1090 x_{t-3} + a_t,$$

where $\hat{\sigma}_a = 0.054$. The actual returns of 2008 are also given in the table. Because of the weak serial dependence in the series, the forecasts and standard deviations of forecast errors converge to the sample mean and standard deviation of the data quickly. For the first 984 observations, the sample mean and standard error are 0.0095 and 0.0540, respectively.

Figure 2.13 shows the corresponding out-of-sample prediction plot for the monthly simple return series of the value-weighted index. The forecast origin $t = 984$ corresponds to December 2007. The prediction plot includes the two standard error limits of the forecasts and the actual observed returns for 2008. The forecasts and actual returns are marked by "o" and "•," respectively. From the plot, except for the return of October 2008, all actual returns are within the 95% prediction intervals.

2.5   SIMPLE MOVING AVERAGE MODELS

We now turn to the class of MA models. These models are useful in modeling asset returns in finance. As is shown in Chapter 6, the bid−ask bounce in stock trading may

TABLE 2.2. Multistep Ahead Forecasts of an AR(3) Model For The Monthly Simple Returns of CRSP Value-Weighted Index[a]

Step	1	2	3	4	5	6
Forecast	0.0076	0.0161	0.0118	0.0099	0.0089	0.0093
Standard Error	0.0534	0.0537	0.0537	0.0540	0.0540	0.0540
Actual	−0.0623	−0.0220	−0.0105	0.0511	0.0238	−0.0786
Step	7	8	9	10	11	12
Forecast	0.0095	0.0097	0.0096	0.0096	0.0096	0.0096
Standard Error	0.0540	0.0540	0.0540	0.0540	0.0540	0.0540
Actual	−0.0132	0.0110	−0.0981	−0.1847	−0.0852	0.0215

[a]The forecast origin is 984.

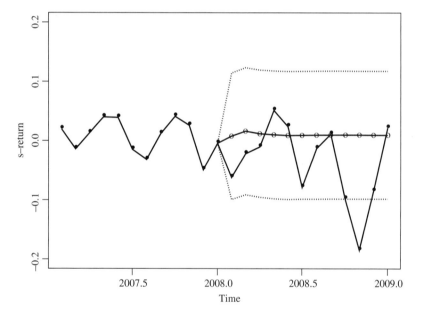

Figure 2.13. Plot of 1-step to 12-step ahead out-of-sample forecasts for the monthly simple return of the CRSP value-weighted index. The forecast origin is $t = 984$, which is December 2007. The forecasts are denoted by "o" and the actual observations by "•." The two dashed lines denote two standard error limits of the forecasts.

introduce the structure of an MA(1) model in a return series. There are several ways to introduce MA models. One approach is to treat the model as a simple extension of white noise series. Another approach is to treat the model as an infinite-order AR model, with some parameter constraints. We adopt the second approach.

There is no particular reason, but simplicity, to assume a priori that the order of an AR model is finite. We may entertain, at least in theory, an AR model with infinite order as

$$x_t = \phi_0 + \phi_1 x_{t-1} + \phi_2 x_{t-2} + \cdots + a_t.$$

However, such an AR model is not realistic because it has infinite parameters. One way to make the model practical is to assume that the coefficients ϕ_is satisfy some constraints so that they are determined by a finite number of parameters. A special case of this idea is

$$x_t = \phi_0 - \theta_1 x_{t-1} - \theta_1^2 x_{t-2} - \theta_1^3 x_{t-3} - \cdots + a_t, \tag{2.17}$$

where the coefficients depend on a single parameter θ_1 via $\phi_i = -\theta_1^i$ for $i \geq 1$. For the model in Equation (2.17) to be stationary, θ_1 must be less than 1 in absolute value; otherwise, θ_1^i and the series will explode. Because $|\theta_1| < 1$, we have $\theta_1^i \to 0$ as $i \to \infty$. Thus, the contribution of x_{t-i} to x_t decays exponentially as i increases. This is reasonable as the dependence of a stationary series x_t on its lagged value x_{t-i}, if any, should decay over time.

The model in Equation (2.17) can be rewritten in a rather compact form. To see this, rewrite the model as

$$x_t + \theta_1 x_{t-1} + \theta_1^2 x_{t-2} + \cdots = \phi_0 + a_t. \tag{2.18}$$

The model for x_{t-1} is then

$$x_{t-1} + \theta_1 x_{t-2} + \theta_1^2 x_{t-3} + \cdots = \phi_0 + a_{t-1}. \tag{2.19}$$

Multiplying Equation (2.19) by θ_1 and subtracting the result from Equation (2.18), we obtain

$$x_t = \phi_0(1 - \theta_1) + a_t - \theta_1 a_{t-1},$$

which says that except for the constant term, x_t is a weighted average of shocks a_t and a_{t-1}. Therefore, the model is called an *MA model* of order 1 or MA(1) model for short. The general form of an MA(1) model is

$$x_t = c_0 + a_t - \theta_1 a_{t-1} \quad \text{or} \quad x_t = c_0 + (1 - \theta_1 B)a_t, \tag{2.20}$$

where c_0 is a constant and $\{a_t\}$ is a white noise series. Similarly, an MA(2) model is in the form

$$x_t = c_0 + a_t - \theta_1 a_{t-1} - \theta_2 a_{t-2}, \tag{2.21}$$

and an MA(q) model is

$$x_t = c_0 + a_t - \theta_1 a_{t-1} - \cdots - \theta_q a_{t-q}, \tag{2.22}$$

or $x_t = c_0 + (1 - \theta_1 B - \cdots - \theta_q B^q)a_t$, where $q > 0$.

2.5.1 Properties of MA Models

Again, we focus on the simple MA(1) and MA(2) models. The results of MA(q) models can easily be obtained by the same techniques.

Stationarity. MA models are always weakly stationary because they are finite linear combinations of a white noise sequence for which the first two moments are time invariant. For example, consider the MA(1) model in Equation (2.20). Taking expectation of the model, we have

$$E(x_t) = c_0,$$

which is time invariant. Taking the variance of Equation (2.20), we have

$$\text{Var}(x_t) = \sigma_a^2 + \theta_1^2 \sigma_a^2 = (1 + \theta_1^2)\sigma_a^2,$$

where we use the fact that a_t and a_{t-1} are uncorrelated. Again, $\text{Var}(x_t)$ is time invariant. The prior discussion applies to general MA(q) models, and we obtain two general properties. First, the constant term of an MA model is the mean of the series (i.e., $E(x_t) = c_0$). Second, the variance of an MA(q) model is

$$\text{Var}(x_t) = (1 + \theta_1^2 + \theta_2^2 + \cdots + \theta_q^2)\sigma_a^2.$$

Autocorrelation Function. Assume for simplicity that $c_0 = 0$ for an MA(1) model. Multiplying the model by $x_{t-\ell}$, we have

$$x_{t-\ell}x_t = x_{t-\ell}a_t - \theta_1 x_{t-\ell}a_{t-1}.$$

Taking expectation, we obtain

$$\gamma_1 = -\theta_1 \sigma_a^2 \quad \text{and} \quad \gamma_\ell = 0, \quad \text{for} \quad \ell > 1.$$

Using the prior result and the fact that $\text{Var}(x_t) = (1 + \theta_1^2)\sigma_a^2$, we have

$$\rho_0 = 1, \quad \rho_1 = \frac{-\theta_1}{1 + \theta_1^2}, \quad \text{and} \quad \rho_\ell = 0, \quad \text{for} \quad \ell > 1.$$

Thus, for an MA(1) model, the lag-1 ACF is not 0, but all higher-order ACFs are 0. In other words, the ACF of an MA(1) model cuts off at lag 1. For the MA(2) model in Equation (2.21), the autocorrelation coefficients are

$$\rho_1 = \frac{-\theta_1 + \theta_1\theta_2}{1 + \theta_1^2 + \theta_2^2}, \quad \rho_2 = \frac{-\theta_2}{1 + \theta_1^2 + \theta_2^2}, \quad \text{and} \quad \rho_\ell = 0, \quad \text{for} \quad \ell > 2. \quad (2.23)$$

Here, the ACF cuts off at lag 2. This property generalizes to other MA models. For an MA(q) model, the lag-q ACF is not 0, but $\rho_\ell = 0$ for $\ell > q$. Consequently, an MA(q) series is only linearly related to its first q lagged values and hence is a "finite memory" model.

Invertibility. Rewriting a zero-mean MA(1) model as $a_t = x_t + \theta_1 a_{t-1}$, one can use repeated substitutions to obtain

$$a_t = x_t + \theta_1 x_{t-1} + \theta_1^2 x_{t-2} + \theta_1^3 x_{t-3} + \cdots.$$

This equation expresses the current shock a_t as a linear combination of the present and past values of x_t. Intuitively, θ_1^j should go to 0 as j increases because the remote return x_{t-j} should have very little impact on the current shock, if any. Consequently, for an MA(1) model to be plausible, we require $|\theta_1| < 1$. Such an MA(1) model is said to be *invertible*. If $|\theta_1| = 1$, then the MA(1) model is noninvertible. See Tsay (2010, Chapter 2) for further discussion on invertibility.

2.5.2 Identifying MA Order

The ACF is useful in identifying the order of an MA model. For a time series x_t with ACF ρ_ℓ, if $\rho_q \neq 0$, but $\rho_\ell = 0$ for $\ell > q$, then x_t follows an MA(q) model.

Figure 2.14 shows the time plot of monthly simple returns of the CRSP equal-weighted index from January 1926 to December 2008 and the sample ACF of the series. The two dashed lines shown on the ACF plot denote the two standard error

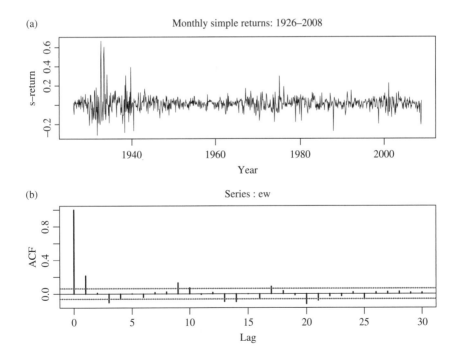

Figure 2.14. Time plot and sample autocorrelation function of monthly simple returns of the CRSP equal-weighted index from January 1926 to December 2008.

limits. It is seen that the series has significant ACF at lags 1, 3, and 9. There are some marginally significant ACF at higher lags, but we do not consider them here. On the basis of the sample ACF, the following MA(9) model

$$x_t = c_0 + a_t - \theta_1 a_{t-1} - \theta_3 a_{t-3} - \theta_9 a_{t-9}$$

is identified for the series. Note that, unlike the sample PACF, sample ACF provides information on the nonzero MA lags of the model. To see this, consider, for example, a simple MA(2) model with $\theta_1 = 0$. The model is $x_t = c_0 + a_t - \theta_2 a_{t-2}$. Using Equation (2.23) or via direct evaluation, the ACF of the model is

$$\rho_0 = 1, \quad \rho_1 = 0, \quad \rho_2 = \frac{-\theta_2}{1 + \theta_2^2}, \quad \text{and} \quad \rho_j = 0 \quad \text{for} \quad j > 2.$$

Therefore, for this particular case, ACF provides the exact information on the structure of the model.

2.5.3 Estimation

Maximum likelihood estimation is commonly used to estimate MA models. There are two approaches for evaluating the likelihood function of an MA model. The first approach assumes that the initial shocks (i.e., a_t for $t \leq 0$) are 0. As such, the shocks needed in likelihood function calculation are obtained recursively from the model, starting with $a_1 = x_1 - c_0$ and $a_2 = x_2 - c_0 + \theta_1 a_1$. This approach is referred to as the *conditional likelihood method* and the resulting estimates the conditional maximum likelihood estimates. The second approach treats the initial shocks $a_t, t \leq 0$ as additional parameters of the model and estimate them jointly with other parameters. This approach is referred to as the *exact likelihood method*. The exact likelihood estimates are preferred over the conditional ones, especially when the MA model is close to being noninvertible. The exact method, however, requires more intensive computation. If the sample size is large, then the two types of maximum likelihood estimates are close to each other. For details of conditional and exact likelihood estimates of MA models, readers are referred to Box et al. (1994) or Tsay (2010, Chapter 8).

For illustration, consider the monthly simple return series of the CRSP equal-weighted index and the specified MA(9) model. The conditional maximum likelihood method produces the fitted model

$$x_t = 0.012 + a_t + 0.189a_{t-1} - 0.121a_{t-3} + 0.122a_{t-9}, \quad \hat{\sigma}_a = 0.0714, \quad (2.24)$$

where standard errors of the coefficient estimates are 0.003, 0.031, 0.031, and 0.031, respectively. The Ljung–Box statistics of the residuals give $Q(12) = 17.5$ with p-value 0.041, which is based on an asymptotic chi-squared distribution with 9 degrees of freedom. The model needs some refinements in modeling the linear dynamic dependence of the data. The p-value would be 0.132 if 12 degrees of freedom are used. The exact maximum likelihood method produces the fitted model

$$x_t = 0.012 + a_t + 0.191a_{t-1} - 0.120a_{t-3} + 0.123a_{t-9}, \quad \hat{\sigma}_a = 0.0714, \quad (2.25)$$

where standard errors of the estimates are $0.003, 0.031, 0.031$, and 0.031, respectively. The Ljung–Box statistics of the residuals gives $Q(12) = 17.6$. The corresponding p-values are 0.040 and 0.128, respectively, when the degrees of freedom are 9 and 12. Again, this fitted model is only marginally adequate. Comparing models (Eq. 2.24) and (Eq. 2.25), we see that for this particular instance, the difference between the conditional and exact likelihood methods is negligible.

Remark. R uses the exact likelihood method in estimation. In addition, the MA polynomial is written as $1 + \theta_1 B + \cdots + \theta_q B^q$ instead of the conventional parameterization $1 - \theta_1 B - \cdots - \theta_q B^q$. More specifically, the ARMA(p,q) model under the R command `arima` is in the form

$$(1 - \phi_1 B - \cdots - \phi_p B^p)(x_t - \mu) = (1 + \theta_1 B + \cdots + \theta_q B^q)a_t,$$

where μ is referred to as the `intercept`. See the attached R output. □

2.5.4 Forecasting Using MA Models

Forecasts of an MA model can easily be obtained. Because the model has finite memory, its point forecasts go to the mean of the series quickly. To see this, assume that the forecast origin is h, and let F_h denote the information available at time h. For the 1-step ahead forecast of an MA(1) process, the model says

$$x_{h+1} = c_0 + a_{h+1} - \theta_1 a_h.$$

Taking the conditional expectation, we have

$$\hat{x}_h(1) = E(x_{h+1}|F_h) = c_0 - \theta_1 a_h$$
$$e_h(1) = x_{h+1} - \hat{x}_h(1) = a_{h+1}.$$

The variance of the 1-step ahead forecast error is $\text{Var}[e_h(1)] = \sigma_a^2$. In practice, the quantity a_h can be obtained in several ways. For instance, assume that $a_0 = 0$, then $a_1 = x_1 - c_0$, and we can compute a_t for $2 \le t \le h$ recursively by $a_t = x_t - c_0 + \theta_1 a_{t-1}$. Alternatively, it can be computed by the AR representation of the MA(1) model (Section 2.6.5). Of course, a_t is the residual series of a fitted MA(1) model. Thus, a_h is readily available from the estimation.

For the 2-step ahead forecast from the equation

$$x_{h+2} = c_0 + a_{h+2} - \theta_1 a_{h+1},$$

we have

$$\hat{x}_h(2) = E(x_{h+2}|F_h) = c_0,$$
$$e_h(2) = x_{h+2} - \hat{r}_h(2) = a_{h+2} - \theta_1 a_{h+1}.$$

The variance of the forecast error is $\mathrm{Var}[e_h(2)] = (1 + \theta_1^2)\sigma_a^2$, which is the variance of the model and is greater than or equal to that of the 1-step ahead forecast error. The prior result shows that for an MA(1) model, the 2-step ahead forecast of the series is simply the unconditional mean of the model. This is true for any forecast origin h. More generally, $\hat{x}_h(\ell) = c_0$ for $\ell \geq 2$. In summary, for an MA(1) model, the 1-step ahead point forecast at the forecast origin h is $c_0 - \theta_1 a_h$ and the multistep ahead forecasts are c_0, which is the unconditional mean of the model. If we plot the forecasts $\hat{x}_h(\ell)$ versus ℓ, we see that the forecasts form a horizontal line after one step. Thus, for MA(1) models, mean reverting only takes one time period.

Similarly, for an MA(2) model, we have

$$x_{h+\ell} = c_0 + a_{h+\ell} - \theta_1 a_{h+\ell-1} - \theta_2 a_{h+\ell-2},$$

from which we obtain

$$\hat{x}_h(1) = c_0 - \theta_1 a_h - \theta_2 a_{h-1}$$
$$\hat{x}_h(2) = c_0 - \theta_2 a_h$$
$$\hat{x}_h(\ell) = c_0, \quad \text{for} \quad \ell > 2.$$

Thus, the multistep ahead forecasts of an MA(2) model go to the mean of the series after two steps. The variances of forecast errors go to the variance of the series after two steps. In general, for an MA(q) model, multistep ahead forecasts go to the mean after the first q steps.

Table 2.3 gives some out-of-sample forecasts of an MA(9) model in the form of Equation (2.25) for the monthly simple returns of the equal-weighted index at the

TABLE 2.3. Out-of-Sample Forecasts of an MA(9) Model For Monthly Simple Returns of CRSP Equal-Weighted Index[a].

Step	1	2	3	4	5
Forecast	0.0043	0.0136	0.0150	0.0144	0.0120
Standard Error	0.0712	0.0724	0.0729	0.0729	0.0729
Actual	−0.0260	0.0312	0.0322	−0.0871	−0.0010

Step	6	7	8	9	10
Forecast	0.0019	0.0122	0.0056	0.0085	0.0128
Standard Error	0.0729	0.0729	0.0729	0.0729	0.0734
Actual	0.0141	−0.1209	−0.2060	−0.1366	0.0431

[a]The Forecast Origin is February 2008 With $h = 986$. The Model is Estimated by the Exact Maximum Likelihood Method.

forecast origin $h = 986$ (February, 2008). The model parameters are reestimated using the first 986 observations. The sample mean and standard error of the estimation subsample are 0.0128 and 0.0736, respectively. As expected, the table shows that (i) the 10-step ahead forecast is the sample mean, and (ii) the standard deviations of the forecast errors converge to the standard deviation of the series as the forecast horizon increases. In this particular case, the point forecasts deviate substantially from the observed returns because of the worldwide financial crisis caused by the subprime mortgage problem and the collapse of Lehman Brothers.

```
> da=read.table("m-ibm3dx2608.txt",header=T)
> head(da)
        date      ibmrtn       vwrtn       ewrtn       sprtn
1 19260130 -0.010381    0.000724   0.023174   0.022472
 ....
> ew=da$ewrtn
> m1=arima(ew,order=c(0,0,9))  % unrestricted model
> m1
arima(x = ew, order = c(0, 0, 9))
Coefficients:
        ma1      ma2      ma3      ma4      ma5      ma6      ma7      ma8
      0.2144 0.0374 -0.1203 -0.0425 0.0232 -0.0302 0.0482 -0.0276
s.e.  0.0316 0.0321   0.0328   0.0336 0.0319   0.0318 0.0364   0.0354
          ma9 intercept
      0.1350    0.0122
s.e.  0.0323    0.0028

sigma^2 estimated as 0.005043: log likelihood=1220.86, aic=-2419.72
%% Refined model
> m1=arima(ew,order=c(0,0,9),fixed=c(NA,0,NA,0,0,0,0,0,NA,NA))
> m1
arima(x=ew, order=c(0,0,9), fixed=c(NA,0,NA,0,0,0,0,0,NA,NA))
Coefficients:
        ma1 ma2    ma3 ma4 ma5 ma6 ma7 ma8    ma9   intercept
      0.1909  0 -0.1199   0   0   0   0   0 0.1227      0.0122
s.e.  0.0293  0  0.0338   0   0   0   0   0 0.0312      0.0027

sigma^2 estimated as 0.005097: log likelihood=1215.61, aic=-2421.22
> sqrt(0.005097)
[1] 0.07139328
>
> Box.test(m1$residuals,lag=12,type='Ljung')   % model checking
        Box-Ljung test
data:  m1$residuals
X-squared = 17.604, df = 12, p-value = 0.1283

> pv=1-pchisq(17.6,9)   % compute p-value after adjusting the d.f.
> pv
[1] 0.04010828
%% To perform out of sample prediction at forecast origin 986.
> m1=arima(ew[1:986],order=c(0,0,9),fixed=c(NA,0,NA,0,0,0,0,0,NA,NA))
> m1
arima(x = ew[1:986], order=c(0,0,9), fixed=c(NA,0,NA,0,0,0,0,0,NA,NA))
```

```
Coefficients:
         ma1  ma2      ma3 ma4 ma5 ma6 ma7 ma8     ma9  intercept
      0.1844    0  -0.1206   0   0   0   0   0 0.1218     0.0128
s.e.  0.0295    0   0.0338   0   0   0   0   0 0.0312     0.0027

sigma^2 estimated as 0.005066: log likelihood =1206.44, aic--2402.88
> predict(m1,10) % prediction
$pred  % point forecast
Time Series:
Start = 987
End = 996
Frequency = 1
 [1] 0.0042826 0.0135589 0.0150242 0.0144534 0.0120463 0.0018056
 [7] 0.0122115 0.0055148 0.0085135 0.0127918
$se   % standard error of prediction
Time Series:
Start = 987
End = 996
Frequency = 1
 [1] 0.071175 0.072375 0.072375 0.072882 0.072882 0.072882
 [7] 0.072882 0.072882 0.072882 0.073396
```

Summary. A brief summary of AR and MA models is in order. We have discussed the following properties:

- for MA models, ACF is useful in specifying the order because ACF cuts off at lag q for an MA(q) series;
- for AR models, PACF is useful in order determination because PACF cuts off at lag p for an AR(p) process;
- an MA series is always stationary, but for an AR series to be stationary, all of its characteristic roots must be less than 1 in modulus;
- for a stationary series, the multistep ahead forecasts converge to the mean of the series and the variances of forecast errors converge to the variance of the series as the forecast horizon increases.

2.6   SIMPLE ARMA MODELS

In some applications, the AR or MA models discussed in the previous sections become cumbersome because one may need a high order model with many parameters to adequately describe the dynamic structure of the data. To overcome this difficulty, the ARMA models are introduced (Box et al.,1994). Basically, an ARMA model combines the ideas of AR and MA models into a compact form so that the number of parameters used is kept small, achieving parsimony in parameterization. The model is useful in modeling business, economic, and engineering time series. For the return series in finance, the chance of using ARMA models is low. However, the concept of ARMA models is highly relevant in volatility modeling. As a matter of fact, the generalized autoregressive conditional heteroscedastic (GARCH) model can be regarded as an

ARMA model, albeit nonstandard, for the a_t^2 series; see Chapter 4 for details. In this section, we study the simplest ARMA(1,1) model.

A time series x_t follows an ARMA(1,1) model if it satisfies

$$x_t - \phi_1 x_{t-1} = \phi_0 + a_t - \theta_1 a_{t-1}, \tag{2.26}$$

where $\{a_t\}$ is a white noise series. The left-hand side of the Equation (2.26) is the AR component of the model and the right-hand side gives the MA component. The constant term is ϕ_0. For this model to be meaningful, we need $\phi_1 \neq \theta_1$; otherwise, there is a cancellation in the equation and the process reduces to a white noise series.

2.6.1 Properties of ARMA(1,1) Models

Properties of ARMA(1,1) models are generalizations of those of AR(1) models with some minor modifications to handle the impact of the MA(1) component. We start with the stationarity condition. Taking expectation of Equation (2.26), we have

$$E(x_t) - \phi_1 E(x_{t-1}) = \phi_0 + E(a_t) - \theta_1 E(a_{t-1}).$$

Because $E(a_i) = 0$ for all i, the mean of x_t is

$$E(x_t) = \mu = \frac{\phi_0}{1 - \phi_1}$$

provided that the series is weakly stationary. This result is exactly the same as that of the AR(1) model in Equation (2.8).

Next, assuming for simplicity that $\phi_0 = 0$, we consider the autocovariance function of x_t. First, multiplying the model by a_t and taking expectation, we have

$$E(x_t a_t) = E(a_t^2) - \theta_1 E(a_t a_{t-1}) = E(a_t^2) = \sigma_a^2. \tag{2.27}$$

Rewriting the model as

$$x_t = \phi_1 x_{t-1} + a_t - \theta_1 a_{t-1}$$

and taking the variance of the prior equation, we have

$$\mathrm{Var}(x_t) = \phi_1^2 \mathrm{Var}(x_{t-1}) + \sigma_a^2 + \theta_1^2 \sigma_a^2 - 2\phi_1 \theta_1 E(x_{t-1} a_{t-1}).$$

Here we make use of the fact that x_{t-1} and a_t are uncorrelated. Using Equation (2.27), we obtain

$$\mathrm{Var}(x_t) - \phi_1^2 \mathrm{Var}(x_{t-1}) = (1 - 2\phi_1 \theta_1 + \theta_1^2)\sigma_a^2.$$

Therefore, if the series x_t is weakly stationary, then $\mathrm{Var}(x_t) = \mathrm{Var}(x_{t-1})$, and we have

$$\mathrm{Var}(x_t) = \frac{(1 - 2\phi_1 \theta_1 + \theta_1^2)\sigma_a^2}{1 - \phi_1^2}.$$

Because the variance is positive, we need $\phi_1^2 < 1$ (i.e., $|\phi_1| < 1$). Again, this is precisely the same stationarity condition as that of the AR(1) model.

To obtain the autocovariance function of x_t, we assume that $\phi_0 = 0$ and multiply the model in Equation (2.26) by $x_{t-\ell}$ to obtain

$$x_t x_{t-\ell} - \phi_1 x_{t-1} x_{t-\ell} = a_t x_{t-\ell} - \theta_1 a_{t-1} x_{t-\ell}.$$

For $\ell = 1$, taking expectation and using Equation (2.27) for $t - 1$, we have

$$\gamma_1 - \phi_1 \gamma_0 = -\theta_1 \sigma_a^2,$$

where $\gamma_\ell = \mathrm{Cov}(x_t, x_{t-\ell})$. This result is different from that of the AR(1) case for which $\gamma_1 - \phi_1 \gamma_0 = 0$. However, for $\ell = 2$ and taking expectation, we have

$$\gamma_2 - \phi_1 \gamma_1 = 0,$$

which is identical to that of the AR(1) case. In fact, the same technique yields

$$\gamma_\ell - \phi_1 \gamma_{\ell-1} = 0, \quad \text{for} \quad \ell > 1. \tag{2.28}$$

In terms of ACF, the previous results show that for a stationary ARMA(1,1) model

$$\rho_1 = \phi_1 - \frac{\theta_1 \sigma_a^2}{\gamma_0}, \quad \rho_\ell = \phi_1 \rho_{\ell-1}, \quad \text{for} \quad \ell > 1.$$

Thus, the ACF of an ARMA(1,1) model behaves very much similar to that of an AR(1) model except that the exponential decay starts with lag 2. Consequently, the ACF of an ARMA(1,1) model does not cut off at any finite lag.

Turning to PACF, one can show that the PACF of an ARMA(1,1) model does not cut off at any finite lag either. It behaves very much similar to that of an MA(1) model except that the exponential decay starts with lag 2 instead of lag 1.

In summary, the stationarity condition of an ARMA(1,1) model is the same as that of an AR(1) model, and the ACF of an ARMA(1,1) exhibits a pattern similar to that of an AR(1) model except that the pattern starts at lag 2.

2.6.2   General ARMA Models

A general ARMA(p, q) model is in the form

$$x_t = \phi_0 + \sum_{i=1}^{p} \phi_i x_{t-i} + a_t - \sum_{i=1}^{q} \theta_i a_{t-i},$$

where $\{a_t\}$ is a white noise series and p and q are nonnegative integers. The AR and MA models are special cases of the ARMA(p, q) model. Using the backshift operator, the model can be written as

$$(1 - \phi_1 B - \cdots - \phi_p B^p)x_t = \phi_0 + (1 - \theta_1 B - \cdots - \theta_q B^q)a_t. \qquad (2.29)$$

The polynomial $1 - \phi_1 B - \cdots - \phi_p B^p$ is the AR polynomial of the model. Similarly, $1 - \theta_1 B - \cdots - \theta_q B^q$ is the MA polynomial. We require that there are no common factors between the AR and MA polynomials; otherwise, the order (p, q) of the model can be reduced. Similar to a pure AR model, the AR polynomial introduces the characteristic equation of an ARMA model. If all of the solutions of the characteristic equation are less than 1 in absolute value, then the ARMA model is weakly stationary. In this case, the unconditional mean of the model is $E(x_t) = \phi_0/(1 - \phi_1 - \cdots - \phi_p)$.

2.6.3 Identifying ARMA Models

The ACF and PACF are not informative in determining the order of an ARMA model. Tsay and Tiao (1984) propose a new approach that uses the extended autocorrelation function (EACF) to specify the order of an ARMA process. The basic idea of EACF is relatively simple. If we can obtain a consistent estimate of the AR component of an ARMA model, then we can derive the MA component. From the derived MA series, we can use ACF to identify the order of the MA component.

The derivation of EACF is relatively involved; see Tsay and Tiao (1984) for details. Yet the function is easy to use. The output of EACF is a two-way table, where the rows correspond to AR order p and the columns to MA order q. The theoretical version of EACF for an ARMA(1,1) model is given in Table 2.4. The key feature of the table is that it contains a triangle of "O" with the upper left vertex located at the order (1,1). This is the characteristic we use to identify the order of an ARMA process. In general, for an ARMA(p, q) model, the triangle of "O" will have its upper left vertex at the (p, q) position.

For illustration, consider the monthly log stock returns of the 3M Company from February 1946 to December 2008. There are 755 observations. The return series and its sample ACF are shown in Figure 2.15. The ACF indicates that there are no significant

TABLE 2.4. Theoretical EACF Table For an ARMA(1,1) Model, Where "X" Denotes Nonzero, "O" Denotes Zero, and "*" Denotes Either Zero or Nonzero[a]

	MA							
AR	0	1	2	3	4	5	6	7
0	X	X	X	X	X	X	X	X
1	X	O	O	O	O	O	O	O
2	*	X	O	O	O	O	O	O
3	*	*	X	O	O	O	O	O
4	*	*	*	X	O	O	O	O
5	*	*	*	*	X	O	O	O

[a]This latter category does not play any role in identifying the order (1,1).

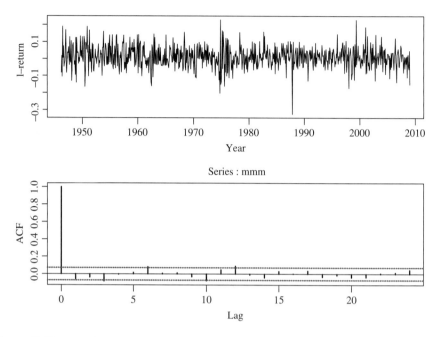

Figure 2.15. Time plot and sample autocorrelation function of monthly log stock returns of 3M Company from February 1946 to December 2008.

serial correlations in the data at the 1% level. Table 2.5 shows the sample EACF and a corresponding simplified table for the series, obtained via the SCA package. One can also use the TSA package in R to compute EACF; see Remark below. The simplified table is constructed using the following notation:

1. "X" denotes that the absolute value of the corresponding EACF is greater than or equal to twice of its asymptotic standard error;
2. "O" denotes that the corresponding EACF is less than twice of its standard error in modulus.

The standard error of EACF can be computed using either the Bartlett's formula of Section 2.2 or simply $2/\sqrt{T}$ with T being the sample size. The simplified table exhibits a triangular pattern of "O" with its upper left vertex at the order $(p, q) = (0,0)$. A few exceptions of "X" appear when $q = 2, 5, 9$, and 11. However, the EACF table shows that the values of sample ACF corresponding to those "X" are around 0.08 or 0.09. These ACFs are only slightly greater than $2/\sqrt{755} = 0.073$. Indeed, if 1% critical value is used, those "X" would become "O" in the simplified EACF table. Consequently, the EACF suggests that the monthly log returns of 3M stock follow an ARMA(0,0) model (i.e., a white noise series). This is in agreement with the result suggested by the sample ACF in Figure 2.15.

Remark. EACF is also available in R with the package TSA of Cryer and Chan (2010). SCA uses the Bartlett's formula to construct the simplified table, whereas TSA uses $2/\sqrt{T}$ as the standard error of all EACF. For the 3M returns, R output is given below: □

```
> da=read.table("m-3m4608.txt",header=T)
> head(da)
        date         rtn
1 19460228 -0.077922
> mmm=log(da$rtn+1)
> library(TSA)      % Load the package
> m1=eacf(mmm,6,12)        % Simplified table
AR/MA
   0 1 2 3 4 5 6 7 8 9 10 11 12
0  o o x o o x o o o x o  x  o
1  x o x o o x o o o o o  x  o
2  x x x o o x o o o o o  o  o
3  x x x o o o o o o o o  o  o
4  x o x o o o o o o o o  o  o
5  x x x o x o o o o o o  o  o
6  x x x x x o o o o o o  o  o
> names(m1)
[1] "eacf"    "ar.max" "ma.ma"   "symbol"
> print(m1$eacf,digits=2)
          [,1]    [,2]    [,3]    [,4]    [,5]     [,6]     [,7]    [,8]     [,9]
[1,] -0.056 -0.038 -0.082 -0.005  0.018  0.0821  0.0080 0.0127 -0.0301
[2,] -0.474  0.010 -0.074 -0.021  0.002  0.0772 -0.0288 0.0026 -0.0068
[3,] -0.383 -0.348 -0.074  0.016 -0.006  0.0772  0.0269 0.0120  0.0004
[4,] -0.177  0.138  0.384 -0.022  0.002  0.0419 -0.0232 0.0154 -0.0044
[5,]  0.421  0.029  0.454 -0.008  0.001  0.0025 -0.0140 0.0305  0.0116
[6,] -0.114  0.214  0.449  0.010  0.202 -0.0063 -0.0038 0.0403 -0.0129
[7,] -0.208 -0.250  0.243  0.311  0.168 -0.0388 -0.0034 0.0429 -0.0101
         [,10]   [,11]   [,12]   [,13]
[1,] -0.078 0.0488  0.0909 -0.011
[2,] -0.069 0.0372  0.0938 -0.024
[3,] -0.027 0.0221  0.0428  0.042
[4,] -0.025 0.0185  0.0100  0.043
[5,]  0.004 0.0191 -0.0043  0.013
[6,] -0.012 0.0315  0.0117  0.028
[7,] -0.026 0.0078  0.0106  0.037
```

The information criteria discussed earlier can also be used to select the order of an ARMA model. Typically, for some prespecified positive integers P and Q, one computes AIC (or BIC) for ARMA(p, q) models, where $0 \le p \le P$ and $0 \le q \le Q$, and selects the model that gives the minimum AIC (or BIC). This approach requires maximum likelihood estimation of many models, and in some cases may encounter the difficulty of overfitting in estimation.

Once an ARMA(p, q) model is specified, its parameters can be estimated by either the conditional or exact likelihood method. In addition, the Ljung–Box statistics of the residuals can be used to check the adequacy of a fitted model. If the model is correctly specified, then $Q(m)$ follows asymptotically a chi-squared distribution with

$m - g$ degrees of freedom, where g denotes the number of AR or MA coefficients fitted in the model.

2.6.4 Forecasting Using an ARMA Model

Similar to the behavior of ACF, forecasts of an ARMA(p, q) model have similar characteristics as those of an AR(p) model after adjusting for the impacts of the MA component on the lower horizon forecasts. Denote the forecast origin by h and the available information by F_h. The 1-step ahead forecast of x_{h+1} can be easily obtained from the model as

$$\hat{x}_h(1) = E(x_{h+1}|F_h) = \phi_0 + \sum_{i=1}^{p} \phi_i x_{h+1-i} - \sum_{i=1}^{q} \theta_i a_{h+1-i},$$

and the associated forecast error is $e_h(1) = x_{h+1} - \hat{x}_h(1) = a_{h+1}$. The variance of 1-step ahead forecast error is $\text{Var}[e_h(1)] = \sigma_a^2$. For the ℓ-step ahead forecast, we have

$$\hat{x}_h(\ell) = E(x_{h+\ell}|F_h) = \phi_0 + \sum_{i=1}^{p} \phi_i \hat{x}_h(\ell - i) - \sum_{i=1}^{q} \theta_i a_h(\ell - i),$$

where it is understood that $\hat{x}_h(\ell - i) = x_{h+\ell-i}$ if $\ell - i \le 0$ and $a_h(\ell - i) = 0$ if $\ell - i > 0$ and $a_h(\ell - i) = a_{h+\ell-i}$ if $\ell - i \le 0$. Thus, the multistep ahead forecasts of an ARMA model can be computed recursively. The associated forecast error is

$$e_h(\ell) = x_{h+\ell} - \hat{x}_h(\ell),$$

which can be computed easily via a formula to be given below in Equation (2.35).

2.6.5 Three Model Representations for an ARMA Model

In this section, we briefly discuss three model representations for a stationary ARMA(p, q) model. The three representations serve three different purposes. Knowing these representations can lead to a better understanding of the model. The first representation is the ARMA(p, q) model in Equation (2.29). This representation is compact and useful in parameter estimation. It is also useful in computing recursively multistep ahead forecasts of x_t; see the discussion in the previous section.

For the other two representations, we use long division of two polynomials. Given two polynomials $\phi(B) = 1 - \sum_{i=1}^{p} \phi_i B^i$ and $\theta(B) = 1 - \sum_{i=1}^{q} \theta_i B^i$, we can obtain, by long division, that

$$\frac{\theta(B)}{\phi(B)} = 1 + \psi_1 B + \psi_2 B^2 + \cdots \equiv \psi(B) \tag{2.30}$$

and

$$\frac{\phi(B)}{\theta(B)} = 1 - \pi_1 B - \pi_2 B^2 - \cdots \equiv \pi(B). \tag{2.31}$$

For instance, if $\phi(B) = 1 - \phi_1 B$ and $\theta(B) = 1 - \theta_1 B$, then

$$\psi(B) = \frac{1 - \theta_1 B}{1 - \phi_1 B} = 1 + (\phi_1 - \theta_1)B + \phi_1(\phi_1 - \theta_1)B^2 + \phi_1^2(\phi_1 - \theta_1)B^3 + \cdots$$

$$\pi(B) = \frac{1 - \phi_1 B}{1 - \theta_1 B} = 1 - (\phi_1 - \theta_1)B - \theta_1(\phi_1 - \theta_1)B^2 - \theta_1^2(\phi_1 - \theta_1)B^3 - \cdots.$$

From the definition, $\psi(B)\pi(B) = 1$. Making use of the fact that $Bc = c$ for any constant (because the value of a constant is time invariant), we have

$$\frac{\phi_0}{\theta(1)} = \frac{\phi_0}{1 - \theta_1 - \cdots - \theta_q} \quad \text{and} \quad \frac{\phi_0}{\phi(1)} = \frac{\phi_0}{1 - \phi_1 - \cdots - \phi_p}.$$

AR Representation. Using the result of long division in Equation (2.31), the ARMA(p, q) model can be written as

$$x_t = \frac{\phi_0}{1 - \theta_1 - \cdots - \theta_q} + \pi_1 x_{t-1} + \pi_2 x_{t-2} + \pi_3 x_{t-3} + \cdots + a_t. \tag{2.32}$$

This representation shows the dependence of the current return x_t on the past returns x_{t-i}, where $i > 0$. The coefficients $\{\pi_i\}$ are referred to as the π-*weights* of an ARMA model. To show that the contribution of the lagged value x_{t-i} to x_t is diminishing as i increases, the π_i coefficient should decay to 0 as i increases. An ARMA(p, q) model that has this property is said to be invertible. For a pure AR model, $\theta(B) = 1$ so that $\pi(B) = \phi(B)$, which is a finite-degree polynomial. Thus, $\pi_i = 0$ for $i > p$, and the model is invertible. For other ARMA models, a sufficient condition for invertibility is that all the zeros of the polynomial $\theta(B)$ are greater than unity in modulus. For example, consider the MA(1) model $x_t = (1 - \theta_1 B)a_t$. The zero of the first-order polynomial $1 - \theta_1 B$ is $B = 1/\theta_1$. Therefore, an MA(1) model is invertible if $|1/\theta_1| > 1$. This is equivalent to $|\theta_1| < 1$.

From the AR representation in Equation (2.32), an invertible ARMA(p, q) series x_t is a linear combination of the current shock a_t and a weighted average of the past values. The weights decay exponentially for more remote past values.

MA Representation. Again, using the result of long division in Equation (2.30), an ARMA(p, q) model can also be written as

$$x_t = \mu + a_t + \psi_1 a_{t-1} + \psi_2 a_{t-2} + \cdots = \mu + \psi(B)a_t, \tag{2.33}$$

where $\mu = E(x_t) = \phi_0/(1 - \phi_1 - \cdots - \phi_p)$. This representation shows explicitly the impact of the past shock a_{t-i} $(i > 0)$ on the current return x_t. The coefficients $\{\psi_i\}$ are referred to as the *impulse response function* of the ARMA model. For a weakly stationary series, the ψ_i coefficients decay exponentially as i increases. This is understandable as the effect of shock a_{t-i} on the return x_t should diminish over time. Thus, for a stationary ARMA model, the shock a_{t-i} does not have a permanent impact on the series. If $\phi_0 \neq 0$, then the MA representation has a constant term, which is the mean of x_t (i.e., $\phi_0/(1 - \phi_1 - \cdots - \phi_p)$)).

The MA representation in Equation (2.33) is also useful in computing the variance of a forecast error. At the forecast origin h, we have the shocks $a_h, a_{h-1}, \ldots$ Therefore, the ℓ-step ahead point forecast is

$$\hat{x}_h(\ell) = \mu + \psi_\ell a_h + \psi_{\ell+1} a_{h-1} + \cdots, \qquad (2.34)$$

and the associated forecast error is

$$e_h(\ell) = a_{h+\ell} + \psi_1 a_{h+\ell-1} + \cdots + \psi_{\ell-1} a_{h+1}.$$

Consequently, the variance of ℓ-step ahead forecast error is

$$\mathrm{Var}[e_h(\ell)] = (1 + \psi_1^2 + \cdots + \psi_{\ell-1}^2)\sigma_a^2, \qquad (2.35)$$

which, as expected, is a nondecreasing function of the forecast horizon ℓ.

Finally, the MA representation in Equation (2.33) provides a simple proof of mean reversion of a stationary time series. The stationarity implies that ψ_i approaches 0, as $i \to \infty$. Therefore, by Equation (2.34), we have $\hat{x}_h(\ell) \to \mu$, as $\ell \to \infty$. Because $\hat{x}_h(\ell)$ is the conditional expectation of $x_{h+\ell}$ at the forecast origin h, the result says that in the long term, the return series is expected to approach its mean, that is, the series is mean reverting. Furthermore, using the MA representation in Equation (2.33), we have $\mathrm{Var}(x_t) = (1 + \sum_{i=1}^{\infty} \psi_i^2)\sigma_a^2$. Consequently, by Equation (2.35), we have $\mathrm{Var}[e_h(\ell)] \to \mathrm{Var}(x_t)$, as $\ell \to \infty$. The speed by which $\hat{r}_h(\ell)$ approaches μ determines the speed of mean reverting.

2.7   UNIT-ROOT NONSTATIONARITY

So far, we have focused on the return series that are stationary. In some studies, interest rates, foreign exchange rates, or the price series of an asset are of interest. These series tend to be nonstationary. For a price series, the nonstationarity is mainly due to the fact that there is no fixed level for the price. In the time series literature, such a nonstationary series is called *unit-root nonstationary time series*. The best-known example of unit-root nonstationary time series is the random walk model.

2.7.1   Random Walk

A time series $\{p_t\}$ is a random walk if it satisfies

$$p_t = p_{t-1} + a_t, \qquad (2.36)$$

where p_0 is a real number denoting the starting value of the process and $\{a_t\}$ is a white noise series. If p_t is the log price of a particular stock at date t, then p_0 could be the log price of the stock at its initial public offering (i.e., the logged IPO price). If a_t has a symmetric distribution around 0, then conditional on p_{t-1}, p_t has a 50–50 chance to go up or down, implying that p_t would go up or down at random. If we treat the random walk model as a special AR(1) model, then the coefficient of p_{t-1} is unity, which does not satisfy the weak stationarity condition of an AR(1) model. A random walk series is, therefore, not weakly stationary, and we call it a unit-root nonstationary time series.

The random walk model has been widely considered as a statistical model for the movement of logged stock prices. Under such a model, the stock price is not predictable or mean reverting. To see this, the 1-step ahead forecast of model (Eq 2.36) at the forecast origin h is

$$\hat{p}_h(1) = E(p_{h+1}|p_h, p_{h-1}, \ldots) = p_h,$$

which is the log price of the stock at the forecast origin. Such a forecast has no practical value. The 2-step ahead forecast is

$$\hat{p}_h(2) = E(p_{h+2}|p_h, p_{h-1}, \ldots) = E(p_{h+1} + a_{h+2}|p_h, p_{h-1}, \ldots)$$
$$= E(p_{h+1}|p_h, p_{h-1}, \ldots) = \hat{p}_h(1) = p_h,$$

which again is the log price at the forecast origin. In fact, for any forecast horizon $\ell > 0$, we have

$$\hat{p}_h(\ell) = p_h.$$

Thus, for all forecast horizons, point forecasts of a random walk model are simply the value of the series at the forecast origin. Therefore, the process is not mean reverting.

The MA representation of the random walk model in Equation (2.36) is

$$p_t = a_t + a_{t-1} + a_{t-2} + \cdots.$$

This representation has several important practical implications. First, the ℓ-step ahead forecast error is

$$e_h(\ell) = a_{h+\ell} + \cdots + a_{h+1},$$

so that $\text{Var}[e_h(\ell)] = \ell \sigma_a^2$, which diverges to infinity as $\ell \to \infty$. The length of an interval forecast of $p_{h+\ell}$ will approach infinity as the forecast horizon increases. This result says that the usefulness of point forecast $\hat{p}_h(\ell)$ diminishes as ℓ increases, which again implies that the model is not predictable. Second, the unconditional variance of p_t is unbounded because $\text{Var}[e_h(\ell)]$ approaches infinity as ℓ increases. Theoretically, this means that p_t can assume any real value for a sufficiently large t. For the log price p_t of an individual stock, this is plausible. Yet for market indexes, negative log price is very rare if it happens at all. In this sense, the adequacy of a random walk model for market indexes is questionable. Third, from the representation, $\psi_i = 1$, for all i. Thus, the impact of any past shock a_{t-i} on p_t does not decay over time. Consequently, the series has a strong memory as it remembers all of the past shocks. In economics, the shocks are said to have a permanent effect on the series. The strong memory of a unit-root time series can be seen from the sample ACF of the observed series. The sample ACFs are all approaching 1 as the sample size increases.

2.7.2 Random Walk with Drift

As shown by empirical examples considered so far, the log return series of a market index tends to have a small and positive mean. This implies that the model for the log price is

$$p_t = \mu + p_{t-1} + a_t, \tag{2.37}$$

where $\mu = E(p_t - p_{t-1})$ and $\{a_t\}$ is a zero-mean white noise series. The constant term μ of model (Eq. 2.37) is very important in financial study. It represents the time trend of the log price p_t and is often referred to as the *drift* of the model. To see this, assume that the initial log price is p_0. Then, we have

$$p_1 = \mu + p_0 + a_1$$
$$p_2 = \mu + p_1 + a_2 = 2\mu + p_0 + a_2 + a_1$$
$$\vdots = \vdots$$
$$p_t = t\mu + p_0 + a_t + a_{t-1} + \cdots + a_1.$$

The last equation shows that the log price consists of a time trend $t\mu$ and a pure random walk process $\sum_{i=1}^{t} a_i$. Because $\mathrm{Var}(\sum_{i=1}^{t} a_i) = t\sigma_a^2$, where σ_a^2 is the variance of a_t, the conditional standard deviation of p_t is $\sqrt{t}\sigma_a$, which grows at a slower rate than the conditional expectation of p_t. Therefore, if we graph p_t against the time index t, we have a time trend with slope μ. A positive slope μ implies that the log price eventually goes to infinity. In contrast, a negative μ implies that the log price would converge to $-\infty$ as t increases. On the basis of the above discussion, it is then not surprising to see that the log return series of the CRSP value- and equal-weighted indexes have a small, but statistically significant, positive mean.

    To illustrate the effect of the drift parameter on the price series, we consider the monthly log stock returns of the 3M Company from February 1946 to December 2008. As shown by the sample EACF in Table 2.5, the series has no significant serial correlation. The series thus follows the simple model

$$x_t = 0.0103 + a_t, \quad \hat{\sigma}_a = 0.0637, \tag{2.38}$$

where 0.0103 is the sample mean of x_t and has a standard error 0.0023. The mean of the monthly log returns of 3M stock is, therefore, significantly different from 0 at

TABLE 2.5. Sample Extended Autocorrelation Function and a Simplified Table for the Monthly Log Returns of 3M Stock from February 1946 to December 2008

					(a) *Sample Extended Autocorrelation Function*							
						MA Order: q						
p	0	1	2	3	4	5	6	7	8	9	10	11	12
0	−0.06	−0.04	−0.08	−0.00	0.02	0.08	0.01	0.01	−0.03	−0.08	0.05	0.09	−0.01
1	−0.47	0.01	−0.07	−0.02	0.00	0.08	−0.03	0.00	−0.01	−0.07	0.04	0.09	−0.02
2	−0.38	−0.35	−0.07	0.02	−0.01	0.08	0.03	0.01	0.00	−0.03	0.02	0.04	0.04
3	−0.18	0.14	0.38	−0.02	0.00	0.04	−0.02	0.02	−0.00	−0.03	0.02	0.01	0.04
4	0.42	0.03	0.45	−0.01	0.00	0.00	−0.01	0.03	0.01	0.00	0.02	−0.00	0.01
5	−0.11	0.21	0.45	0.01	0.20	−0.01	−0.00	0.04	−0.01	−0.01	0.03	0.01	0.03
6	−0.21	−0.25	0.24	0.31	0.17	−0.04	−0.00	0.04	−0.01	−0.03	0.01	0.01	0.04

					(b) *Simplified EACF Table*							
						MA Order: q						
p	0	1	2	3	4	5	6	7	8	9	10	11	12
0	O	O	X	O	O	X	O	O	O	X	O	X	O
1	X	O	O	O	O	X	O	O	O	O	O	X	O
2	X	X	O	O	O	X	O	O	O	O	O	O	O
3	X	X	X	O	O	O	O	O	O	O	O	O	O
4	X	O	X	O	O	O	O	O	O	O	O	O	O
5	X	X	X	O	X	O	O	O	O	O	O	O	O
6	X	X	X	X	X	O	O	O	O	O	O	O	O

the 1% level. As a matter of fact, the one-sample test of zero mean shows a t-ratio of 4.44 with p-value close to 0. We use the log return series to construct two log price series, namely,

$$p_t = \sum_{i=1}^{t} x_i \quad \text{and} \quad p_t^* = \sum_{i=1}^{t} a_i,$$

where a_i is the mean-corrected log return in Equation (2.38) (i.e., $a_t = x_t - 0.0103$). The p_t is the log price of 3M stock, assuming that the initial log price is 0 (i.e., the log price of January 1946 was 0). The p_t^* is the corresponding log price if the mean of log returns was 0. Figure 2.16 shows the time plots of p_t and p_t^*, as well as a straight line $y_t = 0.0103 \times t + 1946$, where t is the time sequence of the returns and 1946 is the starting year of the stock. From the plots, the importance of the constant 0.0103 in Equation (2.38) is evident. In addition, as expected, it represents the slope of the upward trend of p_t.

Interpretation of the Constant Term. From the previous discussions, it is important to understand the meaning of a constant term in a time series model. First, for an MA(q) model in Equation (2.22), the constant term is simply the mean of the

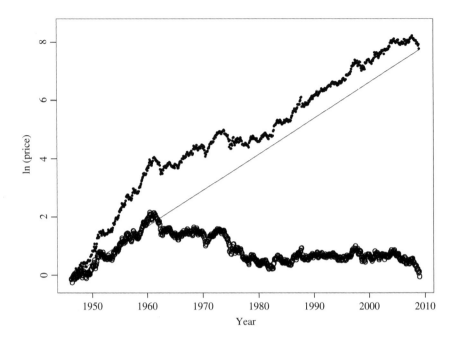

Figure 2.16. Time plots of log prices for 3M stock from February 1946 to December 2008, assuming that the log price of January 1946 was 0. The "o" line is for log price without time trend. The straight line is $y_t = 0.0103 \times t + 1946$.

series. Second, for a stationary AR(p) model in Equation (2.9) or ARMA(p, q) model in Equation (2.29), the constant term is related to the mean via $\mu = \phi_0/(1 - \phi_1 - \cdots - \phi_p)$. Third, for a random walk with drift, the constant term becomes the time slope of the series. These different interpretations for the constant term in a time series model clearly highlight the difference between dynamic and usual linear regression models.

Another important difference between dynamic and regression models is shown by an AR(1) model and a simple linear regression model,

$$x_t = \phi_0 + \phi_1 x_{t-1} + a_t \quad \text{and} \quad y_t = \beta_0 + \beta_1 x_t + a_t.$$

For the AR(1) model to be meaningful, the coefficient ϕ_1 must satisfy $|\phi_1| \leq 1$. However, the coefficient β_1 can assume any fixed real number.

2.7.3 Trend-Stationary Time Series

A closely related model that exhibits linear trend is the trend-stationary time series model,

$$p_t = \beta_0 + \beta_1 t + x_t,$$

where x_t is a stationary time series, for example, a stationary AR(p) series. Here, p_t grows linearly in time with rate β_1 and hence can exhibit behavior similar to that of a random walk model with drift. However, there is a major difference between the two models. To see this, suppose that p_0 is fixed. The random walk model with drift assumes that the mean $E(p_t) = p_0 + \mu t$ and variance $\text{Var}(p_t) = t\sigma_a^2$, both of them are time dependent. On the other hand, the trend-stationary model assumes the mean $E(p_t) = \beta_0 + \beta_1 t$, which depends on time, and variance $\text{Var}(p_t) = \text{Var}(x_t)$, which is finite and time invariant. The trend-stationary series can be transformed into a stationary one by removing the time trend via a simple linear regression analysis. For analysis of trend-stationary time series, see the method described in Section 2.10.

2.7.4 General Unit-Root Nonstationary Models

Consider an ARMA model. If one extends the model by allowing the AR polynomial to have 1 as a characteristic root, then the model becomes the well-known autoregressive integrated moving average (ARIMA) model. An ARIMA model is said to be unit-root nonstationary because its AR polynomial has a unit root. Similar to a random walk model, an ARIMA model has strong memory because the ψ_i coefficients in its MA representation do not decay over time to 0, implying that the past shock a_{t-i} of the model has a permanent effect on the series. A conventional approach for handling unit-root nonstationarity is to use *differencing*.

Differencing. A time series y_t is said to be an ARIMA$(p, 1, q)$ process if the change series $c_t = y_t - y_{t-1} = (1 - B)y_t$ follows a stationary and invertible ARMA(p, q) model. In finance, price series are commonly believed to be nonstationary, but the log return series, $x_t = \ln(P_t) - \ln(P_{t-1})$, is stationary. In this case, the log price series is unit-root nonstationary and hence can be treated as an ARIMA process. The idea of transforming a nonstationary series into a stationary one by considering its change series is called *differencing* in the time series literature. More formally, $c_t = y_t - y_{t-1}$ is referred to as the first differenced series of y_t. In some scientific fields, a time series y_t may contain multiple unit roots and needs to be differenced multiple times to become stationary. For example, if both y_t and its first differenced series $c_t = y_t - y_{t-1}$ are unit-root nonstationary, but $s_t = c_t - c_{t-1} = y_t - 2y_{t-1} + y_{t-2}$ is weakly stationary, then y_t has double unit roots, and s_t is the second differenced series of y_t. In addition, if s_t follows an ARMA(p, q) model, then y_t is an ARIMA$(p, 2, q)$ process. For such a time series, if s_t has a nonzero mean, then y_t has a quadratic time function and the quadratic time coefficient is related to the mean of s_t. The seasonally adjusted series of US quarterly GDP implicit price deflator might have double unit roots. However, the mean of the second differenced series is not significantly different from 0; see Exercises of the chapter. Box et al. (1994) discuss many properties of general ARIMA models.

2.7.5 Unit-Root Test

To test whether the log price p_t of an asset follows a random walk or a random walk with drift, we employ the models

$$p_t = \phi_1 p_{t-1} + e_t, \tag{2.39}$$

$$p_t = \phi_0 + \phi_1 p_{t-1} + e_t, \tag{2.40}$$

where e_t denotes the error term, and consider the null hypothesis $H_0 : \phi_1 = 1$ versus the alternative hypothesis $H_a : \phi_1 < 1$. This is the well-known unit-root testing problem (Dickey and Fuller, 1979). A convenient test statistic is the t ratio of the LS estimate of ϕ_1 under the null hypothesis. For Equation (2.39), the LS method gives

$$\hat{\phi}_1 = \frac{\sum_{t=1}^{T} p_{t-1} p_t}{\sum_{t=1}^{T} p_{t-1}^2}, \quad \hat{\sigma}_e^2 = \frac{\sum_{t=1}^{T} (p_t - \hat{\phi}_1 p_{t-1})^2}{T - 1},$$

where $p_0 = 0$ and T is the sample size. The t ratio is

$$\text{DF} \equiv t\text{-ratio} = \frac{\hat{\phi}_1 - 1}{\text{std}(\hat{\phi}_1)} = \frac{\sum_{t=1}^{T} p_{t-1} e_t}{\hat{\sigma}_e \sqrt{\sum_{t=1}^{T} p_{t-1}^2}},$$

which is commonly referred to as the *Dickey–Fuller test*. If $\{e_t\}$ is a white noise series with finite moments of order slightly greater than 2, then the DF-statistic converges to a function of the standard Brownian motion as $T \rightarrow \infty$; see Chan and Wei (1988) and Phillips (1987) for more information. If ϕ_0 is 0 but Equation (2.40) is employed anyway, then the resulting t-ratio for testing $\phi_1 = 1$ will converge to another nonstandard asymptotic distribution. In either case, simulation is used to obtain critical values of the test statistics; see Fuller (1995, Chapter 8) for selected critical values. Yet if $\phi_0 \neq 0$ and Equation (2.40) is used, then the t ratio for testing $\phi_1 = 1$ is asymptotically normal. However, large sample sizes are needed for the asymptotic normal distribution to hold.

For many economic time series, ARIMA(p, d, q) models might be more appropriate than the simple model used in Equation (2.40). In the econometric literature, AR(p) models are often used. The series is denoted by x_t. To verify the existence of a unit root in an AR(p) process, one may perform the test $H_0 : \beta = 1$ versus $H_a : \beta < 1$ using the regression

$$x_t = c_t + \beta x_{t-1} + \sum_{i=1}^{p-1} \phi_i \Delta x_{t-i} + e_t, \tag{2.41}$$

where c_t is a deterministic function of the time index t and $\Delta x_j = x_j - x_{j-1}$ is the differenced series of x_t. In practice, c_t can be 0 or a constant or $c_t = \omega_0 + \omega_1 t$. The t-ratio of $\hat{\beta} - 1$,

$$\text{ADF-test} = \frac{\hat{\beta} - 1}{\text{std}(\hat{\beta})},$$

where $\hat{\beta}$ denoting the LS estimate of β is the well-known *augmented Dickey–Fuller* unit-root test. Note that because of the first differencing, Equation (2.41) is equivalent to an AR(p) model with deterministic function c_t. Equation (2.41) can also be rewritten as

$$\Delta x_t = c_t + \beta_c x_{t-1} + \sum_{i=1}^{p-1} \phi_i \Delta x_{t-i} + e_t,$$

where $\beta_c = \beta - 1$. One can then test the equivalent hypothesis $H_0 : \beta_c = 0$ versus $H_a : \beta_c < 0$.

Example 2.4. Consider the log series of US quarterly GDP from 1947.I to 2008.IV. The series exhibits an upward trend, showing the growth of US economy and has high sample serial correlations; see the lower-left panel of Figure 2.17. The first differenced series, representing the growth rate of US GDP and also shown in Figure 2.17, seems to vary around a fixed mean level, even though the variability appears to be smaller in recent years. To confirm the observed phenomenon, we apply the augmented Dicky–Fuller unit-root test to the log series. On the basis of the sample PACF of the differenced series shown in Figure 2.17, we choose $p = 10$. Other values of p are also used, but they do not alter the conclusion of the test. With $p = 10$, ADF-test statistic is -1.611 with p-value 0.457, indicating that the unit-root hypothesis cannot be rejected.   □

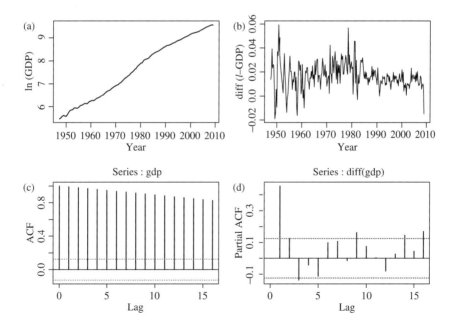

Figure 2.17. Log series of US quarterly GDP from 1947.I to 2008.IV: (a) Time plot the the logged GDP series, (b) sample ACF of the log GDP data, (c) time plot of the first differenced series, and (d) sample PACF of the differenced series.

```
> library(fUnitRoots)
> da=read.table("q-gdp4708.txt",header=T)
> gdp=log(da[,4])
> m1=ar(diff(gdp),method='mle')
> m1$order
[1] 10
> adfTest(gdp,lags=10,type=c("c"))
Title:
 Augmented Dickey--Fuller Test

Test Results:
  PARAMETER:
    Lag Order: 10
  STATISTIC:
    Dickey-Fuller: -1.6109
  P-VALUE:     0.4569
```

As another example, consider the log series of the S&P 500 index from January 3, 1950 to April 16, 2008 for 14,462 observations. The series is shown in Figure 2.18. Testing for a unit root in the index is relevant if one wishes to verify empirically that the Index follows a random walk with drift. To this end, we use $c_t = \omega_0 + \omega_1 t$ in applying the augmented Dickey–Fuller test. Furthermore, we choose $p = 15$ based on the sample PACF of the first differenced series. The resulting test statistic is -1.995 with p-value 0.581. Thus, the unit-root hypothesis cannot be rejected at

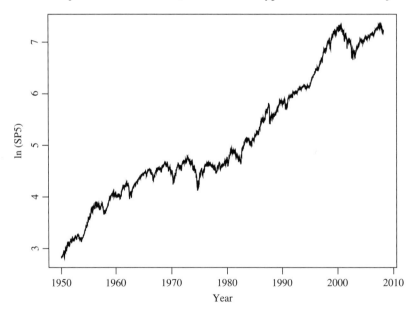

Figure 2.18. Time plot of the logarithm of daily S&P 500 index from January 3, 1950 to April 16, 2008.

any reasonable significance level. Further analysis shows that the constant term is statistically significant, whereas the estimate of the time trend is not at the usual 5% level. In summary, for the period from January 1950 to April 2008, the log series of the S&P 500 index contains a unit root and a positive drift, but there is no strong evidence of a time trend.

```
> library(fUnitRoots)
> da=read.table("d-sp55008.txt",header=T)
> sp5=log(da[,7])
> m2=ar(diff(sp5),method='mle') % Based on AIC
> m2$order
[1] 2
> adfTest(sp5,lags=2,type=("ct"))
Title:
 Augmented Dickey-Fuller Test

Test Results:
  PARAMETER:
    Lag Order: 2
  STATISTIC:
    Dickey-Fuller: -2.0179
  P-VALUE:   0.5708

> adfTest(sp5,lags=15,type=("ct")) % Based on PACF
Title:
 Augmented Dickey-Fuller Test

Test Results:
  PARAMETER:
    Lag Order: 15
  STATISTIC:
    Dickey-Fuller: -1.9946
  P-VALUE:    0.5807

> dsp5=diff(sp5)
> tdx=c(1:length(dsp5))
> m3=arima(dsp5,order=c(2,0,0),xreg=tdx)
> m3
Call: arima(x = dsp5,order=c(2,0,0),xreg = tdx)
Coefficients:
          ar1       ar2   intercept   tdx
       0.0721   -0.0387        4e-04     0
s.e.   0.0083    0.0083        2e-04     0
sigma^2 estimated as 8.1e-05:log likelihood=48287,aic=-96564
> m3$coef
           ar1               ar2         intercept                tdx
 7.214122e-02  -3.868823e-02    3.513995e-04  -7.165372e-09
> sqrt(diag(m3$var.coef))
          ar1             ar2      intercept                tdx
```

```
8.307510e-03 8.285669e-03 1.537309e-04 8.349685e-06
> tratio=m3$coef/sqrt(diag(m3$var.coef))
> tratio
       ar1        ar2  intercept        tdx
 8.6838549 -4.6692947  2.2858091 -0.0008582
```

2.8 EXPONENTIAL SMOOTHING

A commonly used method in forecasting is the exponential smoothing. Consider the 1-step ahead forecast of a time series x_t. Suppose that the forecast origin is h and all past data are available. Under the general belief that the serial dependence of x_t decays exponentially, one can use a weighted average of the past data to predict x_{h+1} with weights decaying exponentially. Specifically, one employs a quantity as

$$\hat{x}_{h+1} \propto wx_h + w^2 x_{h-1} + w^3 x_{h-2} + \cdots$$

$$= \sum_{j=1}^{\infty} w^j x_{h+1-j},$$

where w is a positive real number in $(0,1)$ referred to as the *discounting rate*. However, as one does not want to change the scale of x_t, it is necessary that the sum of the weights is equal to 1. Using properties of a geometric series, it is easy to see that $\sum_{j=1}^{\infty} w^i = \frac{1}{1-w}$. Therefore, a proper way to use weighted average is

$$\hat{x}_h(1) = (1-w)[wx_h + w^2 x_{h-1} + w^3 x_{h-2} + \cdots]. \tag{2.42}$$

This technique to produce forecasts is called the *exponential smoothing method*. It has been widely used in practice because the technique says that more recent data contribute more in predicting x_{h+1}.

It turns out that the exponential smoothing is a special case of the ARIMA models. Specifically, consider the ARIMA(0,1,1) model

$$(1 - B)x_t = (1 - \theta B)a_t,$$

where $\theta \in (0,1)$. Using the AR representation in Section 2.6.5, this model implies that

$$x_{h+1} = (1 - \theta)[\theta x_h + \theta^2 x_{h-1} + \theta^3 x_{h-2} + \cdots] + a_{h+1}.$$

Therefore, the 1-step ahead forecast is

$$\hat{x}_h(1) = (1 - \theta)[\theta x_h + \theta^2 x_{h-1} + \theta^3 x_{h-2} + \cdots].$$

This is precisely the exponential smoothing in Equation (2.42), with $\theta = w$.

Treating exponential smoothing as a special ARIMA(0,1,1) model has several advantages. First, the discounting parameter θ can be estimated via the maximum likelihood method. Second, one can identify and check the adequacy of the exponential smoothing method via the model building procedure of ARIMA models.

Example 2.5. Consider the daily volatility index (VIX) of Chicago Board Options Exchange (CBOE) from May 1, 2008 to April 19, 2010. The data are obtained from the CBOE web. Figure 2.19 shows the time plot of the logarithm of the VIX index and the sample ACFs of the differenced series. Since only the lag-1 ACF is significantly different from 0 at the 5% level, an MA(1) model is identified for the differenced series. Let $x_t = \ln(\text{VIX}_t)$. The fitted model is

$$(1 - B)x_t = (1 - 0.163B)a_t, \quad \tilde{\sigma}_a^2 = 0.0044.$$

The Ljung–Box statistics of the residuals shows that the fitted ARIMA(0,1,1) model is adequate. For instance, we have $Q(10) = 14.25$ with p-value 0.11, based on a chi-squared distribution with 9 degrees of freedom. Consequently, in this particular instance, one can employ the exponential smoothing to predict the log series of daily VIX index. □

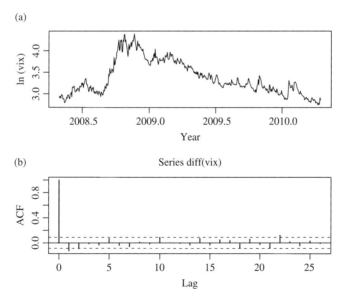

Figure 2.19. Log series of daily VIX index of Chicago Board Options Exchange from May 1, 2008 to April 19, 2010: (a) time plot of log VIX and (b) sample ACFs of the differenced series of log VIX.

R Demonstration

```
> da=read.table("d-vix0810.txt",header=T)
> vix=log(da$Close)
> length(vix)
[1] 496
> m1=arima(vix,order=c(0,1,1))
> m1
Call:
arima(x = vix, order = c(0, 1, 1))

Coefficients:
          ma1
       -0.1629
s.e.    0.0497

sigma^2 estimated as 0.004429: log likelihood=638.97, aic=-1273.94
>
> Box.test(m1$residuals,lag=10,type='Ljung')
        Box-Ljung test

data:  m1$residuals
X-squared = 14.2536, df = 10, p-value = 0.1617

> pp=1-pchisq(14.25,9)
> pp
[1] 0.1137060
```

Finally, one can extend the argument used in this section to show that the double exponential smoothing method for forecasting is a special case of the ARIMA(0,2,2) model.

2.9   SEASONAL MODELS

Some financial time series such as quarterly earnings per share of a company exhibits certain cyclical or periodic behavior. This type of series is called a *seasonal time series*. Recall that Figure 2.2 shows the time plot of quarterly earnings per share of the Coca-Cola Company from the first quarter of 1983 to the third quarter of 2009. The seasonal pattern is clearly seen. A careful examination of the plot shows that the quarterly earnings had a strong seasonality, grew exponentially during the sample period, and were subject to some disturbance in the late 1990s. The seasonal pattern repeats itself every year so that the periodicity of the series is 4. If monthly data are considered (e.g., monthly sales of Wal-Mart Stores), then the periodicity is 12.

Seasonal time series is also highly relevant in empirical studies of the prices of weather-related derivatives and energy futures. It is well known that most environmental time series exhibits strong seasonal behavior.

Analysis of seasonal time series has a long history. In some applications, seasonality is of secondary importance and is removed from the data, resulting in a seasonally adjusted time series that is then used to make inference. The procedure to remove seasonality from a time series is referred to as *seasonal adjustment*. Several methods are available. See, for instance, Peña et al. (2001, Chapter 8) and the references therein. Most economic data published by the US government are seasonally adjusted (e.g., the growth rate of GDP and the unemployment rate). In other applications such as forecasting, seasonality is as important as other characteristics of the data and must be handled accordingly. Because forecasting is a major objective of financial time series analysis, we focus on analyzing directly the seasonal time series. Our goal is to discuss some econometric models and methods that are useful in modeling seasonal time series.

2.9.1 Seasonal Differencing

Figure 2.20 shows the time plot of log earnings per share of the Coca-Cola Company. We took the log transformation for two reasons. First, it is used to handle the exponential growth of the series. Indeed, the plot shows that the log earnings increased linearly and the linear trend continued even after the 1998 disturbance, albeit with a different rate. Second, the transformation is used to stabilize the variability of the series. Compared with Figure 2.2, the increasing pattern in variability of quarterly

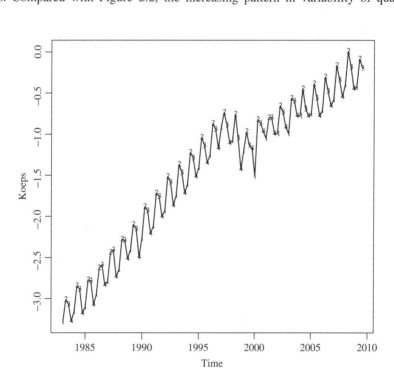

Figure 2.20. Time plot of quarterly earnings per share of the Coca-Cola from 1983.I to 2009.III: log earnings.

earnings disappears after the transformation. In fact, log transformation is commonly used in analysis of financial and economic time series. In this particular instance, all earnings are positive so that no adjustment is needed before taking the transformation. In some cases, one may need to add a positive constant to every data point before taking the transformation.

The log earnings are denoted by x_t. The upper left panel of Figure 2.21 shows the sample ACF of x_t, which suggests that the series of quarterly log earnings has strong serial correlations. A conventional method to handle such strong serial correlations is to consider the first differenced series of x_t (i.e., $\Delta x_t = x_t - x_{t-1} = (1 - B)x_t$). Differencing is a commonly used method to induce stationarity in time series analysis. The top panel of Figure 2.22 shows the time plot of Δx_t. From the plot, we see that (a) the differencing successfully removes the upward trend of the data and (b) the series now shows a very strong seasonal pattern. The lower-left plot of Figure 2.21 gives the sample ACF of Δx_t. The autocorrelations are large at lags, which are multiples of the periodicity 4. In addition, these seasonal autocorrelations decay slowly. What we observe here is a well-documented behavior of the sample ACF of a seasonal time series. Following the procedure of Box et al. (1994, Chapter 9), we take a seasonal difference of Δx_t to handle the strong seasonal pattern. Specifically, we consider

$$\Delta_4(\Delta x_t) = (1 - B^4)\Delta x_t = \Delta x_t - \Delta x_{t-4} = x_t - x_{t-1} - x_{t-4} + x_{t-5}.$$

The operation $\Delta_4 = (1 - B^4)$ is called a *seasonal differencing*. In general, for a seasonal time series y_t with periodicity s, seasonal differencing means

$$\Delta_s y_t = y_t - y_{t-s} = (1 - B^s)y_t.$$

Seasonal differencing is commonly used in business and finance. For example, in reporting quarterly earnings of a company, news media often compare the earnings with that of the same quarter one year earlier. The conventional difference $\Delta y_t = y_t - y_{t-1} = (1 - B)y_t$ is referred to as the *regular differencing*.

The lower panel of Figure 2.22 shows the time plot of $\Delta_4 \Delta x_t$. The strong seasonal pattern of Δx_t disappears. The lower right plot of Figure 2.21 shows the sample ACF of $\Delta_4 \Delta x_t$. From the ACF plot, we make the following observations:

1. The ACFs are negative and statistically significant at lags 1 and 4. That is, the serial correlations are different from 0 at the regular and seasonal lags.
2. The ACF is positive and statistically significant at lag 5.
3. The ACF is positive and marginally significant at lag 3.

These observed characteristics are common among empirical seasonal time series and lead to the development of multiplicative seasonal models introduced in the next section. For completeness, Figure 2.21 also provides the sample ACF of the seasonally differenced series $\Delta_4 x_t$, and the middle panel of Figure 2.22 shows associated time plot.

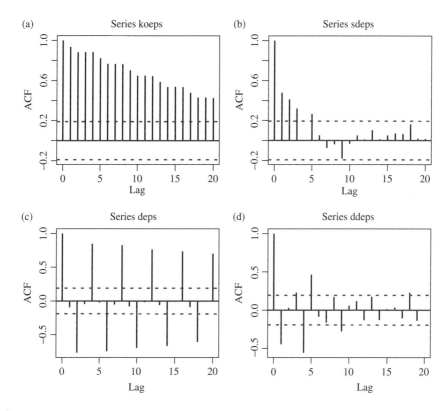

Figure 2.21. Sample ACF of the log series of quarterly earnings per share of the Coca-Cola Company from 1983.I to 2009.III, where `koeps` is the log earnings, `deps` is the first differenced series, `sdeps` is the seasonally differenced series, and `ddeps` denotes series with regular and seasonal differencing. (a) Series koeps, (b) series sdeps, (c) series deps, and (d) series ddeps.

2.9.2 Multiplicative Seasonal Models

A statistical model whose autocorrelations possess the behavior shown by the sample ACF of $(1 - B^4)(1 - B)x_t$ in Figure 2.21 is the multiplicative seasonal model. A simple multiplicative seasonal model assumes the form

$$(1 - B)(1 - B^s)x_t = (1 - \theta B)(1 - \Theta B^s)a_t, \qquad (2.43)$$

where s is the periodicity of the series, a_t is a white noise series, $|\theta| < 1$, and $|\Theta| < 1$. This model is referred to as the *airline model* in the literature (Box et al., 1994, Chapter 9). It has been found to be widely applicable in modeling seasonal time series. The AR part of the model simply consists of the regular and seasonal differences, whereas the MA part involves two parameters. Focusing on the MA part (i.e., on the model),

$$w_t = (1 - \theta B)(1 - \Theta B^s)a_t = a_t - \theta a_{t-1} - \Theta a_{t-s} + \theta \Theta a_{t-s-1},$$

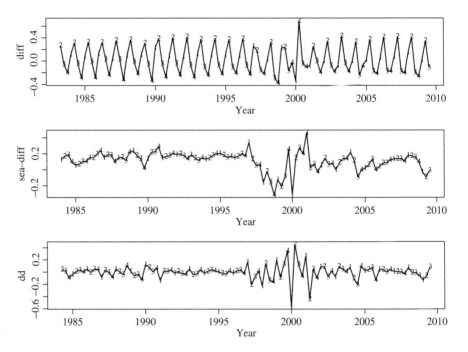

Figure 2.22. Time plots of quarterly earnings per share of the Coca-Cola Company from 1983.I to 2009.III. The top panel is for the first differenced series, the middle panel is for the seasonally differenced series, and the lower panel is for the regularly and seasonally differenced series.

where $w_t = (1 - B^s)(1 - B)x_t$ and $s > 1$. It is easy to obtain that $E(w_t) = 0$ and

$$\text{Var}(w_t) = (1 + \theta^2)(1 + \Theta^2)\sigma_a^2$$
$$\text{Cov}(w_t, w_{t-1}) = -\theta(1 + \Theta^2)\sigma_a^2$$
$$\text{Cov}(w_t, w_{t-s+1}) = \theta\Theta\sigma_a^2$$
$$\text{Cov}(w_t, w_{t-s}) = -\Theta(1 + \theta^2)\sigma_a^2$$
$$\text{Cov}(w_t, w_{t-s-1}) = \theta\Theta\sigma_a^2$$
$$\text{Cov}(w_t, w_{t-\ell}) = 0, \quad \text{for} \quad \ell \neq 0, 1, s-1, s, s+1.$$

Consequently, the ACF of the w_t series is given by

$$\rho_1 = \frac{-\theta}{1 + \theta^2}, \quad \rho_s = \frac{-\Theta}{1 + \Theta^2}, \quad \rho_{s-1} = \rho_{s+1} = \rho_1 \rho_s = \frac{\theta\Theta}{(1 + \theta^2)(1 + \Theta^2)},$$

and $\rho_\ell = 0$ for $\ell > 0$ and $\ell \neq 1, s-1, s, s+1$. For example, if w_t is a quarterly time series, then $s = 4$ and for $\ell > 0$, the ACF ρ_ℓ is nonzero at lags 1, 3, 4, and 5 only. This is indeed the case for the log quarterly earnings of the Coca-Cola Company.

It is interesting to compare the prior ACF with those of the MA(1) model $y_t = (1 - \theta B)a_t$ and the MA(s) model $z_t = (1 - \Theta B^s)a_t$. The ACF of y_t and z_t series are

$$\rho_1(y) = \frac{-\theta}{1 + \theta^2}, \quad \text{and} \quad \rho_\ell(y) = 0, \quad \ell > 1,$$

$$\rho_s(z) = \frac{-\Theta}{1 + \Theta^2}, \quad \text{and} \quad \rho_\ell(z) = 0, \quad \ell > 0, \quad \neq s.$$

We see that (i) $\rho_1 = \rho_1(y)$, (ii) $\rho_s = \rho_s(z)$, and (iii) $\rho_{s-1} = \rho_{s+1} = \rho_1(y) \times \rho_s(z)$. Therefore, the ACF of w_t at lags $(s - 1)$ and $(s + 1)$ can be regarded as the *interaction* between lag-1 and lag-s serial dependence, and the model of w_t is called a *multiplicative* seasonal MA model. In practice, a multiplicative seasonal model says that the dynamics of the regular and seasonal components of the series are approximately orthogonal.

The usefulness of the airline model in Equation (2.43) can be obtained by rewriting the model as

$$\frac{1 - B}{1 - \theta B} \left(\frac{1 - B^s}{1 - \Theta B^s} x_t \right) = a_t.$$

Let $y_t = (1 - B^s)/(1 - \Theta B^s)x_t$. Then, we have

$$(1 - B)y_t = (1 - \theta B)a_t, \quad (1 - B^s)x_t = (1 - \Theta B^s)y_t.$$

Here, y_t is the exponential smoothing model of Section 2.8 and x_t is another exponential smoothing model, but for the seasonal component. Thus, the airline model can be regarded as an exponential smoothing model on top of another exponential smoothing model. One exponential smoothing is for the usual serial dependence, whereas the other one is for the seasonal dependence.

Example 2.6. In this example, we apply the airline model to the log series of quarterly earnings per share of Coca-Cola from 1983 to 2009. On the basis of the exact likelihood method, the fitted model is

$$(1 - B)(1 - B^4)x_t = (1 - 0.4096B)(1 - 0.8203B^4)a_t, \quad \hat{\sigma}_a^2 = 0.00724,$$

where standard errors of the two MA parameters are 0.0866 and 0.0743, respectively. The Ljung–Box statistics of the residuals show $Q(12) = 13.20$ with p-value 0.35 when the degrees of freedom is 12. With adjustment to 10 degrees of freedom, the p-value becomes 0.21. Figure 2.23 shows the model checking plots for the fitted airline model. It consists of three plots. The top plot shows the standardized residuals. This plot can be used to examine the iid assumption of the residuals and to spot possible outliers in the data. The middle plot is the autocorrelations of the residuals. Ideally, all the ACF of the residuals should be within the limit of two standard errors. The bottom plot provides the p-values of the Ljung–Box statistics for several values of m.

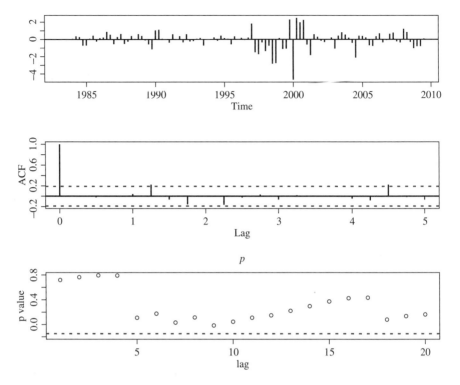

Figure 2.23. Model checking for the fitted Airline model to the log quarterly earnings of the Coca-Cola Company from 1983 to 2009: (a) time plot of the standard residuals, (b) ACF of the standardized residuals, and (c) plot of p-values of the Ljung–Box statistics of the standardized residuals.

If the fitted model is adequate in describing the serial dependence of the data, then all p-values should be greater than the type I error. The dashed line of the plot gives the default type I error of 0.05. The number of m can be specified by the subcommand gof of the command tsdiag. On the basis of the three plots, except for a possible outlier at the end of 1999, the model appears to be adequate for the log quarterly earnings of the Coca-Cola Company.   □

To illustrate the forecasting performance of the fitted seasonal model, we reestimate the model using the first 100 observations, that is, from 1983 to 2007, and reserve the last seven data points for forecasting evaluation. The fitted model becomes $(1 - B)(1 - B^4)x_t = (1 - 0.4209B)(1 - 0.8099B^4)a_t$ with $\sigma_a^2 = 0.00743$. We compute 1-step to 7-step ahead forecasts and their standard errors of the fitted model at the forecast origin $h = 100$. An antilog transformation is taken to obtain forecasts of earnings per share using the relationship between normal and log-normal distributions given in Chapter 1. Figure 2.24 shows the forecast performance of the model, where the data are represented by solid line, observations during the forecasting period are

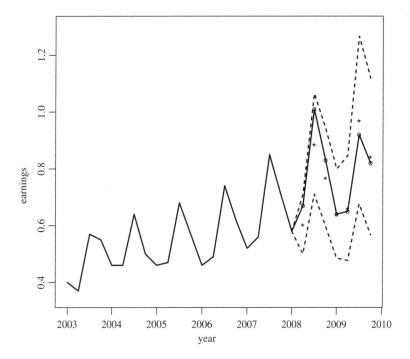

Figure 2.24. Out-of-sample point and interval forecasts for the quarterly earnings of the Coca-Cola Company. The forecast origin is the fourth quarter of 2007. In the plot, circles represent the actual earnings in the forecasting period, asterisks represent point forecasts, and dashed lines represent a 95% interval forecasts.

marked by "o," point forecasts are indicated by "*," and the dashed lines represent 95% interval forecasts. The forecasts show a strong seasonal pattern and are close to the observed data. The actual earnings are all in the interval forecasts.

R Demonstration

```
> da=read.table("q-ko-earns8309.txt",header=T)
> head(da)
      pends   anntime   value
1 19830331 19830426 0.0375
...
6 19840630 19840720 0.0583
> eps=log(da$value)
> koeps=ts(eps,frequency=4,start=c(1983,1))
> plot(koeps,type='l')
> points(koeps,pch=c1,cex=0.6)
% Obtain ACF plot
> par(mfcol=c(2,2))
> koeps=log(da$value)
> deps=diff(koeps)
```

```
> sdeps=diff(koeps,4)
> ddeps=diff(sdeps)
> acf(koeps,lag=20)
> acf(deps,lag=20)
> acf(sdeps,lag=20)
> acf(ddeps,lag=20)
% Obtain time plots
> c1=c("2","3","4","1")
> c2=c("1","2","3","4")
> par(mfcol=c(3,1))
> plot(deps,xlab='year',ylab='diff',type='l')
> points(deps,pch=c1,cex=0.7)
> plot(sdeps,xlab='year',ylab='sea-diff',type='l')
> points(sdeps,pch=c2,cex=0.7)
> plot(ddeps,xlab='year',ylab='dd',type='l')
> points(ddeps,pch=c1,cex=0.7)
%  Estimation
> m1=arima(koeps,order=c(0,1,1),seasonal=list(order=c(0,1,1),period=4))
> m1
Call:
arima(x=koeps,order=c(0,1,1),seasonal=list(order=c(0,1,1),period=4))
Coefficients:
          ma1      sma1
      -0.4096   -0.8203
s.e.   0.0866    0.0743

sigma^2 estimated as 0.00724:  log likelihood = 104.25,  aic = -202.5

> tsdiag(m1,gof=20)  % model checking
> Box.test(m1$residuals,lag=12,type='Ljung')
        Box-Ljung test
data:  m1$residuals
X-squared = 13.3034, df = 12, p-value = 0.3474
> pp=1-pchisq(13.30,10)
> pp
[1] 0.2073788    % p-value
% Out-of-sample forecasting
> koeps=log(da$value)
> length(koeps)
[1] 107
> y=koeps[1:100]
> m1=arima(y,order=c(0,1,1),seasonal=list(order=c(0,1,1),period=4))
> m1
Coefficients:
          ma1      sma1
      -0.4209   -0.8099
s.e.   0.0874    0.0767

sigma^2 estimated as 0.007432:  log likelihood = 95.78,  aic = -185.57
% Prediction
> pm1=predict(m1,7)
> names(pm1)
[1] "pred" "se"
> pred=pm1$pred
```

```
> se=pm1$se
% Anti-log transformation
> ko=da$value
> fore=exp(pred+se^2/2)
> v1=exp(2*pred+se^2)*(exp(se^2)-1)
> s1=sqrt(v1)
> eps=ko[80:107]
> length(eps)
[1] 28
> tdx=(c(1:28)+3)/4+2002
> upp=c(ko[100],fore+2*s1)
> low=c(ko[100],fore-2*s1)
> min(low,eps)
[1] 0.37
> max(upp,eps)
[1] 1.267623
> plot(tdx,eps,xlab='year',ylab='earnings',type='l',ylim=c(0.35,1.3))
> points(tdx[22:28],fore,pch='*')
> lines(tdx[21:28],upp,lty=2)
> lines(tdx[21:28],low,lty=2)
> points(tdx[22:28],ko[101:107],pch='o',cex=0.7)
```

In some applications, the multiplicative model

$$(1 - B)(1 - B^s)x_t = (1 - \theta_1 B - \theta_2 B^2)(1 - \Theta B^s)a_t$$

is needed, especially when $s > 4$. Here, the nonzero ACFs of the differenced series $w_t = (1 - B)(1 - B^s)x_t$ may occur at lags 1, 2, $s - 2, s - 1, s, s + 1$, and $s + 2$. The sample ACF of w_t can be used to identify such a model.

The model

$$w_t = (1 - \theta B - \Theta B^s)a_t, \tag{2.44}$$

where $|\theta| < 1$ and $|\Theta| < 1$, is a nonmultiplicative seasonal MA model. It is easy to see that for the model in Equation (2.44), $\rho_{s+1} = 0$. A multiplicative model is more parsimonious than its nonmultiplicative counterpart because both models use the same number of parameters, but the multiplicative model has more nonzero ACFs.

2.9.3 Seasonal Dummy Variable

When the seasonal pattern of a time series is stable over time (e.g., close to a deterministic function), dummy variables may be used to handle the seasonality. By seasonal dummy variables, we mean the indicator variables for the seasons within a year. For quarterly data, the dummy variables represent spring, summer, autumn and winter, respectively, and three of them are used in an analysis. This approach is adopted by some analysts. However, deterministic seasonality is a special case of the multiplicative seasonal model discussed earlier. Specifically, if $\Theta = 1$, then model (Eq. 2.43) contains a deterministic seasonal component. Consequently, the same forecasts are obtained by using either dummy variables or a multiplicative seasonal model when the seasonal pattern is deterministic. Yet, the use of dummy variables can lead to inferior forecasts if the seasonal pattern is not deterministic. In practice, we recommend that the exact likelihood method should be used to estimate a multiplicative seasonal model, especially when the sample size is small or when there is the possibility of having a deterministic seasonal component.

Example 2.7. To demonstrate the deterministic seasonal behavior, consider the monthly simple returns of the CRSP Decile 1 Index from January 1970 to December 2008 for 468 observations. The series is shown in Figure 2.25a, and the time plot does not show any clear pattern of seasonality. However, the sample ACF of the return series shown in Figure 2.25b contains significant lags at 12, 24, and 36, as well as lag 1. If seasonal ARMA models are entertained, a model in the form

$$(1 - \phi_1 B)(1 - \phi_{12} B^{12})X_t = (1 - \theta_{12} B^{12})a_t$$

is identified, where X_t denotes the monthly simple return. After removing the insignificant parameter, the fitted model becomes

$$(1 - 0.179B)(1 - 0.989B^{12})X_t = (1 - 0.913B^{12})a_t, \quad \tilde{\sigma}_a^2 = 0.00472.$$

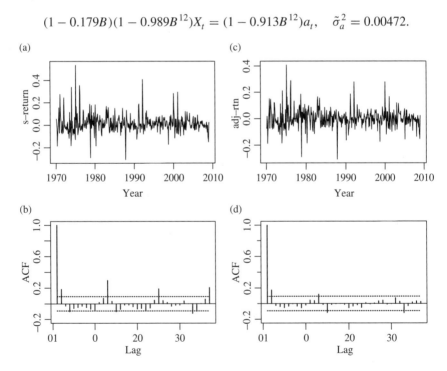

Figure 2.25. Monthly simple returns of CRSP Decile 1 index from January 1970 to December 2008: (a) time plot of the simple returns, (b) sample ACF of the simple returns, (c) time plot of the simple returns after adjusting for January effect, and (d) sample ACF of the adjusted simple returns.

The near cancellation between seasonal AR and MA factors is clearly seen. This highlights the usefulness of using the exact likelihood method, and the estimation result suggests that the seasonal behavior might be deterministic. To further confirm this assertion, we define the dummy variable for January, that is,

$$\text{Jan}_t = \begin{cases} 1 & \text{if } t \text{ is January,} \\ 0 & \text{otherwise,} \end{cases}$$

and employ the simple liner regression

$$X_t = \beta_0 + \beta_1 \, \text{Jan}_t + e_t.$$

The fitted model is $X_t = 0.0029 + 0.1253\text{Jan}_t + e_t$, where the standard errors of the estimates are 0.0033 and 0.0115, respectively. The right panel of Figure 2.25 shows the time plot and sample ACF of the residual series of the prior simple linear regression. From the sample ACF, serial correlations at lags 12, 24, and 36 largely disappear, suggesting that the seasonal pattern of the Decile 1 returns has been successfully removed by the January dummy variable. Consequently, the seasonal behavior in the monthly simple return of Decile 1 is mainly due to the *January effect*.   □

R Demonstration. Output edited and % denotes explanation

```
> da=read.table("m-deciles08.txt",header=T)
> d1=da[,2]
> jan=rep(c(1,rep(0,11)),39) % Create January dummy.
> m1=lm(d1 ~ jan)
> summary(m1)
lm(formula = d1 ~ jan)


Coefficients:
              Estimate Std. Error t value Pr(>|t|)
(Intercept) 0.002864    0.003333   0.859    0.391
jan         0.125251    0.011546  10.848   <2e-16 ***
---

Residual standard error: 0.06904 on 466 degrees of freedom
Multiple R-squared: 0.2016,    Adjusted R-squared: 0.1999

> m2=arima(d1,order=c(1,0,0),seasonal=list(order=c(1,0,1),
+ period=12))
> m2
Coefficients:
         ar1     sar1     sma1   intercept
      0.1769   0.9882  -0.9144    0.0118  % insignificant intercept
s.e.  0.0456   0.0093   0.0335    0.0129

sigma^2 estimated as 0.004717: log likelihood=584.07, aic=-1158.14
> tsdiag(m2,gof=36)   % plot not shown.

> m2=arima(d1,order=c(1,0,0),seasonal=list(order=c(1,0,1),
+ period=12),include.mean=F)
> m2
```

```
Call:
arima(x=d1,order=c(1,0,0),seasonal=list(order=c(1,0,1),period=12),
    include.mean = F)

Coefficients:
          ar1     sar1     sma1
       0.1787   0.9886  -0.9127   % All estimates are significant.
s.e.   0.0456   0.0089   0.0335

sigma^2 estimated as 0.00472: log likelihood=583.68, aic=-1159.36
```

2.10   REGRESSION MODELS WITH TIME SERIES ERRORS

In many applications, the relationship between two time series is of major interest. An obvious example is the *Market Model* in finance that relates the excess return of an individual stock to that of a market index. The term structure of interest rates is another example in which the time evolution of the relationship between interest rates with different maturities is investigated. These examples lead naturally to the consideration of a linear regression in the form

$$y_t = \alpha + \beta x_t + e_t, \tag{2.45}$$

where y_t and x_t are two time series and e_t denotes the error term. The LS method is often used to estimate model (Eq. 2.45). If $\{e_t\}$ is a white noise series, then the LS method produces consistent estimates. In practice, however, it is common to see that the error term e_t is serially correlated. In this case, we have a regression model with time series errors, and the LS estimates of α and β may not be consistent.

A regression model with time series errors is widely applicable in economics and finance. Care must be exercised, however, as not to overlook the serial dependence in e_t. It pays to study the model carefully.

We introduce the model by considering the relationship between two US weekly interest rate series:

1. x_{1t}: The 1-year treasury constant maturity rate,
2. x_{3t}: The 3-year treasury constant maturity rate.

Both series have 2467 observations from January 5, 1962 to April 10, 2009 and are measured in percentages. The series is obtained from the Federal Reserve Bank of St Louis. Strictly speaking, we should model the two interest series jointly using multivariate time series analysis (Tsay, 2010, Chapter 8). However, for simplicity, we focus here on the regression type of analysis and ignore the issue of simultaneity.

Figure 2.26 shows the time plots of the two interest rates with solid line denoting the 1-year rate and dashed line the 3-year rate. Figure 2.27a is the scatter plot between x_{1t} and x_{3t}, indicating that, as expected, the two interest rates are highly correlated. A naive way to describe the relationship between the two interest rates is to use the simple model $x_{3t} = \alpha + \beta x_{1t} + e_t$. This results in a fitted model

$$x_{3t} = 0.832 + 0.930x_{1t} + e_t, \quad \hat{\sigma}_e = 0.523 \tag{2.46}$$

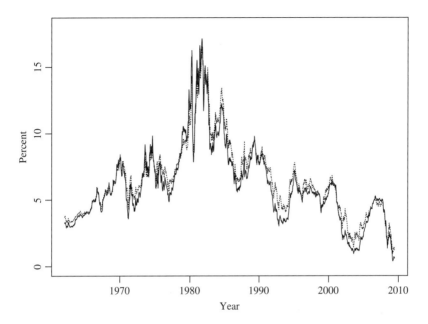

Figure 2.26. Time plots of US weekly interest rates (in percentages) from January 5, 1962 to April 10, 2009. The solid line represents the treasury 1-year constant maturity rate and the dashed line represents the treasury 3-year constant maturity rate.

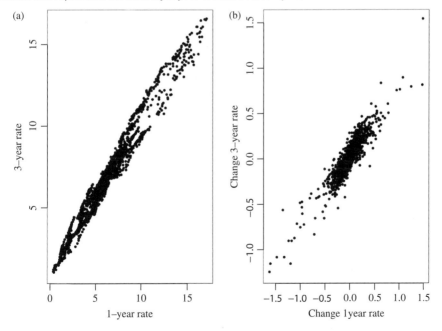

Figure 2.27. Scatter plots of US weekly interest rates from January 5, 1962 to April 10, 2009: (a) 3-year rate versus 1-year rate and (b) changes in 3-year rate versus changes in 1-year rate.

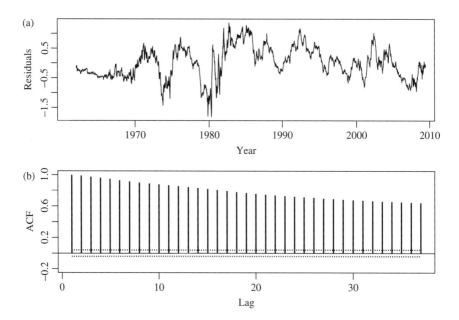

Figure 2.28. Residual series of linear regression (Eq 2.46) for two US weekly interest rates: (a) time plot and (b) sample ACF.

with $R^2 = 96.5\%$, where the standard errors of the two coefficients are 0.024 and 0.004, respectively. Model (Eq. 2.46) confirms the high correlation between the two interest rates. However, the model is seriously inadequate as shown by Figure 2.28, which gives the time plot and ACF of its residuals. In particular, the sample ACF of the residuals is highly significant and decays slowly, showing the pattern of a unit-root nonstationary time series. The behavior of the residuals suggests that marked differences exist between the two interest rates. Using the modern econometric terminology, if one assumes that the two interest rate series are unit-root nonstationary, then the behavior of the residuals of Equation (2.46) indicates that the two interest rates are not *cointegrated*; see Tsay (2010, Chapter 8) for the discussion of cointegration. In other words, the data fail to support the hypothesis that there exists a long-term equilibrium between the two interest rates. In some sense, this is not surprising because the pattern of "inverted yield curve" did occur during the data span. By inverted yield curve, we mean the situation under which interest rates are inversely related to their time to maturities.

The unit-root behavior of both interest rate series and the residuals of Equation (2.46) leads to the consideration of the change series of interest rates. Let

1. $c_{1t} = x_{1t} - x_{1,t-1} = (1 - B)x_{1t}$ for $t \geq 2$: changes in the 1-year interest rate;
2. $c_{3t} = x_{3t} - x_{3,t-1} = (1 - B)x_{3t}$ for $t \geq 2$: changes in the 3-year interest rate,

and consider the linear regression $c_{3t} = \beta c_{1t} + e_t$. Figure 2.29 shows time plots of the two change series, whereas Figure 2.27b provides a scatter plot between them. The

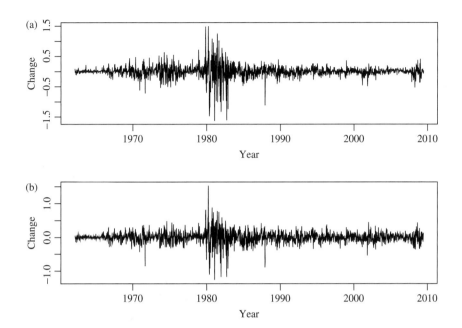

Figure 2.29. Time plots of the change series of US weekly interest rates from January 12, 1962 to April 10, 2009: (a) changes in the treasury 1-year constant maturity rate and (b) changes in the treasury 3-year constant maturity rate.

change series remain highly correlated with a fitted linear regression model given by

$$c_{3t} = 0.792c_{1t} + e_t, \quad \hat{\sigma}_e = 0.0690, \tag{2.47}$$

with $R^2 = 82.5\%$. The standard error of the coefficient is 0.0073. This model further confirms the strong linear dependence between interest rates. Figure 2.30 shows the time plot and sample ACF of the residuals of Equation (2.47). Once again, the ACF shows some significant serial correlations in the residuals, but magnitudes of the correlations are much smaller. This weak serial dependence in the residuals can be modeled by using the simple time series models discussed in the previous sections, and we have a linear regression with time series errors.

The main objective of this section is to discuss a simple approach for building a linear regression model with time series errors. The approach is straightforward. We employ a simple time series model discussed in this chapter for the residual series and estimate the whole model jointly. For illustration, consider the simple linear regression in Equation (2.47). Because residuals of the model are serially correlated, we shall identify a simple ARMA model for the residuals. From the sample ACF of the residuals shown in Figure 2.30, we specify an MA(1) model for the residuals and modify the linear regression model to

$$c_{3t} = \beta c_{1t} + e_t, \quad e_t = a_t - \theta_1 a_{t-1}, \tag{2.48}$$

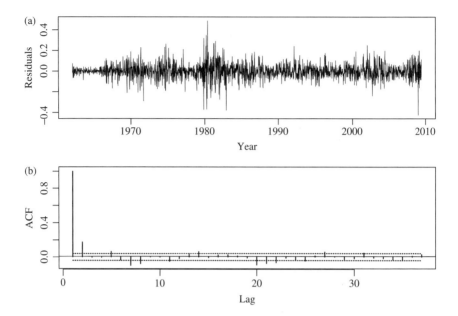

Figure 2.30. Residual series of the linear regression (Eq 2.47) for two change series of US weekly interest rates: (a) time plot and (b) sample ACF.

where $\{a_t\}$ is assumed to be a white noise series. In other words, we simply use an MA(1) model, without the constant term, to capture the serial dependence in the error term of Equation (2.47). The resulting model is a simple example of linear regression with time series errors. In practice, more elaborated time series models can be added to a linear regression equation to form a general regression model with time series errors.

Estimating a regression model with time series errors was not easy before the advent of modern computers. Special methods such as the Cochrane–Orcutt estimator have been proposed to handle the serial dependence in the residuals (Greene, 2003, p. 273). By now, the estimation is as easy as that of other time series models. If the time series model used is stationary and invertible, then one can estimate the model jointly via the maximum likelihood method. This is the approach we take using the command `arima` in R. For the US weekly interest rate data, the fitted version of model (Eq. 2.48) is

$$c_{3t} = 0.794c_{1t} + e_t, \quad e_t = a_t + 0.1823a_{t-1}, \quad \hat{\sigma}_a = 0.0678, \qquad (2.49)$$

with $R^2 = 83.1\%$. The standard errors of the parameters are 0.0075 and 0.0196, respectively. The model no longer has a significant lag-1 residual ACF, even though some minor residual serial correlations remain at lags 4, 6, and 7. The incremental improvement of adding additional MA parameters at lags 4, 6, and 7 to the residual equation is small, and the result is not reported here.

Comparing the models in Equations (2.46), (2.47), and (2.49), we make the following observations. First, the high R^2 96.5% and coefficient 0.930 of model (Eq. 2.46) are misleading because the residuals of the model show strong serial correlations. Second, for the change series, R^2 and the coefficient of c_{1t} of models (Eqs. 2.47 and 2.49) are close. In this particular instance, adding the MA(1) model to the change series provides only a marginal improvement. This is not surprising because the estimated MA coefficient is small numerically, even though it is statistically highly significant. Third, the analysis demonstrates that it is important to check residual serial dependence in linear regression analysis.

From Equation (2.49), the model shows that the two weekly interest rate series are related as

$$x_{3t} = x_{3,t-1} + 0.794(x_{1t} - x_{1,t-1}) + a_t + 0.182a_{t-1}.$$

The interest rates are concurrently and serially correlated.

R Demonstration. Output edited.

```
> r1=read.table("w-gs1yr.txt",header=T)[,4]
> r3=read.table("w-gs3yr.txt",header=T)[,4]
> m1=lm(r3~r1)
> summary(m1)
Call:
lm(formula = r3 ~ r1)
Coefficients:
              Estimate Std. Error t value Pr(>|t|)
(Intercept)   0.83214    0.02417   34.43   <2e-16 ***
r1            0.92955    0.00357  260.40   <2e-16 ***
---
Residual standard error: 0.5228 on 2465 degrees of freedom
Multiple R-squared: 0.9649,    Adjusted R-squared: 0.9649

> plot(m1$residuals,type='l')
> acf(m1$residuals,lag=36)
> c1=diff(r1)
> c3=diff(r3)
> m2=lm(c3 -1+c1)
> summary(m2)
Call:
lm(formula = c3 ~ -1 + c1)
Coefficients:
   Estimate Std. Error t value Pr(>|t|)
c1 0.791935   0.007337   107.9   <2e-16 ***
---
Residual standard error: 0.06896 on 2465 degrees of freedom
Multiple R-squared: 0.8253,    Adjusted R-squared: 0.8253

> acf(m2$residuals,lag=36)

> m3=arima(c3,order=c(0,0,1),xreg=c1,include.mean=F)
> m3
```

```
Call:
arima(x = c3, order = c(0, 0, 1), xreg = c1, include.mean = F)
Coefficients:
          ma1       c1
       0.1823   0.7936
s.e.   0.0196   0.0075

sigma^2 estimated as 0.0046: log likelihood=3136.62, aic=-6267.23
>
> rsq=(sum(c3^2)-sum(m3$residuals^2))/sum(c3^2)
> rsq
[1] 0.8310077
```

Summary. We outline a general procedure for analyzing linear regression models with time series errors as follows:

1. Fit the linear regression model and check serial correlations of the residuals.
2. If the residual series is unit-root nonstationary, take the first difference of both the dependent and explanatory variables. Go to step 1. If the residual series appears to be stationary, identify an ARMA model for the residuals and modify the linear regression model accordingly.
3. Perform a joint estimation via the maximum likelihood method and check the fitted model for further improvement.

To check the serial correlations of residuals, we recommend that the Ljung–Box statistics be used instead of the Durbin–Watson (DW) statistic because the latter only considers the lag-1 serial correlation. There are cases in which serial dependence in residuals appears at higher-order lags. This is particularly so when the time series involved exhibits some seasonal behavior.

Remark. For a residual series e_t with T observations, the Durbin–Watson statistic is

$$\text{DW} = \frac{\sum_{t=2}^{T}(e_t - e_{t-1})^2}{\sum_{t=1}^{T} e_t^2}.$$

Straightforward calculation shows that $\text{DW} \approx 2(1 - \hat{\rho}_1)$, where $\hat{\rho}_1$ is the lag-1 ACF of $\{e_t\}$.   □

Remark. In some application, the structure of the serial dependence in residuals might not be of direct interest, as one focuses on making inference concerning the linear regression model. In this case, it suffices to obtain a consistent estimate of the covariance matrix of the regression coefficient. Several methods are available in the literature to estimate the covariance matrix of the regression coefficients in the presence of serial correlations and conditional heteroscedasticity. Interested readers are referred to Tsay (2010, Section 2.10).   □

2.11 LONG-MEMORY MODELS

We have discussed that for a stationary time series, the ACF decays exponentially to 0 as lag increases. Yet for a unit-root nonstationary time series, it can be shown that the sample ACF converges to 1 for all fixed lags as the sample size increases (Chan and Wei, 1988; Tiao and Tsay, 1983). There exist some time series whose ACF decays slowly to 0 at a polynomial rate as the lag increases. These processes are referred to as *long-memory time series*. One such example is the fractionally differenced process defined by

$$(1 - B)^d x_t = a_t, \quad -0.5 < d < 0.5, \tag{2.50}$$

where $\{a_t\}$ is a white noise series. Properties of model (Eq. 2.50) have been widely studied in the literature (Hosking, 1981). We summarize some of these properties as follows:

1. If $d < 0.5$, then x_t is a weakly stationary process and has the infinite MA representation

$$x_t = a_t + \sum_{i=1}^{\infty} \psi_i a_{t-i}, \quad \text{with}$$

$$\psi_k = \frac{d(1+d)\cdots(k-1+d)}{k!} = \frac{(k+d-1)!}{k!(d-1)!}.$$

2. If $d > -0.5$, then x_t is invertible and has the infinite AR representation

$$x_t = \sum_{i=1}^{\infty} \pi_i x_{t-i} + a_t, \quad \text{with}$$

$$\pi_k = \frac{-d(1-d)\cdots(k-1-d)}{k!} = \frac{(k-d-1)!}{k!(-d-1)!}.$$

3. For $-0.5 < d < 0.5$, the ACF of x_t is

$$\rho_k = \frac{d(1+d)\cdots(k-1+d)}{(1-d)(2-d)\cdots(k-d)}, \quad k = 1, 2, \ldots$$

In particular, $\rho_1 = d/(1-d)$ and

$$\rho_k \approx \frac{(-d)!}{(d-1)!} k^{2d-1}, \quad \text{as} \quad k \to \infty.$$

4. For $-0.5 < d < 0.5$, the PACF of x_t is $\phi_{k,k} = d/(k-d)$ for $k = 1, 2, \ldots$

5. For $-0.5 < d < 0.5$, the spectral density function $f(\omega)$ of x_t, which is the Fourier transform of the ACF of x_t, satisfies

$$f(\omega) \sim \omega^{-2d}, \quad \text{as} \quad \omega \to 0, \tag{2.51}$$

where $\omega \in [0, 2\pi]$ denotes the frequency.

Of particular interest here is the behavior of ACF of x_t when $d < 0.5$. The property says that $\rho_k \sim k^{2d-1}$, which decays at a polynomial, instead of an exponential, rate. For this reason, such an x_t process is called a *long-memory time series*. A special characteristic of the spectral density function in Equation (2.51) is that the spectrum diverges to infinity as $\omega \to 0$. However, the spectral density function of a stationary ARMA process is bounded for all $\omega \in [0, 2\pi]$.

Earlier, we used the binomial theorem for noninteger powers

$$(1-B)^d = \sum_{k=0}^{\infty} (-1)^k \binom{d}{k} B^k, \quad \binom{d}{k} = \frac{d(d-1)\cdots(d-k+1)}{k!}.$$

If the fractionally differenced series $(1-B)^d x_t$ follows an ARMA(p,q) model, then x_t is called an ARFIMA(p,d,q) process, which is a generalized ARIMA model by allowing for noninteger d.

In practice, if the sample ACF of a time series is not large in magnitude, but decays slowly, then the series may have long memory. As an illustration, Figure 2.31 shows the sample ACFs of the absolute series of daily simple returns for the CRSP value- and equal-weighted indexes from January 2, 1970 to December 31, 2008. The ACFs are relatively small in magnitude but decay very slowly; they appear to be significant at the 5% level even after 300 lags. For more information on the behavior of sample ACF of absolute return series, see Ding et al. (1993). For the pure fractionally differenced model in Equation (2.50), one can estimate d using either a maximum likelihood method or a regression method with logged periodogram at the lower frequencies. See, for instance, Geweke and Porter-Hudak (1982). Finally, long-memory models have attracted some attention in the finance literature in part because of the work on fractional Brownian motion in the continuous-time models.

Example 2.8. To illustrate the modeling of long-memory time series, we consider the daily absolute returns of the value-weighted index of CRSP from January 2, 1970 to December 31, 2008. The sample ACF of the series is shown in Figure 2.31a. Treating the series as a purely fractionally differenced process and applying the Geweke-Porter-Hudak method, we obtain $\hat{d} = 0.372$ with standard error 0.070. The estimate is indeed in the stationary and invertible interval (0,0.5). If we employ an AFRIMA(1,d,1) model and use the maximum likelihood method, then we obtain the fitted model

$$(1 - 0.113B)(1-B)^{0.491} y_t = (1 - 0.576B)a_t,$$

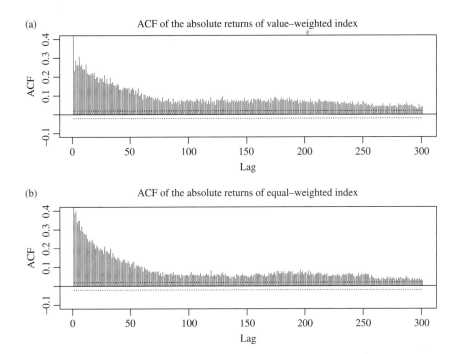

Figure 2.31. Sample autocorrelation function of the absolute series of daily simple returns for the CRSP value- and equal-weighted indexes: (a) the value-weighted index return and (b) the equal-weighted index return. The sample period is from January 2, 1970 to December 31, 2008.

where y_t is the absolute return and the standard errors of the estimates are 0.006, 0.008, and 0.006, respectively, in the order of appearance. The estimates are highly significant. In particular, the estimate of the fractional parameter is close to the nonstationary boundary of 0.5. Finally, the package fracdiff can be used in R to estimate AFRIMA(p, d, q) models. □

R Demonstration

```
> library(fracdiff)
> da=read.table("d-ibm3dx7008.txt",header=T)
> head(da)
      Date        rtn     vwretd     ewretd     sprtrn
1 19700102   0.000686   0.012137   0.033450   0.010211
 ....
6 19700109  -0.001353  -0.002797  -0.002923  -0.003021
> ew=abs(da$vwretd)
% obtain Geweke-Port-Hudak estimate using command fdGPH
```

```
> m3=fdGPH(ew)
> m3
$d
[1] 0.3722260
$sd.as
[1] 0.0698385
$sd.reg
[1] 0.06868857
% Maximum likelihood estimation of an AFRIMA(1,d,1) model.
> m2=fracdiff(ew,nar=1,nma=1)
> summary(m2)
Call:
  fracdiff(x = ew, nar = 1, nma = 1)

Coefficients:
    Estimate Std. Error z value Pr(>|z|)
d   0.490938   0.007997   61.39  <2e-16 ***
ar  0.113389   0.005988   18.94  <2e-16 ***
ma  0.575895   0.005946   96.85  <2e-16 ***
---
[d.tol = 0.0001221, M = 100, h = 0.0003742]
Log likelihood: 3.551e+04 ==> AIC = -71027.02 [1 deg.freedom]
```

2.12   MODEL COMPARISON AND AVERAGING

In applications, there is no *true* model for a given time series. All statistical models are approximations used to describe the dynamic dependence of the data. It is then common to see that several models fit a given data set well, and the question of model comparison arises. In this section, we discuss two methods to compare time series models. These methods are statistical criteria. They are designed to provide guidances in selecting a model. It should also be kept in mind, however, that the objective of data analysis and substantive information of the problem at hand are also important in model selection.

2.12.1   In-sample Comparison

If the objective of data analysis is to gain insight into the dynamic structure of a time series, then one can use in-sample measurement to compare different models. By in-sample, we mean that all data are used in model estimation and comparison. In this case, information criteria, such as AIC and BIC, and the estimate of residual variance can be used for model comparison. For a selected criterion, the model with a smaller value is preferred. To illustrate, consider Example 2.7, in which a seasonal ARMA model and a regression model with January dummy are used to model the CRSP Decile 1 index from January 1970 to December 2008. For the regression model, the residual standard error is 0.0690 while that of the seasonal ARMA model is

$\sqrt{0.00472} = 0.0687$. Therefore, if the residual standard error is used as the criterion for in-sample comparison, then the seasonal ARMA model is selected. In this particular instance, the difference between the two competing models is small.

2.12.2 Out-of-sample Comparison

When the objective of time series modeling is forecasting, one should use the forecasting performance of the models in model comparison. A commonly used measure to quantify the forecasting performance of statistical models is the mean square of forecast errors (MSFE) in an out-of-sample exercise. This model comparison method is known as *backtesting* in the finance literature. We use 1-step ahead forecasts to introduce the method. However, the idea applies to multistep ahead forecasts too.

Backtesting. The procedure of backtesting for a statistical model is as follows:

1. Divide the data set into estimation and forecasting subsamples. There is no specific rule to guide the division, but each subsample should contain sufficient data points so that the estimation and MSFE can be as accurate as possible.
2. Perform model estimation using data in the estimation subsample and use the fitted model to obtain 1-step ahead forecast and its forecast error. Specifically, suppose the estimation subsample is $\{x_t | t = 1, \ldots, h\}$. We estimate the model using the first h data points to compute the 1-step ahead prediction $\hat{x}_h(1)$ and its forecast error $e_h(1) = x_{h+1} - \hat{x}_h(1)$. The data point x_{h+1} is not used in model estimation.
3. Advance the estimation subsample by one data point, that is, $\{x_t | t = 1, \ldots, h + 1\}$. Reestimate the model using $h + 1$ data points and compute the 1-step ahead forecast and its forecast error. That is, compute $e_{h+1}(1) = x_{h+2} - \hat{x}_{h+1}(1)$, where $\hat{x}_{h+1}(1)$ is the 1-step ahead prediction of the newly fitted model at the forecast origin $h + 1$.
4. Repeat step 3 until we have the 1-step ahead forecast error $e_{T-1}(1) = x_T - \hat{x}_{T-1}(1)$, where T is the sample size.

The MSFE of the model is then given by

$$\text{MSFE}(m) = \frac{\sum_{j=h}^{T-1} [e_j(1)]^2}{T - h},$$

where m denotes the model used. One selects the model with the smallest MSFE as the best model for the data. In practice, one often uses the square root ot MSFE instead of MSFE itself. Other measurements of forecasting performance include the mean absolute forecast errors and bias, that is,

$$\text{MAFE}(m) = \frac{\sum_{j=h}^{T-1} |e_j(1)|}{T - h}, \quad \text{Bias}(m) = \frac{\sum_{j=h}^{T-1} e_j(1)}{T - h}.$$

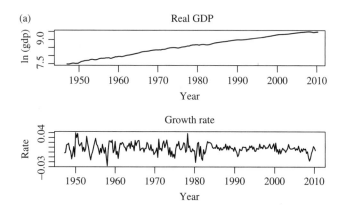

Figure 2.32. Time plots of the US quarterly real gross domestic product from the first quarter of 1947 to the second quarter of 2010: (a) real GDP in logarithm and (b) the growth rates of GDP.

Example 2.9. Consider the US quarterly real GDP from the first quarter of 1947 to the second quarter of 2010. The GDP data are obtained from the Federal Reserve Bank at St Loius, in billions of chained 2005 dollars, and seasonally adjusted. Figure 2.32 shows the time plots of the quarterly real GDP, in logarithm, and its growth rate series. We shall focus on the growth rate series, denoted by x_t. Figure 2.33 shows the sample ACF and PACF of x_t. From the plots, both ACF and PACF decay quickly. The PACF suggests an AR(3) model for the data. Using AIC via the `ar` command in R, we also obtain an AR(3) model. The fitted model is

$$(1 - 0.346B - 0.130B^2 + 0.123B^3)(x_t - 0.0079) = a_t, \quad \hat{\sigma}_a^2 = 8.32 \times 10^{-5},$$
$$(2.52)$$

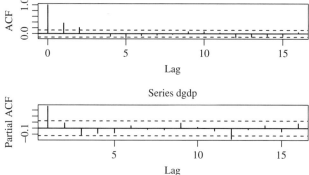

Figure 2.33. Sample ACF and PACF of the growth rates of US quarterly real GDP from 1947.II to 2010.II: (a) sample ACF and (b) sample PACF. Dashed lines represent two standard error limits.

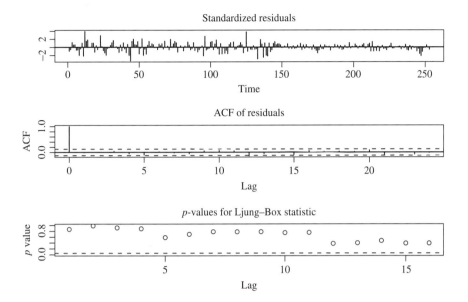

Figure 2.34. Model checking for an AR(3) model fitted to the growth rates of US quarterly real GDP from 1947.II to 2010.II: (a) standardized residuals, (b) residual sample ACF, and (c) p-values of the Ljung–Box statistics for residuals.

where the standard errors of the estimates are, in order, 0.062, 0.066, 0.062, and 0.0009, respectively. The AIC of the fitted AR(3) model is -1648.45. Figure 2.34 shows the model checking statistics of the AR(3) model. These plots indicate that the model is adequate.

Since the data are seasonally adjusted and for the purpose of model comparison, we also entertain a seasonal model,

$$(1 - 0.331B - 0.152B^2 + 0.110B^3)(1 - 0.497B^4)(x_t - 0.0079) = (1 - 0.587B^4)a_t,$$
$$(2.53)$$

where the standard errors of the estimates are 0.063, 0.067, 0.064, 0.258, 0.0008, and 0.236, respectively, and the residual variance is $\tilde{\sigma}_a^2 = 8.24 \times 10^{-5}$. The seasonal AR parameter is only marginally significant. Model checking fails to indicate any inadequacy of this fitted seasonal model. The AIC of model (Eq. 2.53) is -1646.93.

Turn to model comparison between the two models in Equations (2.52) and (2.53). For in-sample comparison, AIC selects the AR(3) model. For out-sample comparison, we apply the backtesting procedure with initial forecast origin being the fourth quarter of 2000, so that there are 38 quarters in the forecasting subperiod. For 1-step ahead prediction, the root mean square of forecast errors for the AR(3) and seasonal models are 0.00615 and 0.00632, respectively. Again, the AR(3) model is preferred. The mean absolute forecast errors are 0.00443 and 0.00455, respectively, and is also in favor of the AR(3) model.

Finally, in this particular instance, both in-sample and out-sample comparisons select the same AR(3) model for the real GDP growth rates. It is possible in a real application that different models are selected for in-sample and out-sample comparisons. It is also possible that the choice of the initial forecast origin affects the result of model selection. We recommend that one uses several initial forecast origins to obtain a better understanding of the comparison.

Remark. The backtesting of Example 2.9 was done by a R script backtest, which is available on the web page of the book.   □

R Demonstration

```
> da=read.table("q-gdpc96.txt",header=T)
> head(da)
  Year Mon Day       gdp
1 1947   1   1 1772.204
  ....
6 1948   4   1 1856.930
> gdp=log(da$gdp)
> dgdp=diff(gdp)

> m1=ar(dgdp,method='mle')
> m1$order
[1] 3
> m2=arima(dgdp,order=c(3,0,0))
> m2
Coefficients:
         ar1     ar2      ar3  intercept
      0.3462  0.1299  -0.1225     0.0079
s.e.  0.0623  0.0655   0.0624     0.0009

sigma^2 estimated as 8.323e-05: log likelihood=829.23, aic=-1648.45

> m3=arima(dgdp,order=c(3,0,0),season=list(order=c(1,0,1),period=4))
> m3
Coefficients:
         ar1     ar2      ar3    sar1     sma1  intercept
      0.3305  0.1521  -0.1103  0.4966  -0.5865     0.0079
s.e.  0.0633  0.0668   0.0635  0.2578   0.2357     0.0008

sigma^2 estimated as 8.24e-05: log likelihood=830.47, aic=-1646.93
>
> source("backtest.R")     % Perform backtest
> mm2=backtest(m2,dgdp,215,1)
[1] "RMSE of out-of-sample forecasts"
[1] 0.006153102
[1] "Mean absolute error of out-of-sample forecasts"
[1] 0.004430387
```

```
> mm3=backtest(m3,dgdp,215,1)
[1] "RMSE of out-of-sample forecasts"
[1] 0.006322009
[1] "Mean absolute error of out-of-sample forecasts"
[1] 0.004553896
```

2.12.3 Model Averaging

When several models fit a given time series well, instead of selecting a single model, one can use all the models to produce a combined forecast. This technique is referred to as *model averaging* in the statistical literature. Suppose that there are m models available and they all produce unbiased forecasts for a time series. By unbiased forecast, we mean that the expectation of the associated forecast error is 0. Let $\hat{x}_{i,h+1}$ be the 1-step ahead forecast of model i at the forecast origin h. Then, a combined forecast is

$$\hat{x}_{h+1} = \sum_{i=1}^{m} w_i \hat{x}_{i,h+1},$$

where w_i is a nonnegative real number denoting the weight for model i and satisfies $\sum_{i=1}^{m} w_i = 1$. The weights w_i can be determined in various ways. For example, in Bayesian inference, w_i is the posterior probability of model i. Here, we use the simple average, namely, $w_i = \frac{1}{m}$. Limited experience shows that this simple average works well in practice.

EXERCISES

If not specifically specified, use 5% significance level to draw conclusions in the exercises.

1. Consider the monthly US unemployment rate from January 1948 to November 2011 in the file m-unrate-4811.txt. The data are seasonally adjusted and obtained from the Federal Reserve Bank at St Louis.

   (a) Does the monthly unemployment rate have a unit root? Why?

   (b) Build a time series model for the monthly unemployment rates. Check the fitted model for adequacy. Then, use the model to forecast the unemployment rate for the December 2011 and the first three months of 2012. (Note that there are more than one model that fits the data well. You only need an adequate model.)

   (c) Does the fitted model imply the existence of business cycles? Why?

2. Consider the monthly simple returns of CRSP Decile 1, 2, 5, 9, and 10 portfolios based on the market capitalization of NYSE/AMEX/NASDAQ. The data span is from January 1961 to September 2011.

   (a) For the return series of Decile 2 and Decile 10, test the null hypothesis that the first 12 lags of autocorrelations are 0 at the 5% level. Draw your conclusion.

(b) Build an ARMA model for the return series of Decile 2. Perform model checking and write down the fitted model.

(c) Use the fitted ARMA model to produce 1- to 12-step ahead forecasts of the series and the associated standard errors of forecasts.

3. Consider the daily range (daily high–daily low) of Apple stock from January 2, 2007 to December 23, 2011. One can obtain the data by the package quantmod from Yahoo. Compute the first 100 lags of ACF of the series. Is there evidence of long-range dependence? Why? If the range series has long memory, build an AFRIMA model for the data.

4. Consider the monthly yields of Moody's Aaa & Baa seasoned bonds from January 1919 to November, 2011. The data are obtained from FRED of Federal Reserve Bank of St. Louis. Consider the log series of monthly Aaa bond yields. Build a time series model for the series, including model checking.

5. Consider again the monthly log series of Moody's Aaa bound yield. Use the exponential smoothing method to produce 1- to 12-step ahead out-of-sample forecasts at the forecast origin November 2010.

6. Consider the two bond yield series of the previous exercise. What is the relationship between the two series? To answer this question, take the log transformation of the data to build a time series model for the Aaa yields using Baa yields as an explanatory variable. Write down the fitted model, including model checking.

7. Consider the quarterly earnings per share of the Johnson & Johnson from the first quarter of 1992 to the second quarter of 2011. The data are in the file q-jnj-earns-9211.txt and are obtained from the First Call Historical Database of Thomson Reuters. Take log transformation of the data if necessary. Build a time series model for the data. Perform model checking to assess the adequacy of the fitted model. Write down the model. Refit the model using data from 1992 to 2008. Perform 1-step to 10-step ahead forecasts of the quarterly earnings and obtain a forecast plot.

8. Consider the US quarterly real gross national product from the first quarter of 1947 to the third quarter of 2011. The data are in the file q-GNPC96.txt, seasonally adjusted, and in billions of chained 2005 dollars. Let x_t be the growth rate series of the real GDP.

(a) The ar command identifies an AR(4) model for x_t via the AIC criterion. Fit the model. Is the model adequate? Why?

(b) The sample PACF of x_t specifies an AR(3) model. Fit the model. Is it adequate? Why?

(c) What is the model for x_t if one uses in-sample model comparison? Why?

(d) Divide the data into estimation and forecasting subsamples using the fourth quarter of 2000 as the initial forecast origin and apply the backtesting procedure with MSFE as the criterion. Select a model for x_t. Justify the choice.

REFERENCES

Akaike H. Information theory and an extension of the maximum likelihood principle. In: Petrov BN, Csaki F, editors. 2nd International Symposium on Information Theory. Budapest: Akademia Kiado; 1973. p 267–281.

Box GEP, Jenkins GM, Reinsel GC. Time Series Analysis: Forecasting and Control. 3rd ed. Englewood Cliffs (NJ): Prentice Hall; 1994.

Box GEP, Pierce D. Distribution of residual autocorrelations in autoregressive-integrated moving average time series models. J Am Stat Assoc 1970; 65: 1509–1526.

Brockwell PJ, Davis RA. Introduction to Time Series and Forecasting. 2nd ed. New York: Springer; 2002.

Brockwell PJ, Davis RA. Time Series: Theory and Methods. 2nd ed. New York: Springer; 2009.

Chan NH, Wei CZ. Limiting distributions of least squares estimates of unstable autoregressive processes. Ann Stat 1988; 16: 367–401.

Cryer JD, Chan KS. Time Series Analysis: With Applications in R. New York: Springer-Verlag; 2010.

Dickey DA, Fuller WA. Distribution of the estimates for autoregressive time series with a unit root. J Am Stat Assoc 1979; 74: 427–431.

Ding Z, Granger CWJ, Engle RF. A long memory property of stock returns and a new model. J Empir Finance 1993; 1: 83–106.

Fuller WA. Introduction to Statistical Time Series. 2nd ed. Hoboken (NJ): John Wiley & Sons; 1995.

Geweke J, Porter-Hudak S. The estimation and application of long memory time series models. J Time Anal 1982; 4: 221–238.

Greene WH. Econometric Analysis. 5th ed. Upper Saddle River (NJ): Prentice-Hall; 2003.

Hosking JRM. Fractional differencing. Biometrika 1981; 68: 165–176.

Ljung G, Box GEP. On a measure of lack of fit in time series models. Biometrika 1978; 66: 67–72.

Peña D, Tiao GC, Tsay RS. A Course in Time Series Analysis. Hoboken (NJ): Wiley & Sons; 2001.

Phillips PCB. Time series regression with a unit root. Econometrica 1987; 55: 277–301.

Shumway RH, Stoffer DS. Time Series Analysis and its Applications. New York: Springer; 2000.

Tiao GC, Tsay RS. Consistency properties of least squares estimates of autoregressive parameters in ARMA models. Ann Stat 1983; 11: 856–871.

Tsay RS. Analysis of Financial Time Series. 3rd ed. Hoboken (NJ): Wiley & Sons; 2010.

Tsay RS, Tiao GC. Consistent estimates of autoregressive parameters and extended sample autocorrelation function for stationary and nonstationary ARMA models. J Am Stat Assoc 1984; 79: 84–96.

Woodward WA, Gray HL, Elliott AC. Applied Time Series Analysis. Baca Raton (FL): CRC Press, Taylor & Francis Group; 2012.

3

CASE STUDIES OF LINEAR TIME SERIES

In this chapter, we consider three case studies of linear time series analysis. Our goals are (i) to demonstrate applications of the methods discussed in Chapter 2; (b) to show the usefulness and limitations of linear time series models; and (c) to gain further experience in analyzing time series data with R. The three cases considered are (i) the monthly global temperature anomalies from January 1880 to August 2010, (ii) the monthly US unemployment rate with or without the weekly initial jobless claims, and (iii) the weekly US regular gasoline price from January 6, 1997, to September 27, 2010, and the crude oil price from January 3, 1997, to September 24, 2010. We chose these three cases because they are timely, have important implications to the US economy, and are informative in achieving the goals of the chapter.

A main difficulty for the beginners of time series analysis is finding an adequate model for a given series. This is particularly so when the dynamic dependence of the data is complex or when many models seem to fit the data well. In this chapter, we tackle this difficulty by working through real examples. Our goal is that the three case studies are helpful to the reader.

Let us start by accepting Professor George Box's dictum concerning statistical models: *All models are wrong, but some are useful* (Box, 1976). Our goal then is to find an appropriate model that is useful to the objective of data analysis. Thus, the

An Introduction to Analysis of Financial Data with R, First Edition. Ruey S. Tsay.
© 2013 John Wiley & Sons, Inc. Published 2013 by John Wiley & Sons, Inc.

choice of a final model often depends on the objective of data analysis. It would not surprise me that a reader can find alternative models for the three time series considered in the chapter.

There are some general guidelines available for time series modeling. First, data are only part of information available in an application. We may have some prior knowledge about the problem at hand. In this situation, it is important to make use of the substantive information in model selection. Combination and cross-validation between prior knowledge and data often lead to an improved model selection. Second, in some cases, many models are available and the distinction between these competing models is small. The issue of model selection then becomes less important and one can comfortably use one of the models. Third, in some applications, especially in forecasting, one may want to combine several competing models. This results in pooling or combining forecasts. Fourth, a general principle of modeling is to start with a simple model. We can refine the model by using the iterative model-building process of Box and Jenkins, namely identification, estimation, and model checking, see Box et al.(2008). Finally, another general principle of statistical modeling is simplicity, a concept often referred to as *keeping it sophisticatedly simple* (KISS) by my late colleague Professor Arnold Zellner. Readers will find via examples that simple models can often capture the basic information embedded in the data.

3.1   WEEKLY REGULAR GASOLINE PRICE

Since the oil crisis of early 1970s, oil price has shown marked impacts on the global economy. The high gasoline price in 2008 further demonstrated the effects of oil price on daily life. High gasoline price results in high transportation and heating costs and, hence, higher prices for foods and services. This in turn leads to inflation and lower disposable income for consumers on other items. In some cases, it can even lead to recession in an economy. Thus, it is of interest and importance to analyze the gasoline price. In this case study, we analyze the weekly retail regular gasoline price of the United States. We also study the dependence of gasoline price on the crude oil price and use the latter to improve the forecast of the former. The data used are obtained from the web site of US Energy Information Administration at http://www.eia.gov. The data span is from January 1997 to September 2010. More specifically, the weekly data of regular all formulations retail gasoline prices (dollars per gallon) are from January 06, 1997, to September 27, 2010. The weekly crude oil data are the US spot price FOB weighted by estimated import volume (dollars per barrel) and are from January 03, 1997, to September 24, 2010. Thus, the crude oil prices are available three days prior to the gasoline prices, that is, from Fridays to Mondays. Here, FOB stands for free on board, denoting a transaction whereby the seller makes the product available within an agreed on period at a given port at a given price. Because the prices vary substantially, we use log prices in our analysis.

Figure 3.1 shows the time plots of weekly log prices of US regular gasoline and crude oil. As expected, the two series are highly related and move in unison. The log prices also show an increasing trend. To render the series stationary, we consider the

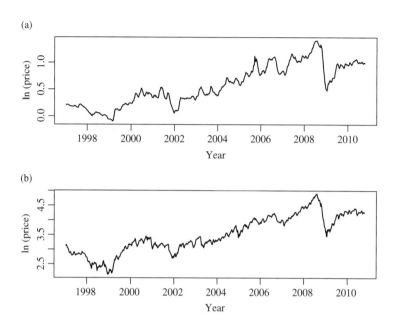

Figure 3.1. Weekly log prices of (a) regular gasoline (dollars per gallon) and (b) crude oil (dollars per barrel) of the United States from January 1997 to September 2010. The data are from US Energy Information Administration.

first differenced data. In other words, we focus on the log returns (or growth rates) of the weekly regular gasoline and crude oil prices.

3.1.1 Pure Time Series Model

We begin our analysis with a pure time series model. Let x_t be the weekly growth rates of US regular gasoline price. Figure 3.2 gives the time plot of x_t. It shows that there was a price jump in 2005 and the price increases in the first half of 2008 result in some big drops in the gasoline price. These features often lead to some large outliers in the subsequent analysis. Figure 3.3 shows the sample ACFs and PACFs of x_t. Both ACFs and PACFs decay quickly, confirming that the series is weakly stationary. The plot shows that the first five lags of PACF are significantly different from zero, suggesting that an AR(5) model might be appropriate for x_t. Indeed, the ar command in R selects an AR(5) model. Furthermore, the one-sample t-test gives a t-ratio of 1.306 with p value 0.192 for x_t, implying that the mean of the growth rate is not significantly different from zero. Therefore, we specify an AR(5) model without the constant for the growth rate series of weekly regular gasoline prices.

The fitted model is

$$(1 - 0.507B - 0.079B^2 - 0.136B^3 + 0.036B^4 + 0.086B^5)x_t = a_t,$$
$$\sigma_a^2 = 3.26 \times 10^{-4}.$$

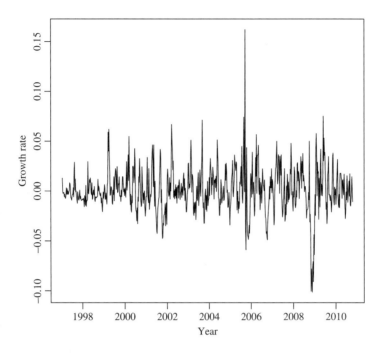

Figure 3.2. Weekly growth rates of the US regular gasoline price from January 1997 to September 2010.

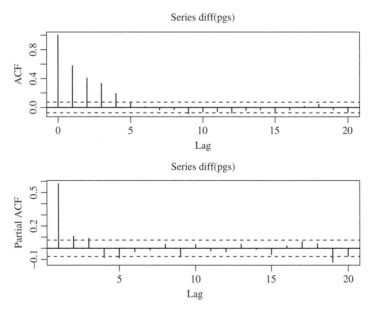

Figure 3.3. Sample autocorrelations and partial autocorrelations of the growth rates of weekly US regular gasoline prices from January 6, 1997 to September 27, 2010.

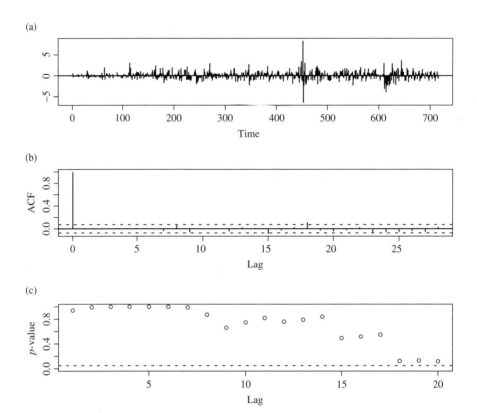

Figure 3.4. Model checking of AR(5) model in Equation (3.1) for the growth rate of weekly US regular gasoline prices from January 6, 1997 to September 27, 2010. (a) Standardized residuals, (b) ACF of residuals, (c) p-values for Ljung–Box statistic.

Since the lag-4 coefficient is insignificant with t-ratio less than 1, we refine the model as

$$(1 - 0.504B - 0.079B^2 - 0.122B^3 + 0.101B^5)x_t = a_t,$$
$$\sigma_a^2 = 3.265 \times 10^{-4}, \tag{3.1}$$

where the standard errors of the coefficients are 0.037, 0.042, 0.039, and 0.033, respectively. The t-ratio of lag-2 coefficient is 1.89 and the AIC of the model is -3704.96. One can further remove the lag-2 coefficient, but the AIC of the resulting model is -3703.4, which is greater than that of model (3.1). Consequently, we prefer the AR(5) model in Equation (3.1). Figure 3.4 shows the diagnostic plots of the fitted model in Equation (3.1). As expected, the plot of standardized residuals indicates some possible outliers, that is, big residuals, in the data. The sample ACF of the residuals and the p values of Ljung–Box statistics show that the fitted AR(5) model is adequate.

The sample PACF of x_t in Figure 3.3 shows a dominating correlation at lag 1. The ACF of the same figure indicates that the autocorrelations decay exponentially. These two features suggest that $p = 1$. On the other hand, the significance of ACFs and PACFs at higher order lags indicates $p = 1$ is not sufficient. Thus, another possibility is to entertain an ARMA model. To this end, we consider an ARMA(1,3) model. After removing an insignificant coefficient, we obtain the model

$$(1 - 0.633B)x_t = (1 - 0.127B + 0.141B^3)a_t,$$
$$\sigma_z^2 = 3.276 \times 10^{-4},$$

where the standard errors of the coefficient estimates are 0.051, 0.060, and 0.041, respectively. Model checking, not shown, fails to suggest any major inadequacy of the model. Thus, this ARMA(1,3) model is also adequate for the weekly growth rates of regular gasoline price. However, AIC of the model is -3704.6, which is larger than that of an AR(5) model in Equation (3.1). Consequently, we select the AR(5) model in Equation (3.1) as the pure time series model for x_t.

3.1.2 Use of Crude Oil Prices

Next, as gasoline prices depend heavily on spot prices of crude oil, we employ a regression model with time series errors to improve the accuracy in forecasting weekly gasoline price. Let z_t be the weekly growth rates of the US crude oil price. A simple linear regression gives

$$x_t = 0.287z_t + \epsilon_t, \tag{3.2}$$

where the standard error of the coefficient is 0.015. The adjusted R-squared of the model in Equation (3.2) is 33.57%. The sample ACF and PACF of the residual ϵ_t of Equation (3.2) are similar to those of x_t. Thus, the use of z_t does not alter the model specification of x_t. As a matter of fact, the `ar` command in R specifies an AR(6) model for the residual ϵ_t of Equation (3.2). The fitted model becomes

$$(1 - 0.40B - 0.16B^2 - 0.09B^3 - 0.03B^4 + 0.09B^5 + 0.05B^6)(x_t - 0.193z_t) = a_t,$$

where the coefficients at lag-4 and lag-6 are insignificant at the 5% level. Therefore, we simplify the model and obtain

$$(1 - 0.404B - 0.164B^2 - 0.096B^3 + 0.101B^5)(x_t - 0.191z_t) = a_t,$$
$$\sigma_a^2 = 2.53 \times 10^{-4}, \tag{3.3}$$

where the standard errors of the coefficient estimates are 0.039, 0.040, 0.039, 0.035, and 0.014, respectively. Figure 3.5 shows the diagnostic plots of this regression model with time series errors. The large standardized residuals remain, but the sample ACF

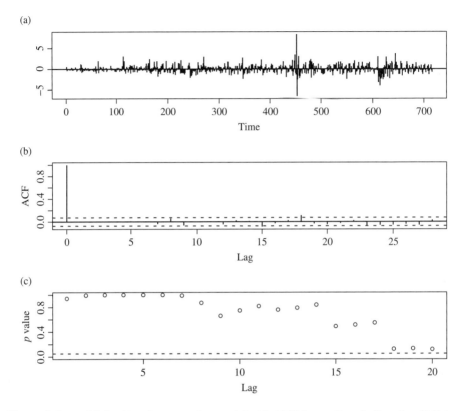

Figure 3.5. Model checking for regression model with AR(5) innovations in Equation (3.3) for the growth rates of US weekly regular gasoline price from January 6, 1997 to September 27, 2010. (a) Standardized residuals, (b) ACF of residuals, (c) p-values for Ljung–Box statistic.

and p values of the Ljung–Box statistics suggest that the model is adequate. The AIC of the model is -3884.95.

It is tempting to compare the models in Equations (3.1) and (3.3). However, this is an unfair comparison because the regression model uses additional information of the growth rates of crude oil price. Given the high dependence between the two oil prices, one would expect the regression model with time series errors in Equation (3.3) to be a better model. This is indeed the case. The information on crude oil prices reduces the residual variance from 3.265×10^{-4} to 2.532×10^{-4}, a 22.5% reduction. Similarly, the AIC drops from -3704.96 to -3884.95. As it is seen later, the regression model also produces more accurate out-of-sample1 forecasts.

3.1.3   Use of Lagged Crude Oil Prices

The usefulness of the improved model in Equation (3.3) is limited to 3 days, because it uses the growth rate of crude oil price 3 days earlier. To increase the lead time in forecasting, one can use the lagged growth rate of crude oil price. For instance, the model

$$x_t = \beta z_{t-1} + \epsilon_t$$

would give the analysts 10 days in advance to predict the weekly gasoline price. The fitted model is

$$x_t = 0.186 z_{t-1} + \epsilon_t, \quad \sigma_\epsilon = 0.0209, \tag{3.4}$$

where the standard error of the coefficient estimate is 0.0172. The adjusted R-squared of this model is 14.1%, which, as expected, is much lower than that of the regression model in Equation (3.2). This is understandable because the correlation between x_t and z_{t-1} is smaller than that between x_t and z_t.

The residuals of the model in Equation (3.4) are serially correlated. Indeed, the sample ACFs and PACFs of the residuals are similar to those of the residuals of Equation (3.2), albeit a little bit more complicated. The `ar` command of R identifies an AR(9) model for the residuals. After removing all insignificant estimates, we obtain the model

$$(1 - 0.454B - 0.088B^2 - 0.142B^3 + 0.083B^5 + 0.064B^9)(x_t - 0.041 z_{t-1}) = a_t, \tag{3.5}$$

where $\sigma_a^2 = 3.23 \times 10^{-4}$ and the standard errors of the estimates are, in order, 0.043, 0.041, 0.039, 0.035, 0.032, and 0.018, respectively. The AIC of the model is -3703.4, which is close to that of the pure time series model in Equation (3.1), but much higher than that of the model in Equation (3.3). Strictly speaking, the AIC of this model is not directly comparable with those of the previous models because the model has one less observation owing to lagging. Nevertheless, the contribution of z_{t-1} to x_t is small. The coefficient 0.041 is statistically significant, but small. Diagnostic checking statistics of model (3.5) are similar to those of model (3.1) and, hence, are omitted. The model is also adequate.

3.1.4 Out-of-Sample Predictions

In this section we consider the performance of the three models built for the growth rates of regular gasoline price in out-of-sample prediction. To this end, we employ the backtesting method of Chapter 2. The method divides the data into modeling and forecasting subsamples and uses an iterative procedure to compute prediction. Specifically, the iterative procedure consists of an estimation–prediction cycle, and starts with the last data point of the modeling subsample as the first forecast origin. Once a forecast is produced, the procedure advances the forecast origin by 1 and repeats the estimation-forecasting cycle. The recursive 1-step ahead forecast errors in the forecasting subsample are then used to measure the accuracy of prediction. Two most widely used measures of forecasting accuracy are the root mean square of forecast errors (RMSFE) and the mean absolute forecast errors (MAFE).

For the growth rates of weekly gasoline price, we divide the data into modeling and forecasting subsamples with the latter consisting of the last 400 data points. In other words, we start the forecast origin on January 24, 2003. The 400 observations in the forecasting subsample should provide reliable measures of RMSFE and MAFE. The results are given below:

Model	RMSFE	MAFE
AR(5) model in Equation (3.1)	0.02171	0.01538
Regression model in Equation (3.3)	0.01926	0.01285
Regression model in Equation (3.5)	0.02166	0.01548

From the table, we make the following observations. First, the regression model with crude oil price 3-day earlier performs best. This is consistent with the in-sample comparison and understandable. It shows that the gasoline price reflects the crude oil price quickly. Second, the other two models fare similarly in out-of-sample prediction. The contribution of crude oil price 10-day earlier is small, if any. If one wants to predict the gasoline price more than 10 days in advance, one can simply use the pure time series model in Equation (3.1). On the other hand, one should use the regression model in Equation (3.3) if the forecast horizon is less than 3 days.

Finally, Figure 3.6 shows the time plot of recursive 1-step ahead predictions in the forecasting subsample based on the pure time series model in Equation (3.1). The solid line denotes the observed value, whereas the asterisk the prediction. Figure 3.7 provides the same for the regression model in Equation (3.3). These two plots provide a direct comparison of the two models at each prediction.

R Demonstration. Output edited.

```
> da=read.table("w-petroprice.txt",header=T)
> da1=read.table("w-gasoline.txt")
> pgs=log(da1[,1])
> pus=log(da$US)
> tdx=c(1:717)/52+1997  % calendar time
> par(mfcol=c(2,1))
> plot(tdx,pgs,xlab='year',ylab='ln(price)',type='l')
> title(main='(a) Gasoline')
> plot(tdx,pus,xlab='year',ylab='ln(price)',type='l')
> title(main='(b) Crude oil')
> dpgs=diff(pgs)
> acf(dpgs,lag=20)
> pacf(dpgs,lag=20)
> m1=ar(diff(pgs),method='mle')
> m1$order
[1] 5
> t.test(dpgs)
        One Sample t-test
data:  dpgs
```

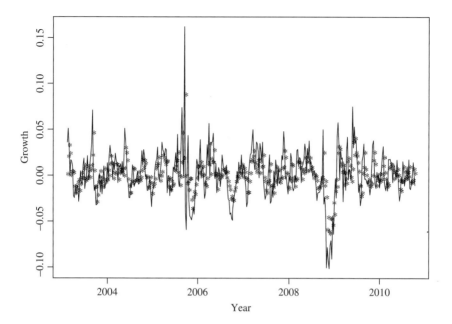

Figure 3.6. 1-step ahead out-of-sample predictions of model (3.1) for the growth rates of weekly US regular gasoline price from January 24, 2003, to September 27, 2010. The data are in solid line and predictions in asterisk.

```
t = 1.3062, df = 715, p-value = 0.1919
alternative hypothesis: true mean is not equal to 0
> m1=arima(dpgs,order=c(5,0,0),include.mean=F)
> m1
arima(x = dpgs, order = c(5, 0, 0), include.mean = F)
Coefficients:
         ar1      ar2      ar3      ar4      ar5
      0.5073   0.0788   0.1355  -0.0360  -0.0862
s.e.  0.0372   0.0417   0.0415   0.0417   0.0372

sigma^2 estimated as 0.0003262: log likelihood=1857.85,aic=-3703.71
> m1=arima(dpgs,order=c(5,0,0),include.mean=F,fixed=c(NA,NA,NA,0,NA))
> m1
arima(x=dpgs,order=c(5,0,0),include.mean=F,fixed=c(NA,NA,NA,0,NA))

Coefficients:
         ar1      ar2      ar3   ar4       ar5
      0.5036   0.0789   0.1220     0   -0.1009
s.e.  0.0370   0.0418   0.0385     0    0.0330

sigma^2 estimated as 0.0003265:  log likelihood = 1857.48,  aic = -3704.96
> tsdiag(m1,gof=20)
> dpus=diff(pus)
> m3=lm(dpgs~-1+dpus)
> summary(m3)
Call:
lm(formula = dpgs ~ -1 + dpus)
```

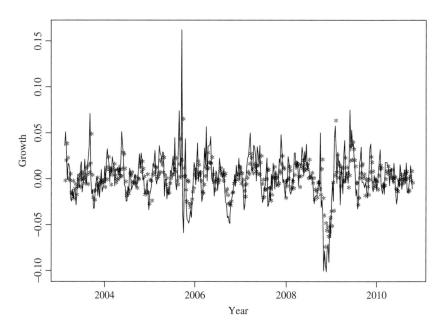

Figure 3.7. 1-step ahead out-of-sample predictions of model (3.3) for the growth rates of weekly US regular gasoline prices from January 24, 2003, to September 27, 2010. The data are in solid line and predictions in asterisk.

```
Coefficients:
      Estimate Std. Error t value Pr(>|t|)
dpus   0.28703    0.01507   19.05   <2e-16 ***
---
Residual standard error: 0.01839 on 715 degrees of freedom
Multiple R-squared: 0.3366,     Adjusted R-squared: 0.3357
F-statistic: 362.8 on 1 and 715 DF,   p-value: < 2.2e-16

> acf(m3$residuals,lag=20)
> pacf(m3$residuals,lag=20)
> m4=ar(m3$residuals,method='mle')
> m4$order
[1] 6
> m4=arima(dpgs,order=c(6,0,0),include.mean=F,xreg=dpus)
> m4
arima(x=dpgs,order=c(6, 0, 0), xreg = dpus, include.mean = F)

Coefficients:
          ar1      ar2      ar3      ar4      ar5      ar6     dpus
       0.3953   0.1634   0.0946   0.0297  -0.0873  -0.0525   0.1927
s.e.   0.0389   0.0400   0.0404   0.0405   0.0400   0.0373   0.0136

sigma^2 estimated as 0.0002524: log likelihood=1949.6,aic=-3883.21
>
> m4=arima(dpgs,order=c(5,0,0),include.mean=F,xreg=dpus)
> m4
arima(x=dpgs,order=c(5,0,0),xreg=dpus,include.mean=F)
```

```
Coefficients:
         ar1      ar2      ar3      ar4      ar5      dpus
      0.4022   0.1621   0.0899   0.0209  -0.1086   0.1914
s.e.  0.0387   0.0401   0.0403   0.0400   0.0371   0.0136

sigma^2 estimated as 0.0002531: log likelihood=1948.6,aic=-3883.23
>
> m4=arima(dpgs,order=c(5,0,0),include.mean=F,xreg=dpus,fixed=c(NA,NA,NA,0,NA,NA))
> m4
arima(x=dpgs,order=c(5,0,0),xreg=dpus,include.mean=F,fixed=c(NA,NA,NA,0,NA,NA))

Coefficients:
         ar1      ar2      ar3   ar4      ar5      dpus
      0.4037   0.1642   0.0961     0  -0.1014   0.1911
s.e.  0.0386   0.0399   0.0386     0   0.0345   0.0136

sigma^2 estimated as 0.0002532: log likelihood=1948.5, aic=-3884.95
> tsdiag(m4,gof=20)

> c1=c(NA,NA,NA,0,NA)
> pm1=backtest(m1,dpgs,316,1,fixed=c1,inc.mean=F)
[1] "RMSE of out-of-sample forecasts"
[1] 0.02171235
[1] "Mean absolute error of out-of-sample forecasts"
[1] 0.01537881
> c4=c(NA,NA,NA,0,NA,NA)
> pm4=backtest(m4,dpgs,316,1,xre=dpus,inc.mean=F,fixed=c4)
[1] "RMSE of out-of-sample forecasts"
[1] 0.01925732
[1] "Mean absolute error of out-of-sample forecasts"
[1] 0.01285104
> tdx=tdx[2:717]
> pm4fit=dpgs[317:716]-pm4$error
> pm1fit=dpgs[317:716]-pm1$error
> plot(tdx[317:716],dpgs[317:716],xlab='year',ylab='growth',type='l')
> points(tdx[317:716],pm1fit,pch='*')
> plot(tdx[317:716],dpgs[317:716],xlab='year',ylab='growth',type='l')
> points(tdx[317:716],pm4fit,pch='*')
 % Use lagged growth rate of crude oil price
> m6=lm(dpgs[2:716]~-1+dpus[1:715])
> summary(m6)
lm(formula=dpgs[2:716]~-1+dpus[1:715])
Coefficients:
            Estimate Std. Error t value Pr(>|t|)
    dpus[1:715]  0.18560    0.01716    10.81   <2e-16 ***
---
Residual standard error: 0.02093 on 714 degrees of freedom
Multiple R-squared: 0.1408,    Adjusted R-squared: 0.1395
> acf(m6$residuals,lag=20)
> pacf(m6$residuals,lag=20)
> m7=ar(m6$residuals,method='mle')
> m7$order
[1] 9
> m7=arima(dpgs[2:716],order=c(9,0,0),include.mean=F,xreg=dpus[1:715])
> m7
arima(x=dpgs[2:716],order=c(9,0,0),xreg=dpus[1:715],include.mean=F)

Coefficients:
         ar1      ar2      ar3      ar4      ar5      ar6      ar7      ar8
      0.4559   0.0888   0.1679  -0.0468  -0.0653  -0.0195  -0.0362   0.0797
```

```
s.e. 0.0425  0.0410  0.0423  0.0415  0.0416  0.0414  0.0410 0.0408
          ar9   dpus[1:715]
      -0.0882          0.0454
s.e.   0.0373          0.0174

sigma^2 estimated as 0.0003204: log likelihood=1861.55, aic=-3701.1
> m7=arima(dpgs[2:716],order=c(9,0,0),include.mean=F,xreg=dpus[1:715],
  fixed=c(NA,NA,NA,0,NA,0,0,0,NA,NA))
> m7
arima(x=dpgs[2:716],order=c(9,0,0),xreg=dpus[1:715],include.mean=F,
      fixed = c(NA,NA,NA,0,NA,0,0,0,NA,NA))

Coefficients:
         ar1     ar2     ar3 ar4     ar5 ar6 ar7 ar8     ar9 dpus[1:715]
      0.4544 0.0877 0.1415    0 -0.0830   0   0   0 -0.0640      0.0406
s.e. 0.0427 0.0413 0.0393    0  0.0345   0   0   0  0.0318      0.0176

sigma^2 estimated as 0.000323: log likelihood=1858.7, aic=-3703.4
> tsdiag(m7,gof=20)
> c7=c(NA,NA,NA,0,NA,0,0,0,NA,NA)
> pm7=backtest(m7,dpgs[2:716],315,1,xre=dpus[1:715],inc.mean=F,fixed=c7)
[1] "RMSE of out-of-sample forecasts"
[1] 0.0216638
[1] "Mean absolute error of out-of-sample forecasts"
[1] 0.01548401
```

3.2   GLOBAL TEMPERATURE ANOMALIES

Global warming is a topic of considerable importance and has attracted much attention in recent years, ranging from environmental engineers to scientists to economists. If the rise in global temperature continues, it will have a major impact on the global economy. In this section, we analyze the monthly global temperature anomalies from January 1880 to August 2010. Our goal is not to debate the evidence of global warming, but to demonstrate empirical time series analysis. Specifically, our goals are (i) to illustrate the methods discussed in Chapter 2 about time series modeling and forecasting, (ii) to compare different models, (iii) to see the limitation of time series models in long-term prediction, and (iv) to show the difficulty in distinguishing trend-stationarity from unit-root stationarity based purely on the data. Global climate changes involve many other factors.

There are several data sets available for global temperature anomalies. See the web sites http://data.giss.nasa.gov/gistemp/ of the Goddard Institute for Space Studies (GISS), National Aeronautics and Space Administration (NASA) and http://www.ncdc.noaa.gov/cmb-faq/anomalies.html of the National Climatic Data Center (NCDC), National Oceanic and Atmospheric Administration (NOAA). We employ the series of monthly means based on land-surface air temperature anomalies of GISS, NASA. However, we obtained similar results from the data of NOAA. In fact, the same models apply to both series.

Figure 3.8 shows the time plot of the global temperature anomalies from January 1880 to August 2010 for 1568 observations over 131 years. The GISS data are in $0.01°C$. An upward trend is clearly seen from the plot. In particular, the slope of the

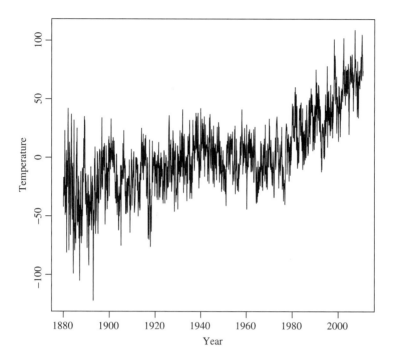

Figure 3.8. Monthly global temperature anomalies from January 1880 to August 2010.

trend seems to increase in the early 1980s. On the other hand, the variability of the temperature is relatively stable over the 131 years.

3.2.1 Unit-Root Stationarity

Let G_t denote the monthly global temperature anomalies. To specify a model for G_t, we start by examining the dynamic dependence of the series. Figure 3.9 gives the sample ACFs of G_t. As expected, the ACFs are high and decay slowly. A careful inspection also shows that the ACFs exhibit a cyclic pattern with peaks occurring around lags 24 and 36. This latter feature is not surprising because temperature often has a seasonal pattern.

Because of the strong serial dependence, we consider the differenced data $x_t = (1 - B)G_t$. Figure 3.10 shows the sample ACFs and PACFs of x_t. Both ACFs and PACFs become small, indicating that x_t can be approximated by a stationary time series model. As the sample size 1568 is large, we can entertain a relatively more complicated model. A careful examination of the sample ACFs of x_t shows that (i) ACF at lags 1, 2, 4, 5, and 8 is either significant or marginally significant, and (ii) ACF at lag 24 is significant. The sample PACF, on the other hand, has several significant values. In particular, the lag-1 PACF is much larger than others and the PACF does not decay exponentially. Putting information together and following the general guidelines

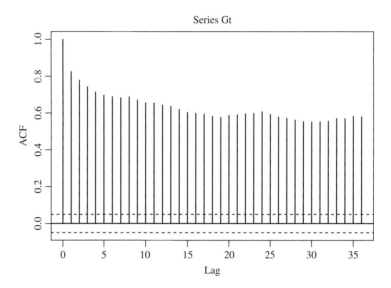

Figure 3.9. Sample autocorrelation function of the monthly global temperature anomalies.

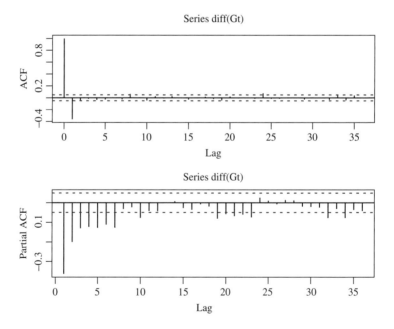

Figure 3.10. Sample autocorrelation and partial autocorrelation functions of the differenced global temperature anomalies.

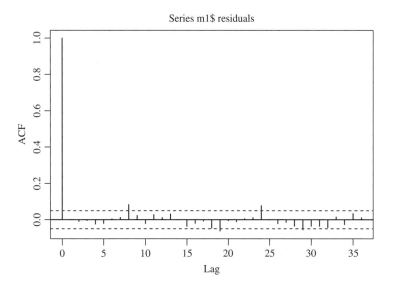

Figure 3.11. Sample autocorrelation function of the residuals of ARIMA(1,1,2) model for global temperature anomalies.

of the previous section, we start with the simple ARIMA(1,1,2) model

$$(1 - \phi B)(1 - B)G_t = (1 - \theta_1 B - \theta_2 B^2)a_t, \tag{3.6}$$

for G_t. Here, we use $p = 1$ because the differenced series x_t has a large lag-1 PACF and $q = 2$ because the first two ACFs of x_t are significant. As the specified MA(2) model can have significant PACFs at lower order lags, we decide to keep $p = 1$. The high order ACFs are temporarily ignored because we like to keep the model simple. The fitted model is

$$(1 - 0.739B)(1 - B)G_t = \left(1 - 1.297B + 0.318B^2\right)a_t, \quad \sigma_a^2 = 272.1. \tag{3.7}$$

All estimates are highly significant. Figure 3.11 shows the sample ACF of the residuals of the ARIMA(1,1,2) model in Equation (3.7). Based on the residual ACFs, the model is inadequate because the ACFs are significant at lags 8 and 24.

Turn to model refinement. The significance of ACF at lag 24 is understandable because of the seasonal nature of temperature. On the other hand, it is not easy to explain the serial correlation at lag 8. Consequently, we refine the model as

$$(1 - \phi B)(1 - B)G_t = \left(1 - \theta_1 B - \theta_2 B^2\right)\left(1 - \theta_{24}B^{24}\right)a_t. \tag{3.8}$$

The fitted model is

$$(1 - 0.761B)(1 - B)G_t = \left(1 - 1.324B + 0.342B^2\right)\left(1 - 0.072B^{24}\right)a_t,$$
$$\sigma_a^2 = 270.6, \tag{3.9}$$

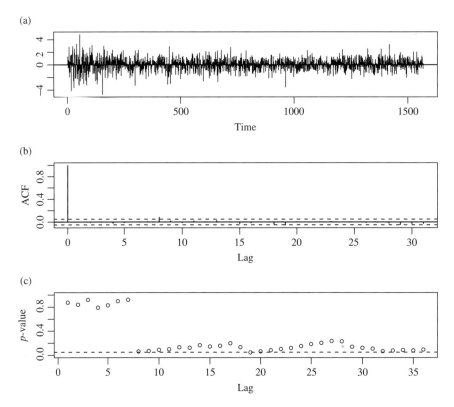

Figure 3.12. Diagnostic checking plots of the model in Equation (3.9) for the monthly global temperature anomalies. (a) Standardized residuals, (b) ACF of residuals, (c) p-values for Ljung–Box statistic.

where the standard errors of the estimates are, in order, 0.038, 0.052, 0.049, and 0.024, respectively. These estimates are statistically significant. Figure 3.12 gives the diagnostic plots for the seasonal model in Equation (3.9). The residual plot looks reasonable and the p values of Ljung–Box statistics are above 0.05 except for $Q(8)$ and $Q(19)$. As expected, the residual ACFs show marginally significant values at lags 8 and 19. As mentioned earlier, it is hard to explain the lag-8 serial correlation and the magnitude of the ACF is small, we terminate the modeling process and treat the model in Equation (3.9) as an adequate model. The AIC of model (3.9) is 13,234.4, which is smaller than 13,241.1 of model (3.7).

The model in Equation (3.9) is called a unit-root nonstationary model because it uses the first difference to transform the global temperature into a stationary series. The time series G_t is said to be difference-stationary.

R Demonstration

```
> Gt=scan(file='m-GLBTs.txt')
```

```
> Gtemp=ts(Gt,frequency=12,start=c(1880,1))
> plot(Gtemp,xlab='year',ylab='temperature',type='l') % Plot the data
> acf(diff(Gt),lag=36)
> pacf(diff(Gt),lag=36)

> m1=arima(Gt,order=c(1,1,2))
> m1
arima(x = Gt, order = c(1, 1, 2))

Coefficients:
          ar1      ma1      ma2
       0.7387  -1.2973   0.3183
s.e.   0.0406   0.0533   0.0492

sigma^2 estimated as 272.1: log likelihood=-6616.6, aic=13241.1
> acf(m1$residuals,lag=36)
> m1=arima(Gt,order=c(1,1,2),seasonal=list(order=c(0,0,1),period=24))
> m1
arima(x=Gt,order=c(1,1,2),seasonal=list(order=c(0,0,1),period=24))

Coefficients:
          ar1      ma1      ma2     sma1
       0.7612  -1.3241   0.3416   0.0717
s.e.   0.0379   0.0519   0.0485   0.0243

sigma^2 estimated as 270.6: log likelihood=-6612.2, aic=13234.4
> tsdiag(m1,gof=36)
```

3.2.2 Trend-Nonstationarity

In the literature, some analysts and scientists use *time trend* to model the global temperature anomalies. By time trend, we mean using time index as an explanatory variable. Consider the model

$$G_t = \beta_0 + \beta_1 t + z_t, \tag{3.10}$$

where z_t is an innovation series denoting the deviation of the global temperature anomalies from a time trend. If z_t is a stationary time series, then G_t is called a *trend-stationary time series*, meaning that it can be transformed into a stationary series by removing the effect of a time trend. In model (3.10), β_1 is the slope of the time trend. A positive β_1 indicates that G_t will increase with time and eventually goes to positive infinity as t approaches infinity. For monthly data, β_1 is the monthly growth rate of G_t. Conceptually, a trend-stationary time series is very different from a difference-stationary one because the latter does not contain a fixed trend. We shall discuss further the difference between the two models when we consider long-term prediction.

For the global temperature anomalies, the fitted linear regression model is

$$G_t = -38.04 + 0.05156t + z_t, \tag{3.11}$$

where the standard errors of the coefficient estimates are 1.135 and 0.0013, respectively, and the standard deviation of z_t is 22.46. The time slope is positive and highly significant. Figure 3.13 shows the sample ACFs and PACFs of the innovation series

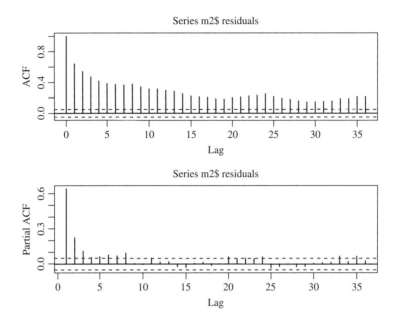

Figure 3.13. Sample autocorrelations and partial autocorrelations of the innovation series z_t in Equation (3.11) for monthly global temperature anomalies.

z_t of Equation (3.11). The PACFs decay quickly and the ACFs do not show any high value. Therefore, it is reasonable to assume that z_t does not have a unit root. That is, z_t is stationary and, hence, G_t is trend-stationary.

Next, we specify a model for the innovation series z_t. Because its ACFs in Figure 3.13 do not cut-off at any finite lag, z_t does not follow a simple MA model. In other words, some AR component is needed. That is, $p > 0$. The PACFs of Figure 3.13 have two discernible features. First, the first eight lags of PACFs are significant, indicating that z_t does not follow a low order AR model. This implies that $q > 0$. Second, the PACFs do not follow a simple exponentially decaying pattern. This means that $p > 1$. Putting information together and keeping the order simple, we start with an ARMA(2,1) model for z_t, that is,

$$\left(1 - \phi_1 B - \phi_2 B^2\right) z_t = (1 - \theta_1 B) a_t.$$

The model for G_t then becomes

$$\left(1 - \phi_1 B - \phi_2 B^2\right) (G_t - \beta_0 - \beta_1 t) = (1 - \theta_1 B) a_t. \qquad (3.12)$$

The fitted model is

$$\left(1 - 1.239 B + 0.272 B^2\right) (G_t + 38.72 - 0.053 t) = (1 - 0.78 B) a_t,$$

$$\sigma_a^2 = 272.9, \qquad (3.13)$$

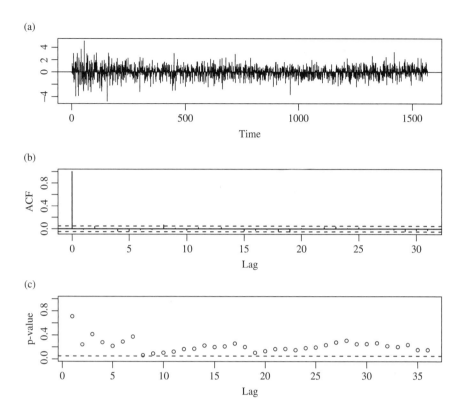

Figure 3.14. Diagnostic checking of the fitted model in Equation (3.14) for the monthly global temperature anomalies. (a) Standardized residuals, (b) ACF of residuals, (c) p-values for Ljung–Box statistic.

where all estimates are highly significant. However, the residual ACFs of the model show a significant value at lag 24. This is not surprising based on the ARIMA model used in the previous section. Consequently, we further refine the model and obtain

$$\left(1 - 1.196B + 0.239B^2\right)\left(G_t + 38.72 - 0.0529t\right)$$
$$= (1 - 0.745B)\left(1 - 0.0856B^{24}\right)a_t, \tag{3.14}$$

where $\sigma_a^2 = 270.8$ and all estimates are statistically significant at the 5% level. The standard errors of the coefficient estimates are, in order, 0.059, 0.048, 5.18, 0.006, 0.049, and 0.024, respectively. Figure 3.14 gives the diagnostic checking plots of the model in Equation (3.14). Except for a minor residual ACF at lag 8, the checking statistics fail to indicate any inadequacy of the fitted model. Consequently, we select the model in Equation (3.14) as the final model for G_t under trend-stationarity.

On the basis of model in Equation (3.14), the global temperature increases on an average 0.0529/100 °C per month. That is, the global temperature increases 0.00635 °C per year. This is very significant because it implies that the global temperature, on an average, will increase 1 °C every 157 years. The AIC of the model in Equation (3.14) is 13,247.5, which is larger than that of the difference-stationary model in Equation (3.9).

R Demonstration

```
> time=c(1:1568) % time index
> m2=lm(Gt~time)
> summary(m2)
lm(formula = Gt ~time)

Coefficients:
              Estimate Std. Error t value Pr(>|t|)
(Intercept) -38.039763   1.134960  -33.52   <2e-16 ***
time          0.051560   0.001253   41.15   <2e-16 ***
---
Residual standard error: 22.46 on 1566 degrees of freedom
Multiple R-squared: 0.5195,    Adjusted R-squared: 0.5192
> par(mfcol=c(2,1))
> acf(m2$residuals,lag=36)
> pacf(m2$residuals,lag=36)
> m2=arima(Gt,order=c(2,0,1),xreg=time)
> m2
arima(x = Gt, order = c(2, 0, 1), xreg = time)

Coefficients:
          ar1       ar2      ma1  intercept     time
       1.2385   -0.2719  -0.7802   -38.8493   0.0530
s.e.   0.0567    0.0477   0.0460     5.3548   0.0059

sigma^2 estimated as 272.9: log likelihood=-6623.0, aic=13257.97
> tsdiag(m2,gof=36)   % Significant ACF at lag 24.
> m2=arima(Gt,order=c(2,0,1),seasonal=list(order=c(0,0,1),
  period=24),xreg=time)
> m2
arima(x=Gt,order=c(2,0,1),seasonal=list(order=c(0,0,1),period=24),
    xreg = time)

Coefficients:
          ar1       ar2      ma1     sma1  intercept     time
       1.1960   -0.2394  -0.7451   0.0856   -38.7150   0.0529
s.e.   0.0587    0.0482   0.0486   0.0241     5.1843   0.0057

sigma^2 estimated as 270.8: log likelihood=-6616.7, aic=13247.5
> tsdiag(m2,gof=36) % model checking
```

3.2.3   Model Comparison

We have obtained two models for the monthly global temperature anomalies from January 1880 to August 2010. The first model in Equation (3.9) is difference-stationary, whereas the second one in Equation (3.14) is trend-stationary. Both models are adequate because they passed the rigorous model checking. A question that arises naturally is which model should one choose. We shall address this question in this section.

In-Sample Comparison. As mentioned in Chapter 2, one approach to compare models for a given time series is to consider their in-sample goodness of fit. A commonly used criterion here is the Akaike information criterion or the Bayesian information criterion. For the global temperature data, the difference-stationary model in Equation (3.9) is selected based on the AIC because it has a smaller value at 13,234. The two models, however, are close in many ways. For instance, Figure 3.15 shows time plots of the residuals of the two competing models with the top panel for the difference-stationary model. The two residuals series are essentially the same.

Another approach to in-sample model comparison is to study the implications of the competing models. Here, the two fitted models for the global temperature anomalies differ dramatically. As mentioned before, the trend-stationary model in Equation (3.14) imposes *a priori* a time trend for the temperature. The estimated time slope is 0.0529, which is positive and highly significant. Thus, the model implies that the global temperature will continue to increase at a pace of 0.00635 °C per year under the current environment. On the other hand, the difference-stationary model in Equation (3.9) does not provide a definite support for the global warming. Similar to the random walk model for stock prices, the future temperature may increase or decrease because a random walk has no fixed direction. What the model implies is that there exists substantial uncertainty in the future global temperature. It can go anywhere from positive infinity to negative infinity. The observed data, which were over 131 years, provide little guidance about the future global temperature. This appears to be unconceivable at the first glance, but 131 years are not sufficiently long when we are making inference about hundreds or thousands of years into the future.

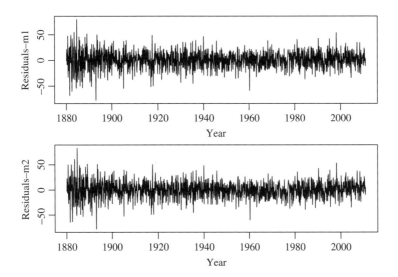

Figure 3.15. Residual plots for the models in Equations (3.9) and (3.14) for the monthly global temperature anomalies. The upper panel is for model (3.9).

In short, the data we have do not contain sufficient evidence to distinguish between trend-stationarity and difference-stationarity.

Out-of-Sample Comparison. If the goal of time series analysis is forecasting, one can use out-of-sample prediction to compare competing models. Again, we use the backtesting of Chapter 2 to evaluate out-of-sample prediction. For the global temperature data, we divide the sample into modeling and forecasting subsamples with the latter consisting of the last 200 observations. We then apply the backtesting method to compute the 1-step ahead prediction of the two competing models in Equations (3.9) and (3.14). For the global temperature data with 200 1-step ahead out-of-sample predictions, we obtain the following results:

Model	RMSFE	MAFE
Difference-stationary model in Equation (3.9)	14.526	11.167
Trend-stationary model in Equation (3.14)	15.341	11.966

Clearly, the difference-stationary model is preferred based on the 1-step ahead prediction. The drop in RMSFE is about $(15.341 - 14.526)/14.526 = 5.6\%$, a moderate amount. This exercise also shows that time series models are useful in short-term prediction because the RMSFEs of the two models are smaller than the unconditional standard error 16.43 of the innovations of the fitted models. For the difference-stationary model, the reduction in RMSFE is approximately $(16.43 - 14.53)/14.53 = 13.1\%$.

R Demonstration

```
> source("backtest.R")
> pm1=backtest(m1,Gt,1368,1)
[1] "RMSE of out-of-sample forecasts"
[1] 14.52598
[1] "Mean absolute error of out-of-sample forecasts"
[1] 11.16746
> time=as.matrix(time)
> pm2=backtest(m2,Gt,1368,1,xre=time)
[1] "RMSE of out-of-sample forecasts"
[1] 15.34131
[1] "Mean absolute error of out-of-sample forecasts"
[1] 11.96595
```

3.2.4 Long-Term Prediction

Global warming is concerned with long-term prediction. In this section, we consider and compare the performance of the two competing models in Equations (3.9) and (3.14) using long-term prediction. Specifically, using August 2010, which gives the

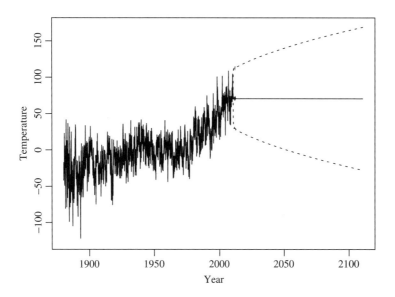

Figure 3.16. Long-term point and interval forecasts of the monthly global temperature anomalies based on the difference-stationary model in Equation (3.9). The forecast origin is August 2010 and the forecast horizon is 100 years.

last data point, as the forecast origin, we compute 1-step to 1200-step ahead predictions of the monthly global temperature anomalies. In other words, we use the models built based on data of the past 131 years to predict the global temperatures for the next 100 years. To compute the predictions of the trend-stationary model in Equation (3.14), the time index is given. Figure 3.16 shows the point predictions and the corresponding 95% interval forecasts of the global temperature anomalies based on the difference-stationary model in Equation (3.9). The plot highlights several features of the model. First, similar to other unit-root models, the long-term forecasts converge to a constant represented by a horizontal line in the plot. The level of this horizontal line depends on the forecast origin. Second, the length of the 95% interval forecasts continues to grow with the forecast horizon. In fact, the length of the interval diverges to infinity eventually. These two features have important implications in forecasting. First, they indicate that the long-term forecasts are rather uncertain. This makes intuitive sense because long-term predictions of the model are dominated by its random walk component and for a random walk the current value contains little information about the future. Second, they demonstrate clearly that the model is only informative in short-term prediction. To a great degree, this is true for most time series models discussed in the book.

Turn to the trend-stationary model in Equation (3.14). Figure 3.17 shows the point predictions and the corresponding 95% interval forecasts of global temperature anomalies of the model. Again, the forecast origin is August 2010 and the forecast horizon is 100 years. The plot also highlights the key features of the model. First,

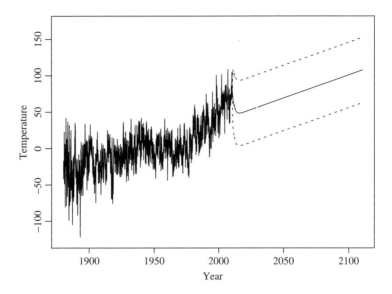

Figure 3.17. Long-term point and interval forecasts of the monthly global temperature anomalies based on the trend-stationary model in Equation (3.14). The forecast origin is August 2010 and the forecast horizon is 100 years.

because of the positive time slope, the predictions grow with the forecast horizon. Second, the lengths of the interval forecasts are stable over time. In fact, the lengths quickly converge to a constant with the constant being approximately $4 \times \sigma_z$, where σ_z is the sample standard error of the innovation series z_t. As stated in Equation (3.11), the innovation series z_t is stationary. As such the variances of the forecasts of z_t converge to its variance, σ_z^2, when the forecast horizon increases. For the trend-stationary model in Equation (3.14), the prediction of the time trend is certain conditioned on the coefficients β_0 and β_1 being fixed. The uncertainty in forecasts is determined by that of z_t. Consequently, the variances of forecast errors of the model in Equation (3.14) converge to that of z_t. As a matter of fact, the lengths of the forecast intervals match well with the range of the data. These two features also have important implications about the model. First, by imposing a time trend, the forecasts of the model diverge to infinity as the forecast horizon increases. Is this reasonable? How certain are we that the model is the true model for the global temperature anomalies? This type of uncertainty is not shown by the model nor by its forecasts. Second, the finite interval forecasts further confirm that the global temperature will continue to rise under the model. This is very different from that of the difference-stationary model in Equation (3.9).

Finally, we provide a direct comparison by plotting the long-term predictions of the two competing models together. See Figure 3.18. The contrast between the two models is clearly seen. On the other hand, the plot contains no information about which model is more appropriate for the data, leading to the assertion that the two models are not useful in long-term prediction.

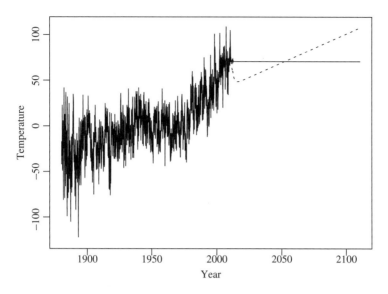

Figure 3.18. Point forecasts of the monthly global temperature anomalies based on two competing models in Equations (3.9) and (3.14). The forecast origin is August 2010 and the forecast horizon is 100 years.

3.2.5 Discussion

In this case study, we employed two models to show several important issues of linear time series analysis. We demonstrated the process of model building and model selection. The models built are useful in short-term prediction because their out-of-sample root mean squares of forecast errors are smaller than their residual standard errors. Both the in-sample AIC criterion and the 1-step ahead backtesting select the difference-stationary model for the global temperature anomalies. On the other hand, the selected models are not informative in long-term prediction. Other important issues raised by the case study are summarized below.

Model Uncertainty. All statistical models are wrong; they simply provide approximation to the underlying process. As such, statistical models contain uncertainty not only in parameter estimates but also in the choice of the models themselves. One should consider model uncertainty in making inference whenever possible. For time series analysis, model uncertainty can be addressed by model averaging, for example, combining forecasts. A simple approach to model averaging in forecasting is to use the simple average among predictions of the competing models. For the monthly global temperature anomalies, we can use the average of two predictions to obtain a combined forecast. Again, this combination is useful mainly in short-term prediction.

Short versus Long-term Forecasts. The difference between short- and long-term predictions shows that in time series analysis the choice of a model may depend on the forecast horizon. The best model for 1-step ahead prediction might be different from that for 2-step ahead prediction. This type of ideas has been explored in the literature and leads to the development of *adaptive forecast*. See Tiao and Tsay (1994) and references therein.

Trend-Shift Model. To demonstrate the arbitrariness (or subjectivity) involved in using linear time trend model, we consider an alternative specification. Recall that the time plot in Figure 3.8 shows an increase in time slope around 1980. This feature has led some analysts to employ a *trend-shift* model. Specifically, consider the model

$$G_t = \beta_0 + \beta_1 t + \beta_2 x_t + n_t, \tag{3.15}$$

where x_t is defined as

$$x_t = \begin{cases} 0 & \text{if } t \le 1212, \\ t & \text{if } 1212 < t \le 1568, \end{cases} \tag{3.16}$$

where $t = 1212$ corresponds to December 1980. This simple model allows the time slope to change at the beginning of 1981. Specifically, the time slope is β_1 before January 1981 and is $\beta_1 + \beta_2$ starting in 1981. The parameter β_2 thus denotes the change in time slope. If β_2 is positive, the time slope jumps from β_1 to $\beta_1 + \beta_2$ in January 1981. The time plot of $\beta_1 t + \beta_2 x_t$ would show a jump and a change of direction in January 1981 so that the model in Equation (3.15) is referred to as a trend-shift model.

For the global temperature anomalies, we obtain the model

$$G_t = -28.92 + 0.0313t + 0.0214x_t + n_t,$$

where all estimates are highly significant. This regression model says that the time slope for the monthly global temperature was 0.0313 before January 1981 and it increased to 0.0527 thereafter. The increase was statistically significant. The sample ACFs and PACFs of the innovation series z_t of the prior regression model are similar to those of the model in Equation (3.11). Therefore, we employ the same ARMA model for z_t. The resulting model for the monthly global temperature anomalies is

$$(1 - 1.122B + 0.197B^2)(1 - B)z_t = (1 - 0.684B)(1 + 0.0823B^{24})a_t, \tag{3.17}$$

where $z_t = G_t + 29.263 - 0.317t - 0.0219x_t$, $\sigma_a^2 = 267.5$, and all coefficient estimates are significantly different from zero. The AIC of the model is 13,230, which is smaller than those of the two competing models built before. The diagnostic checking plots of this trend-shift model are similar to those of Equation (3.14), indicating that model (3.17) is also adequate for the monthly global temperature anomalies. Note that this refined model shows some improvements over that of Equation (3.14) because its residual variance reduces to 267.5.

The significance of the slope change 0.0219 in January 1981 suggests that, under the postulated trend-shift model, the global warming has accelerated in recent years. The monthly temperature change increased from 0.0317 to 0.0536 starting in January 1981.

Finally, on the basis of the AIC, the trend-shift model in Equation (3.17) is preferred over the two competing models discussed before. This is not surprising because we impose a trend shift after examining the data. One can further improve the in-sample fit by employing a more flexible polynomial trend for the data. This type of improvement is not recommended in general because it can easily lead to over-fitting in a real application. Furthermore, long-term forecasts of such models do not take into account the possibility of slope changes in the future.

R Demonstration

```
> Gt=scan(file='m-GLBTs.txt')
> time=c(1:1568)
> time1=c(rep(0,1212),time[1213:1568])
> mm1=lm(Gt~time+time1)
> summary(mm1)
lm(formula = Gt ~time + time1)

Coefficients:
              Estimate Std. Error t value Pr(>|t|)
(Intercept) -28.924419   1.191061  -24.29   <2e-16 ***
time          0.031330   0.001702   18.40   <2e-16 ***
time1         0.021397   0.001318   16.23   <2e-16 ***
---
Residual standard error: 20.79 on 1565 degrees of freedom
Multiple R-squared: 0.5887,     Adjusted R-squared: 0.5882
F-statistic:  1120 on 2 and 1565 DF,  p-value: < 2.2e-16

> x1=cbind(time,time1)
> mm1=arima(Gt,order=c(2,0,1),seasonal=list(order=c(0,0,1),period=24),
   xreg=x1)
> mm1
arima(x=Gt,order=c(2,0,1),seasonal=list(order=c(0,0,1),period=24),xreg=x1)

Coefficients:
          ar1      ar2      ma1    sma1  intercept    time   time1
       1.1220  -0.1973  -0.6835  0.0823   -29.2630  0.0317  0.0219
s.e.   0.0727   0.0542   0.0643  0.0239     4.1411  0.0058  0.0044

sigma^2 estimated as 267.5:  log likelihood = -6607,  aic = 13230
> tsdiag(mm1,gof=36)
> Box.test(mm1$residuals,lag=8,type='Ljung')
        Box-Ljung test

data:  mm1$residuals
X-squared = 15.4598, df = 8, p-value = 0.0508
```

Other Data Set. There exist other time series measuring the monthly global temperature anomalies. See, for instance, the series given in NCDC, NOAA mentioned before. This series measured in degrees of Celsius is close, but not identical, to that of GISS used in our case study. The basic characteristics of the data, however, are the same. As a matter of fact, the same models apply to this new series. Thus, the choice of data sets does not affect the general conclusions of the analysis. For example, consider the trend-stationary model. We have

$$(1 - 1.298B + 0.306B^2)(y_t + 0.495 - 0.000661t) = (1 - 0.934B)(1 - 0.083B^{24})a_t,$$

where y_t denotes the monthly global temperature anomalies, $\sigma_a^2 = 0.0883$ and all coefficient estimates are statistically significant at the 5% level. The time slope is 0.000661, implying that the global temperature increases on an average 0.00793 °C per year. This is slightly greater than 0.00635 °C suggested by the GISS data set. The difference, however, appears to be statistically not significant.

R Demonstration

```
> da=read.table("m-ncdc-noaa-glbtemp.txt")
> head(da)
    V1 V2       V3
1 1880  1 -0.0405
2 1880  2 -0.6112
...
> tail(da)
       V1 V2       V3
1568 2010  8   0.8970
1569 2010  9 -999.0000
....
1572 2010 12 -999.0000
> da=da[1:1568,]
> temp=da[,3]
> m3=arima(temp,order=c(1,1,2),seasonal=list(order=c(0,0,1),period=24))
> m3
arima(x=temp,order=c(1,1,2),seasonal=list(order=c(0,0,1),period=24))

Coefficients:
         ar1      ma1     ma2    sma1
      0.5817  -1.2414  0.2639  0.0854
s.e.  0.0704   0.0827  0.0781  0.0243

sigma^2 estimated as 0.0881:  log likelihood = -321.11,  aic = 652.21
> tsdiag(m3,gof=36)
> m4=arima(temp,order=c(2,0,1),seasonal=list(order=c(0,0,1),period=24),
  xreg=time)
> m4
arima(x=temp,order=c(2,0,1),seasonal=list(order=c(0,0,1),period=24),
    xreg = time)

Coefficients:
         ar1      ar2      ma1    sma1  intercept   time
      1.2975  -0.3057  -0.9344  0.0827    -0.4952  7e-04
s.e.  0.0562   0.0480   0.0430  0.0255     0.1178  1e-04

sigma^2 estimated as 0.08825: log likelihood=-322.2, aic=658.4
> m4$coef
          ar1           ar2          ma1      sma1   intercept
 1.2974716910 -0.3056922789 -0.9343898897 0.0826971  -0.4952302
```

```
        time
 0.0006613375
> sqrt(diag(m4$var.coef))
      ar1         ar2         ma1        sma1   intercept        time
0.0561629   0.048026   0.0429865  0.02545154  0.11775435  0.00013652
> m4$coef/sqrt(diag(m4$var.coef))   % Compute t-ratios
       ar1         ar2         ma1        sma1   intercept        time
 23.101928   -6.365140  -21.736831    3.249198   -4.205621      4.8441
> tsdiag(m4,gof=36)
> %%% Backtesting
> pm3=backtest(m3,temp,1368,1)
[1] "RMSE of out-of-sample forecasts"
[1] 0.3160872
[1] "Mean absolute error of out-of-sample forecasts"
[1] 0.2361009
> pm4=backtest(m4,temp,1368,1,xre=time)
[1] "RMSE of out-of-sample forecasts"
[1] 0.3270271
[1] "Mean absolute error of out-of-sample forecasts"
[1] 0.241655
```

3.3   US MONTHLY UNEMPLOYMENT RATES

The persistently high unemployment rate is a serious problem that many nations have been facing in recent years. The seasonally adjusted monthly rate in the United States remains around 9% in 2011 and declines slowly. In this case study, we consider the problem of modeling and forecasting monthly US unemployment rate with and without using the information of weekly initial jobless claims. The data used are from the US Bureau of Labor Statistics, Department of Labor, and seasonally adjusted. The unemployment rate, in percentage, is the civilian unemployment rate for persons 16 years of age and older and the weekly initial jobless claims denote the number of new unemployment insurance claims. The unemployment rate is from January 1948 to September 2010 and reported on the first day of each month. The weekly initial jobless claims are available from January 7, 1967, to August 28, 2010, and reported on Saturdays.

The objectives of this case study include (i) to demonstrate data analysis with mixed frequencies, that is, monthly and weekly data, (ii) to emphasize the potential model mis-specification in using seasonally adjusted data, and (iii) to illustrate that trial and error based on prior experience can be helpful in model specification.

3.3.1   Univariate Time Series Models

We begin with time series models that use only the dynamic dependence of the unemployment rates. As such, we can employ the data from January 1948 to September 2010 with 753 observations. Figure 3.19 shows the time plot of US monthly unemployment rate. The asterisk of the plot denotes the seasonally adjusted unemployment rate. The plot shows several characteristics of the series. First, as expected, unemployment rates exhibit strong cyclical pattern signifying expansions and contractions of the US economy. The pattern does not have a fixed period because economic

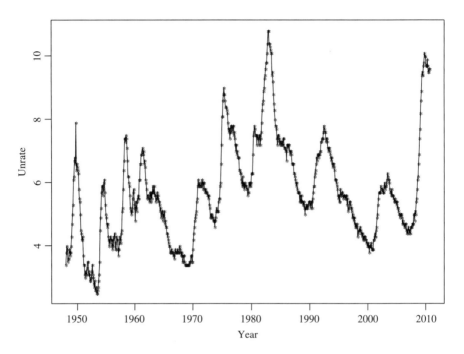

Figure 3.19. US monthly unemployment rates from January 1948 to September 2010. Asterisk denotes observations. Data are seasonally adjusted.

expansions and contractions have no fixed durations. Second, the unemployment rates show an upward trend. There are several possible explanations for the trend, including the increases in labor forces and participations, and advances in technology. Third, the unemployment rate rose quickly and declined slowly. This asymmetric behavior indicates that unemployment rates do not follow a linear time series model. Indeed, several researchers have employed nonlinear models for the unemployment rates. See, for instance, Tsay (2010, Chapter 4) and the references therein. As stated earlier in the introduction, there is no *true* model for a given time series and all statistical models are approximations. We shall not entertain nonlinear models in our analysis. Instead, we focus on linear approximations.

Let x_t be the monthly unemployment rate. Figure 3.20 gives the sample ACF and PACF of x_t and its differenced series $(1 - B)x_t$. The sample ACFs of x_t are high and persistent reflecting the upward trend of the data. On the other hand, the sample ACFs and PACFs of the differenced data, $(1 - B)x_t$, are relatively small and decay quickly. A careful examination shows some basic features of the sample ACFs and PACFs of $(1 - B)x_t$. First, even though the data are seasonally adjusted, the ACFs and PACFs are significant at lags 12, 24, and 36, indicating that seasonality of the data is not completely removed. This is not uncommon for seasonally adjusted data, and a seasonal model is needed for the seasonally adjusted unemployment rates. Second, the sample ACFs and PACFs of $(1 - B)x_t$ at the seasonal lags are relatively small

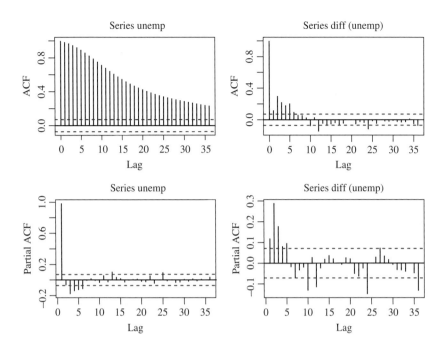

Figure 3.20. Sample and partial autocorrelations of the monthly unemployment rates and their first differenced series.

in magnitude, around 0.15, but they decay slowly. As a matter of fact, the PACFs at the seasonal lags shown in Figure 3.20 do not decay at all. This feature is similar to the behavior of regular sample ACFs and PACFs of a fractionally differenced series, that is, a long-memory series. As stated in Chapter 2, an ARMA model that can approximate long-memory time series is an ARMA(1,1) model with similar AR and MA coefficients. For seasonal data, this means that the model for seasonal component is a seasonal ARMA(1,1) model with period 12. Third, focusing on the regular sample ACFs and PACFs, we see that both ACFs and PACFs have approximately five or six significant lags, and the PACFs show a rough pattern of exponential decay. This latter feature suggests $p = 1$ and $q = 5$. Consequently, we tentatively specify the model

$$(1 - \Phi B^{12})(1 - \phi B)(1 - B)x_t = (1 - \theta_1 B - \cdots - \theta_5 B^5)(1 - \Theta B^{12})a_t,$$

for the monthly unemployment rate. The values of Φ and Θ should be close. This specification is rather complicated and not easy for inexperienced analysts. We shall consider some alternative models later.

On the basis of data from January 1948 to September 2010, we obtain the fitted model

$$(1 - 0.60B^{12})(1 - 0.73B)(1 - B)x_t = (1 - 0.75B + 0.22B^2 + 0.01B^3$$
$$+ 0.04B^4 + 0.08B^5)(1 - 0.85B^{12})a_t,$$

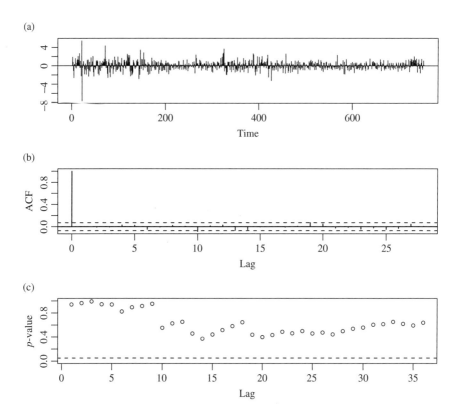

Figure 3.21. Model checking statistics of model 3.18 for monthly unemployment rates. (a) Standardized residuals, (b) ACF of residuals, (c) p-values for Ljung–Box statistic.

where $\sigma_a^2 = 0.0364$. The MA coefficients at lags 3 and 4 are insignificant because their standard errors are 0.05 and 0.047, respectively. Therefore, we remove these two coefficients and refine the model as

$$(1 - 0.61B^{12})(1 - 0.75B)(1 - B)x_t = (1 - 0.77B + 0.24B^2 + 0.099B^5)$$
$$(1 - 0.85B^{12})a_t, \tag{3.18}$$

where $\sigma_a^2 = 0.0365$ and the standard errors of the coefficient estimates are, in the order of appearance, 0.065, 0.057, 0.065, 0.037, 0.039, and 0.046, respectively. All estimates are significantly different from zero. The AIC of model (3.18) is -337.5. Figure 3.21 shows some plots of diagnostic checking for model (3.18). Except for one or two big outliers at the beginning of the series, the fitted model is adequate. The ACFs of the residuals and the associated Ljung–Box statistics fail to indicate any model inadequacy.

R Demonstration

```
> da=read.table("m-unrate.tct",header=T)
> dim(da)
[1] 753    4
> head(da)
  Year mon dd rate
1 1948 1   1  3.4
......
6 1948    6  1  3.6
> unemp=da$rate
> unrate=ts(unemp,frequency=12,start=c(1948,1))
> plot(unrate,xlab='year',ylab='unrate',type='l')
> par(mfcol=c(2,2))
> acf(unemp,lag=36)
> pacf(unemp,lag=36)
> acf(diff(unemp),lag=36)
> pacf(diff(unemp),lag=36)
> m1=arima(unemp,order=c(1,1,5),seasonal=list(order=c(1,0,1),period=12))
> m1
arima(x=unemp,order=c(1,1,5),seasonal=list(order=c(1,0,1),period=12))
Coefficients:
         ar1      ma1     ma2    ma3    ma4    ma5   sar1     sma1
      0.7301 -0.7468 0.2194 0.0073 0.0383 0.0831 0.5978 -0.8469
s.e.  0.0686  0.0776 0.0462 0.0501 0.0467 0.0431 0.0672  0.0477

sigma^2 estimated as 0.03643: log likelihood = 176.43,aic=-334.87
> c1=c(NA,NA,NA,0,0,NA,NA,NA)
> m1=arima(unemp,order=c(1,1,5),seasonal=list(order=c(1,0,1),period=12),
   fixed=c1)
> m1
arima(x=unemp,order=c(1,1,5),seasonal=list(order=c(1,0,1),period=12),
    fixed = c1)

Coefficients:
         ar1      ma1     ma2   ma3 ma4     ma5   sar1     sma1
      0.7536 -0.7744  0.2351     0   0  0.0990 0.6051  -0.8525
s.e.  0.0569  0.0650  0.0365     0   0  0.0386 0.0654   0.0457

sigma^2 estimated as 0.03649: log likelihood = 175.75,aic=-337.5
> tsdiag(m1,gof=36)
> Box.test(m1$3$residuals,lag=24,type='Ljung')
 Box-Ljung test
data:  m3$residuals
X-squared = 23.349, df = 24, p-value = 0.4993

> Box.test(m1$residuals,lag=36,type='Ljung')
  Box-Ljung test
data:  m3$residuals
X-squared = 32.4586, df = 36, p-value = 0.6378
```

3.3.2   An Alternative Model

An alternative model, which is easier to specify, for the monthly unemployment rates can be obtained by a two-step procedure. In the first step, we focus on the seasonal pattern of the data. As stated in the previous section, the behavior of sample ACFs and PACFs of $(1 - B)x_t$ at the seasonal lags suggests a seasonal ARMA(1,1) model with period 12, see Figure 3.20. Therefore, we start with the model

$$(1 - \Phi B^{12})(1 - B)x_t = (1 - \Theta B^{12})a_t.$$

The fitted model is

$$(1 - 0.62B^{12})(1 - B)x_t = (1 - 0.87B^{12})b_t, \quad \sigma_b^2 = 0.043, \tag{3.19}$$

where b_t denotes the residual series. This model is inadequate for the data, but its residual b_t should be approximately free of seasonality. Indeed, this is the case. Figure 3.22 shows the sample ACF and PACF of the residual series b_t. As expected, the ACFs and PACFs at the seasonal lags are no longer significant at the 5% level. In the second step, we specify a regular ARMA model for b_t. On the basis of the sample PACF in Figure 3.22, we can easily identify an AR(5) model for b_t because the PACF cuts off at lag 5. From Equation (3.19), we have

$$b_t = \frac{(1 - 0.62B^{12})(1 - B)}{1 - 0.82B^{12}} x_t.$$

With an AR(5) specification for b_t, we have

$$(1 - \phi_1 B - \cdots - \phi_5 B^5)b_t = a_t.$$

Putting the previous two equations together, we obtain a model for x_t as

$$(1 - \phi_1 B - \cdots - \phi_5 B^5)(1 - \Phi B^{12})(1 - B)x_t = (1 - \Theta B^{12})a_t.$$

The fitted model is

$$(1 + 0.01B - 0.21B^2 - 0.17B^3 - 0.10B^4 - 0.12B^5)(1 - 0.56B^{12})(1 - B)x_t$$
$$= (1 - 0.82B^{12})a_t, \quad \sigma_a^2 = 0.037.$$

Since the lag-1 coefficient is insignificant at the 5% level, we refine the model as

$$(1 - 0.21B^2 - 0.17B^3 - 0.10B^4 - 0.12B^5)(1 - 0.56B^{12})(1 - B)x_t$$
$$= (1 - 0.82B^{12})a_t, \quad \sigma_a^2 = 0.037, \tag{3.20}$$

where the standard errors of the coefficient estimates are 0.037, 0.036, 0.036, 0.036, 0.072, and 0.053, respectively. All estimates are statistically significant at the 5% level and the AIC of the model is -335.02. Figure 3.23 shows some diagnostic plots for the model in Equation (3.20). Again, except for one or two big outliers in the beginning of the series, the model is adequate.

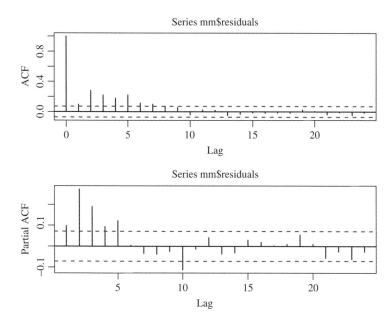

Figure 3.22. Sample ACF and PACF of the residuals of Equation (3.19) for US monthly unemployment rates.

R Demonstration

```
> mm=arima(unemp,order=c(0,1,0),seasonal=list(order=c(1,0,1),period=12))
> mm
arima(x=unemp,order=c(0,1,0),seasonal=list(order=c(1,0,1),period=12))

Coefficients:
         sar1      sma1
       0.6195   -0.8670
s.e.   0.0658    0.0468

sigma^2 estimated as 0.04267: log likelihood = 116.9,aic=-227.8
> par(mfcol=c(2,1))
> acf(mm$residuals,lag=24)
> pacf(mm$residuals,lag=24)
> mm1=arima(unemp,order=c(5,1,0),seasonal=list(order=c(1,0,1),period=12))
> mm1
arima(x=unemp,order=c(5,1,0),seasonal=list(order=c(1,0,1),period=12))

Coefficients:
          ar1      ar2      ar3      ar4      ar5     sar1      sma1
      -0.0124   0.2101   0.1682   0.1024   0.1207   0.5624   -0.8233
s.e.   0.0365   0.0366   0.0366   0.0370   0.0366   0.0723    0.0526

sigma^2 estimated as 0.03663: log likelihood = 174.57, aic=-333.13
> cc1=c(0,NA,NA,NA,NA,NA,NA)
> mm1=arima(unemp,order=c(5,1,0),seasonal=list(order=c(1,0,1),period=12),
  fixed=cc1)
```

(a)

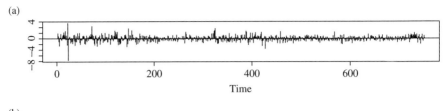

(b)

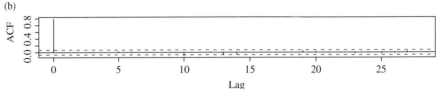

(c)

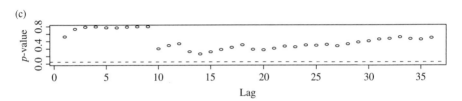

Figure 3.23. Model checking statistics of model (3.20) for US monthly unemployment rates. (a) Standardized residuals, (b) ACF of residuals, (c) p-values for Ljung–Box statistic.

```
> mm1
arima(x=unemp,order=c(5,1,0),seasonal=list(order=c(1,0,1),period=12),
    fixed = cc1)

Coefficients:
      ar1      ar2     ar3     ar4     ar5    sar1     sma1
        0   0.2104  0.1652  0.0996  0.1194  0.5643  -0.8240
s.e.    0   0.0366  0.0355  0.0362  0.0364  0.0724   0.0528

sigma^2 estimated as 0.03664: log likelihood = 174.51,aic=-335.02
> tsdiag(mm1,gof=36)
% Backtesting
> source("backtest.R")
> pm1=backtest(m1,unemp,700,1,fixed=c1,inc.mean=F)
[1] "RMSE of out-of-sample forecasts"
[1] 0.1662391
[1] "Mean absolute error of out-of-sample forecasts"
[1] 0.1349363
>
> pmm1=backtest(mm1,unemp,700,1,fixed=cc1,inc.mean=F)
[1] "RMSE of out-of-sample forecasts"
[1] 0.1679285
[1] "Mean absolute error of out-of-sample forecasts"
[1] 0.1350412
```

3.3.3 Model Comparison

It is interesting to compare the two models in Equations (3.18) and (3.20) for the US monthly unemployment rates. Both models are adequate based on the usual model checking statistics. For in-sample comparison, the AICs of the two models are -337.5 and -335.02, respectively. Thus, the model in Equation (3.18) is preferred according to AIC. For out-of-sample comparison, we divide the data into modeling and forecasting subsamples with the latter consisting of the last 53 observations. The results of forecasting performance are as follows:

Model	RMSFE	MAFE
Model in Equation (3.18)	0.1662	0.1349
Model in Equation (3.20)	0.1679	0.1350

Again, the model in Equation (3.18) slightly outperforms the model in Equation (3.20). But the improvement is so small, there is hardly any difference between the two models.

In short, there is no significant difference between the two models. Either one of them can be used to make inference about US monthly unemployment rates. To further demonstrate the closeness between the two models, we ignore the seasonal components and consider the π-weights of model in Equation (3.18). These weights are

$$\frac{1 - 0.75B}{1 - 0.77B + 0.24B^2 + 0.099B^5} \approx 1 + 0.02B - 0.22B^2$$
$$- 0.174B^3 - 0.081B^4 - 0.119B^5.$$

The π-weights are close to the AR(5) coefficients of the model in Equation (3.20). Consequently, the two models are indeed very close.

3.3.4 Use of Initial Jobless Claims

In this section, we make use of the information embedded in the weekly initial jobless claims to predict the monthly unemployment rates. To this end, we use a shorter data span because the numbers of initial jobless claims were only available starting from January 1967. Again, all data were seasonally adjusted. As the unemployment rates were reported on the first day of each month, we use the weekly initial jobless claims in the previous month as explanatory variables. As such, the actual data span for the unemployment rates was from February 1967 to September 2010 and that for the initial jobless claims was from January 1967 to August 2010. The effective sample size is 524.

Because the data of initial jobless claims are weekly, we consider two categories of explanatory variables. In the first category, we add the weekly numbers within each month to form a monthly number of initial jobless claims. In the second category, we directly use the weekly data of initial jobless claims. Again, let x_t be the monthly unemployment rates. Then, the explanatory variables considered are $w_{1,t-1}$, $w_{2,t-1}$, $w_{3,t-1}$, $w_{4,t-1}$, and c_{t-1}, where $w_{i,t-1}$ is the number of initial jobless claims in week i of month $t-1$ and c_{t-1} is the total number of initial jobless claims in month $t-1$. Note that some months have 5 weeks so that c_{t-1} may not be equal to $\sum_{i=1}^{4} w_{i,t-1}$. Also, for numerical stability, the initial jobless claims were divided by 1,000,000.

Using Monthly Initial Jobless Claims. Figure 3.24 shows the time plots of monthly unemployment rates x_t and numbers of initial jobless claims c_{t-1} from February 1967 to September 2010. As expected, the two series show a general pattern of co-movement. Our analysis starts with a simple linear regression model

$$x_t = 1.52 + 2.905c_{t-1} + n_t, \tag{3.21}$$

where n_t denotes the error term and the standard errors of the two coefficients are 0.179 and 0.110, respectively. The adjusted R-squared of the regression is 57.23%, indicating that the monthly number of jobless claims explains about half of the variation of unemployment rates. As n_t is not a white noise series, the prior linear regression model is inadequate. Figure 3.25 gives the sample ACF and PACF of the residual series n_t. The ACFs decay slowly, indicating that there remains strong serial dependence in the residual series n_t. The PACFs show large values at the first three lags and significant values at the seasonal lags 12 and 36. It is, however, not easy to specify a model for n_t from its sample ACF and PACF.

To overcome the difficulty in model specification, we use trial and error and some prior knowledge. First, the significance of seasonal ACF and PACF indicates that we should continue to employ a seasonal ARMA(1,1) model with period 12. Second, the unemployment rates show a clear cyclical pattern, that is, existence of certain business cycles. This suggests that some characteristic roots of the model for unemployment rates are complex numbers, see Chapter 2. Therefore, the AR order p satisfies $p \geq 2$. For simplicity, we assume $p = 2$. Finally, the ACFs in Figure 3.25 do not show any clear pattern of exponential decay or damping sine function. This indicates that the MA order q is positive. For simplicity, we try $q = 3$. Consequently, we tentatively specify the model

$$(1 - \phi_1 B - \phi_2 B^2)(1 - \Phi B^{12})(x_t - \beta_0 - \beta_1 c_{t-1})$$
$$= (1 - \theta_1 B - \cdots - \theta_3 B^3)(1 - \Theta B^{12})a_t,$$

for the unemployment rates. The fitted model is

$$(1 - 1.900B + -.902B^2)(1 - 0.65B^{12})(x_t - 6.04 - 0.077c_{t-1})$$
$$= (1 - 0.893B + 0.146B^2 + 0.056B^3)(1 - 0.85B^{12})a_t, \quad \sigma_a^2 = 0.0242,$$

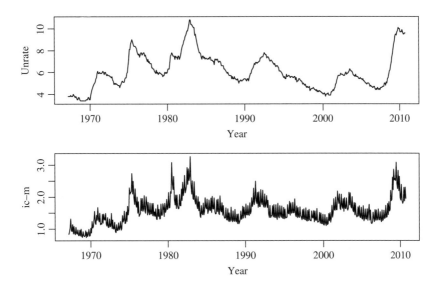

Figure 3.24. Time plots of US monthly unemployment rates and numbers of initial jobless claims from February 1967 to September 2010. Numbers of initial jobless claims were divided by 1,000,000 and shifted by 1 month.

where all estimates, but the MA(3) coefficient, are significant at the 5% level. The model checking statistics indicate that the model is adequate. Therefore, we refine the model as

$$(1 - 1.912B + 0.915B^2)(1 - 0.65B^{12})(x_t - 6.11 - 0.078c_{t-1})$$

$$= (1 - 0.910B + 0.186B^2)(1 - 0.85B^{12})a_t, \quad \sigma_a^2 = 0.0243. \quad (3.22)$$

The standard errors of the coefficient estimates are, in the order of appearance, 0.028, 0.028, 0.082, 0.375, 0.021, 0.053, 0.049, and 0.059, respectively. Figure 3.26 provides some plots of model checking for the model in Equation (3.22). The plots indicate that the model is adequate. The AIC of the model is −435.93.

Remark. For initial model specification, we used an ARMA(2,3) model for the regular component of the unemployment rate. In fact, one can use an ARMA(2,q) model for $2 \le q \le 5$ and obtain the same final model in Equation (3.22). □

R Demonstration

```
> da=read.table("m-unrateic.txt",header=T)
> head(da)
     year mon dd rate w1m1 w2m1 w3m1 w4m1 icm1
1    1967   2  1  3.8  208  207  217  204  836
. . . .
```

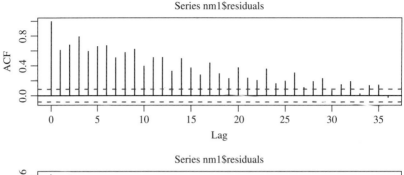

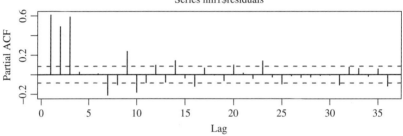

Figure 3.25. Sample ACF and PACF of the residual series of linear regression in Equation (3.21).

```
6    1967   7  1  3.8  248  238  224  218  928
> unrate=da$rate
> x=da[,5:9]/1000
> nm1=lm(unrate~icm1,data=x)
> summary(nm1)
Call: lm(formula = unrate ~icm1, data = x)

Coefficients:
             Estimate Std. Error t value Pr(>|t|)
(Intercept)   1.5202     0.1785    8.518   <2e-16 ***
icm1          2.9047     0.1097   26.475   <2e-16 ***

---
Residual standard error: 1.051 on 522 degrees of freedom
Multiple R-squared: 0.5731,      Adjusted R-squared: 0.5723

> par(mfcol=c(2,1))
> acf(nm1$residuals,lag=36)
> pacf(nm1$residuals,lag=36)
> nm1=arima(unrate,order=c(2,0,3),xreg=x[,5],seasonal=list(order=c(1,0,1),
period=12))
> nm1 arima(x=unrate,order=c(2,0,3),seasonal=list(order=c(1,0,1),period=12),
xreg = x[, 5])

Coefficients:
         ar1      ar2      ma1     ma2     ma3     sar1    sma1   intercept
      1.8997  -0.9021  -0.8932  0.1458  0.0555  0.6501  -0.8520    6.0373
s.e.  0.0332   0.0331   0.0543  0.0565  0.0466  0.0824   0.0586    0.3706
         icm1
```

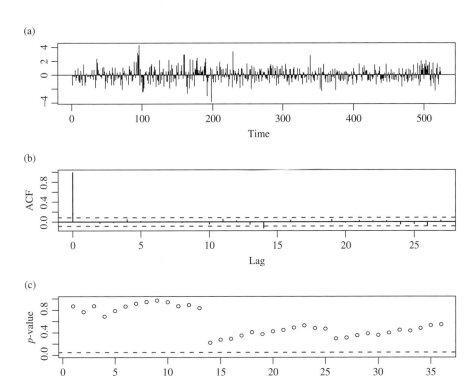

Figure 3.26. Diagnostic plots of model (3.22) for the monthly unemployment rates with monthly initial jobless claims. The data are from February 1967 to September 2010. (a) Standardized residuals, (b) ACF of residuals, (c) p-values for Ljung–Box statistic.

```
          0.0772
s.e.      0.0212

sigma^2 estimated as 0.02419:  log likelihood = 227.7, aic=-435.39
> nm1=arima(unrate,order=c(2,0,2),seasonal=list(order=c(1,0,1),period=12),
  xreg=x[,5])
> nm1 arima(x=unrate,order=c(2,0,2),seasonal=list(order=c(1,0,1),period=12),
          xreg = x[, 5])

Coefficients:
          ar1      ar2      ma1     ma2    sar1    sma1   intercept   x[, 5]
       1.9123  -0.9145  -0.9100  0.1860  0.6465  -0.8483     6.1111   0.0782
s.e.   0.0283   0.0282   0.0527  0.0479  0.0823   0.0591      0.3748   0.0213

sigma^2 estimated as 0.02426:  log likelihood = 226.97, aic=-435.93
> tsdiag(nm1,gof=36)
```

Using Weekly Initial Jobless Claims. Figure 3.27 shows the time plots of monthly unemployment rates x_t and the weekly number of initial jobless claims for the first two weeks of each month. All three plots show a similar cyclical pattern, indicating that weekly data might be helpful in predicting the monthly unemployment rates. An advantage of using the data of initial jobless claims of the first two weeks in a month is that they provide a longer lead time in prediction as compared with monthly initial jobless claims. We use data of the first two weeks of each month based on the following results:

$$x_t = 0.516 + 6.522w_{1,t-1} + 9.671w_{2,t-1} - 2.446w_{3,t-1} + 1.663w_{4,t-1} + e_t, \quad (3.23)$$

where $\sigma_e = 0.884$ and the standard errors of the coefficient estimates are 0.165, 2.115, 2.863, 2.798, and 2.062, respectively. The $w_{i,t-1}$ have high correlations so that the standard errors of their coefficient estimates are relatively large. From the multiple linear regression in Equation (3.23), we only need claim data of the first two weeks. In addition, even with the monthly data of initial jobless claims, we found that only data of the first two weeks are needed:

$$x_t = 0.511 + 6.391w_{1,t-1} + 8.465w_{2,t-1} + 0.131c_{t-1} + e_t,$$

where standard errors of the coefficient estimates are 0.165, 2.075, 2.237, and 0.211, respectively. The coefficient of c_{t-1} is not statistically significant at the 5% level. In summary, with weekly data of initial jobless claims, we employ the linear regression

$$x_t = 0.513 + 6.459w_{1,t-1} + 8.961w_{2,t-1} + n_t, \quad (3.24)$$

where the standard error of n_t is 0.883 and the standard errors of the coefficient estimates are 0.165, 2.071, and 2.087, respectively.

Figure 3.28 shows the sample ACF and PACF of the multiple linear regression model in Equation (3.24). The ACF decays slowly, but the PACF only has a few significant values. The behavior of ACF and PACF in Figure 3.28 is similar to that of ACF and PACF in Figure 3.25. The main difference is that the ACF in Figure 3.28 appears to be smoother. Consequently, we specify a similar model for the unemployment rate:

$$(1 - \phi_1 B - \phi_2 B^2)(1 - \Phi B^{12})(x_t - \beta_0 - \beta_1 w_{1,t-1} - \beta_2 w_{2,t-1})$$
$$= (1 - \theta_1 B - \theta_2 B^2)(1 - \Theta B^{12})a_t.$$

The fitted model is

$$(1 - 1.917B + 0.920B^2)(1 - 0.611B^{12})(x_t - 5.656 - 0.427w_{1,t-1} - 0.969w_{2,t-1})$$
$$= (1 - 0.996B + 0.253B^2)(1 - 0.79B^{12})a_t, \quad \sigma_a^2 = 0.024, \quad (3.25)$$

where the standard errors of the estimates are, in the order of appearance, 0.027, 0.027, 0.112, 0.391, 0.272, 0.321, 0.056, 0.051, and 0.088, respectively. All estimates, but the coefficient of $w_{1,t-1}$, are significant at the 5% level. The t-ratio of the coefficient of $w_{1,t-1}$ is 1.57, which is marginally significant with p-value 0.12. Figure 3.29 shows some diagnostic plots of the model in Equation (3.25). The plots indicate that the model is adequate for the monthly unemployment rates. The AIC of the model is -440.59, which is smaller than that of the model in Equation (3.22).

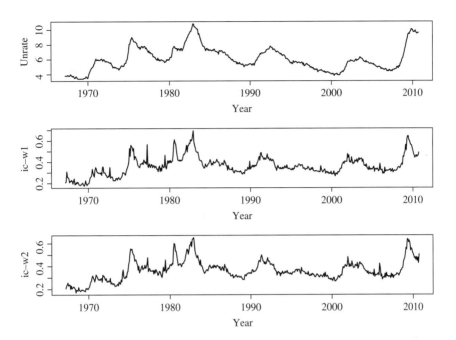

Figure 3.27. Time plots of monthly US unemployment rates and weekly numbers of initial jobless claims from February 1967 to September 2010. The claim data are the numbers of the first two weeks of each month at time $t - 1$.

R Demonstration

```
> nm2=lm(unrate~w1m1+w2m1+w3m1+w4m1,data=x)
> summary(nm2)
lm(formula = unrate ~w1m1 + w2m1 + w3m1 + w4m1, data = x)
Coefficients:
            Estimate Std. Error t value Pr(>|t|)
(Intercept)   0.5160     0.1651   3.125 0.001877 **
w1m1          6.5221     2.1145   3.084 0.002148 **
w2m1          9.6711     2.8630   3.378 0.000785 ***
w3m1         -2.4455     2.7980  -0.874 0.382506
w4m1          1.6626     2.0624   0.806 0.420528
---
Residual standard error: 0.8841 on 519 degrees of freedom

> nm2=lm(unrate~w1m1+w2m1+icm1,data=x)
> summary(nm2)
lm(formula = unrate ~w1m1 + w2m1 + icm1, data = x)
Coefficients:
            Estimate Std. Error t value Pr(>|t|)
(Intercept)   0.5111     0.1648   3.101 0.002033 **
w1m1          6.3906     2.0751   3.080 0.002182 **
w2m1          8.4654     2.2365   3.785 0.000171 ***
icm1          0.1307     0.2110   0.619 0.535950
---
Residual standard error: 0.8837 on 520 degrees of freedom
```

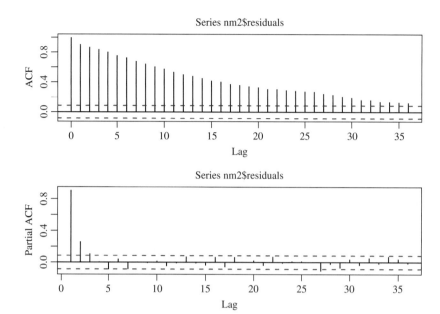

Figure 3.28. Sample ACF and PACF of the residual series of linear regression model in Equation (3.24).

```
> nm2=lm(unrate~w1m1+w2m1,data=x)
> summary(nm2)
lm(formula = unrate ~w1m1 + w2m1, data = x)
Coefficients:
              Estimate Std. Error t value Pr(>|t|)
(Intercept)    0.5127     0.1647    3.113  0.00195 **
w1m1           6.4594     2.0709    3.119  0.00191 **
w2m1           8.9609     2.0872    4.293  2.1e-05 ***
---
Residual standard error: 0.8832 on 521 degrees of freedom
Multiple R-squared: 0.6993,     Adjusted R-squared: 0.6981
> acf(nm2$residuals,lag=36)
> pacf(nm2$residuals,lag=36)
> nm2=arima(unrate,order=c(2,0,2),seasonal=list(order=c(1,0,1),
  period=12),xreg=x[,1:2])
> nm2
arima(x=unrate,order=c(2,0,2),seasonal=list(order= c(1,0,1),period=12),
    xreg = x[, 1:2])

Coefficients:
         ar1      ar2      ma1     ma2    sar1     smal intercept    w1m1
      1.9172  -0.9197  -0.9958  0.2532  0.6111  -0.7915    5.6555  0.4265
s.e   0.0269   0.0268   0.0563  0.0507  0.1119   0.0883    0.3912  0.2721
         w2m1
       0.9693
s.e.   0.3206

sigma^2 estimated as 0.024:  log likelihood = 230.29, aic=-440.59
> tsdiag(nm2,gof=36)
```

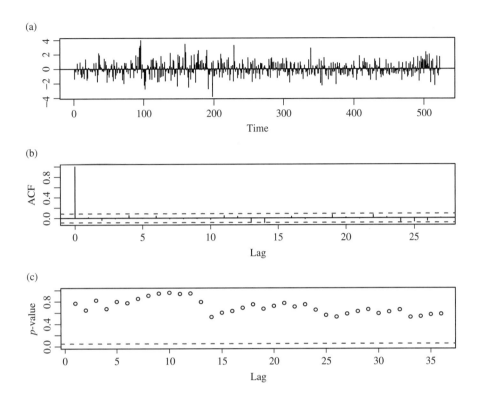

Figure 3.29. Model checking plots of model (3.25) for the monthly US unemployment rates from February 1967 to September 2010. (a) Standardized residuals, (b) ACF of residuals, (c) p-values for Ljung–Box statistic.

3.3.5 Comparison

We have entertained two relatively complicated models for the monthly unemployment rates. These two models made use of the information in initial jobless claims. For comparison purpose, we also built a time series for the unemployment rates from February 1967 to September 2010. The fitted model is

$$(1 - 0.901B)(1 - B)(1 - 0.625B^{12})x_t = (1 - 0.868B + 0.170B^2)(1 - 0.83B^{12})a_t, \tag{3.26}$$

where $\sigma_a^2 = 0.0252$ and the standard errors of the coefficient estimates are 0.031, 0.084, 0.053, 0.047, and 0.062, respectively. All estimates are statistically significant at the 5% level. Model checking, not shown, also fails to reject the adequacy of the model. The AIC of this pure time series model is -426.17.

It is interesting to compare the three models. For in-sample comparison, the AIC selects the model in Equation (3.25), indicating that numbers of initial jobless claims of the first two weeks of the previous month are helpful in explaining the variability of monthly unemployment rates. Between the models in Equations (3.22) and (3.26), AIC also supports the usefulness of monthly numbers of initial jobless claims. For out-of-sample comparison, we, again, divided the data into modeling and forecasting subsamples with the latter consisting of the last 53 observations. The results of backtesting are given below:

Model	RMSFE	MAFE
Equation (3.22) with monthly initial jobless claims	0.1706	0.1454
Equation (3.25) with weekly initial jobless claims	0.1683	0.1372
Pure time series model in Equation (3.26)	0.1684	0.1370

Clearly, there is no major difference in out-of-sample forecasting between the three models. Granted that out-of-sample forecasting performance depends on the choice of forecasting subsamples, our simple exercise indicates that numbers of initial jobless claims are of limited value in predicting the seasonally adjusted monthly unemployment rate. This case study, thus, highlights the difference between in-sample and out-of-sample model comparisons.

The similarity of the three competing models in Equations (3.22), (3.25), and (3.26) can also be seen as follows: First, the seasonal components of the three models are close. They all show the inadequacy of the seasonal adjustment method used. Second, even though we did not impose a unit root on the models in Equations (3.22) and (3.25), the estimation results hint strongly the presence of a unit root. The regular AR polynomials of the two models can be written approximately as $1 - 0.912B + 0.915B^2 \approx (1 - 0.912B)(1 - B)$ and $1 - 1.917 + 0.920B^2 \approx (1 - 0.917B)(1 - B)$. These factorizations are close to $(1 - 0.901B)(1 - B)$ of the model in Equation (3.26). It is then understandable that the 1-step ahead predictions of the three models are similar.

EXERCISES

1. Consider the monthly unemployment rates of the State of California and United States from January 1976 to September 2011. The data are in the file m-CAUS-7611.txt (year, mon, CA, US).

   (a) Build a pure time series model for the monthly unemployment rates of California. Perform model checking and write down the fitted model.

   (b) Build a time series model for the monthly unemployment rates of California using the lag-1 US monthly unemployment rate as an explanatory variable. Perform model checking and write down the fitted model.

   (c) Let the forecasting period be from January 2008 to September 2011. Compare the two models using out-of-sample forecasts.

2. Consider the US monthly 30-year conventional mortgage rates from April 1971 to November 2011. The data are available from FRED and are in the file m-morgfed-7111.txt (year, mon, day, morg, fed).

(a) Build a pure time series model for the monthly mortgage rate. Perform model checking and write down the fitted model.

(b) Mortgage rate is known to depend on the Federal Funds rate. Build a time series model for the mortgage rate using the lag-1 effective Federal Funds rate as an explanatory variable. Perform model checking and write down the fitted model. On the basis of the fitted model, does mortgage rate depend on the Federal Funds rate at the 5% significance level?

(c) Let the forecasting period be from January 2007 to November 2011. Use out-of-sample predictions to compare the two models.

REFERENCES

Box GEP. Science and statistics. J Am Stat Assoc 1976;33526–536.

Box GEP, Jenkins GM, Reinsel GC. Time Series Analysis: Forecasting and Control. 4th ed. Hoboken (NJ): John Wiley & Sons; 2008.

Tiao GC, Tsay RS. Some advances in non-linear and adaptive modeling in time series. J Forecast 1994;13109–131.

Tsay RS. Analysis of Financial Time Series. 3rd ed. Hoboken (NJ): John Wiley & Sons; 2010.

4

ASSET VOLATILITY AND VOLATILITY MODELS

An important measure in finance is the risk associated with an asset and asset volatility is perhaps the most commonly used risk measure. There are several types of asset volatility, however. We define them in the next section. The goal of this chapter is to understand the characteristics of asset volatility, to study volatility models, and to gain experience in modeling and using asset volatility.

Volatility has many important applications in finance. It is a key factor in options pricing and asset allocation. It plays an important role in value at risk calculation for risk management. More recently, some volatility indices have become a financial instrument. For example, the VIX index of the Chicago Board Options Exchange (CBOE) started to trade in futures on March 26, 2004.

Although asset volatility is well defined, it is not directly observable in practice. What we observe are the prices of an asset and its derivatives. One must estimate the volatility from these observed prices. The fact that volatility is not directly observable has several important implications in studying and modeling volatility. We shall discuss these implications throughout the chapter.

There are many volatility models available in the literature. The univariate models discussed in this chapter include the autoregressive conditional heteroscedastic (ARCH) model of Engle (1982), the generalized autoregressive conditional

An Introduction to Analysis of Financial Data with R, First Edition. Ruey S. Tsay.
© 2013 John Wiley & Sons, Inc. Published 2013 by John Wiley & Sons, Inc.

heteroscedastic (GARCH) model of Bollerslev (1986), the exponential generalized autoregressive conditional heteroscedastic (EGARCH) model of Nelson (1991), the threshold generalized autoregressive conditional heteroscedastic (TGARCH) model of Glosten et al. (1993) and Zakoian (1994), the nonsymmetric generalized autoregressive conditional heteroscedastic (NGARCH) model of Engle and Ng (1993) and Duan (1995), and the stochastic volatility (SV) models of Melino and Turnbull (1990), Taylor (1994), Harvey et al. (1994), and Jacquier et al. (1994). We discuss advantages and weaknesses of each volatility model and consider some applications of volatility. The chapter also studies alternative approaches to volatility modeling, including the use of daily high and low prices of an asset and realized volatility based on high frequency data.

4.1 CHARACTERISTICS OF VOLATILITY

Although volatility is not directly observable, it has some characteristics that are commonly seen in asset returns. First, there exist volatility clusters (i.e., volatility is high for certain time periods and low for other periods). Second, volatility evolves over time in a continuous manner—that is, volatility jumps are rare. Third, volatility does not diverge to infinity—that is, volatility varies within some fixed range. Statistically speaking, this means that volatility is often stationary. Fourth, volatility seems to react differently to a big price increase and a big price drop with the latter having a greater impact. This phenomenon is referred to as the *leverage* effect. These properties play an important role in the development of volatility models. Some volatility models were proposed specifically to correct the weaknesses of the existing ones for their inability to capture the characteristics mentioned earlier. For example, the EGARCH and TGARCH models were developed to capture the asymmetry in volatility induced by big "positive" and "negative" asset returns.

In practice, we typically estimate the volatility of an asset using the prices of its stock or derivatives or both. Consider the daily volatility of IBM stock. What we observe are (i) the daily return for each trading day, (ii) tick-by-tick data for intraday transactions and quotes, and (iii) the prices of options contingent on IBM stock. These three data sources give rise to three types of volatility measures for IBM stock. They are as follows:

- Volatility as the conditional standard deviation of daily returns: This is the usual definition of volatility and is the focus of volatility models discussed in this chapter.
- Implied volatility: Using prices from options markets, one can use a pricing formula, for example, the Black–Scholes pricing formula, to deduce the volatility of the stock price. This volatility measure is called the *implied* volatility. Because implied volatility is derived under certain assumptions that relate the price of an option to that of the underlying stock, it is often criticized to be model dependent. Experience shows that implied volatility of an asset return tends to be larger than that obtained by using daily returns and a volatility

model. This might be due to the risk premium for volatility in options markets or to the way daily returns are calculated. The VIX of CBOE is an implied volatility.

- Realized volatility: With the availability of high frequency financial data, one can use intraday returns, for example, 5-min returns, to estimate the daily volatility. This volatility measure is called *realized volatility*, which is a subject of some intensive research in recent years. We shall discuss realized volatility in Chapter 6.

In real applications, the time interval used to measure volatility is a year. Thus, volatility is often annualized. If daily returns are used to estimate volatility, one can obtain the annualized volatility by multiplying daily volatility by $\sqrt{252}$, for there are typically 252 trading days in the United States.

4.2 STRUCTURE OF A MODEL

Let r_t be the log return of an asset at time index t. The basic idea behind volatility study is that the series $\{r_t\}$ is either serially uncorrelated or with minor lower order serial correlations, but it is a dependent series. For illustration, consider the monthly log stock returns of Intel Corporation from January 1973 to December 2009 for 444 observations. Figure 4.1 shows the time plot of the returns. From the plot, the return series appears to be stationary and random. Figure 4.2a shows the sample ACF of the log returns, which, as expected, suggests no significant serial correlations except for some minor ones at lags 7 and 14. Indeed, the Ljung–Box statistics show that $Q(12) = 18.68$ with p value 0.10 for r_t. On the other hand, Figure 4.2b shows the sample ACF of the absolute log returns (i.e., $|r_t|$). It is clear that $|r_t|$ has serial correlations. The Ljung–Box statistics of $|r_t|$ show that $Q(12) = 124.91$ with p value close to zero. Consequently, the monthly log returns of Intel stock are serially uncorrelated, but dependent. This is the feature that a univariate volatility model is designed to capture.

To put the volatility models in proper perspective, it is informative to consider the conditional mean and variance of r_t given F_{t-1}; that is,

$$\mu_t = E(r_t|F_{t-1}), \qquad \sigma_t^2 = \text{Var}(r_t|F_{t-1}) = E[(r_t - \mu_t)^2|F_{t-1}], \qquad (4.1)$$

where F_{t-1} denotes the information set available at time $t-1$. Typically, F_{t-1} consists of all linear functions of the past returns. As shown by the empirical examples of Chapter 2 and Figure 4.2, serial dependence in a stock return series r_t is weak if it exists at all. Therefore, the equation for μ_t in Equation (4.1) should be simple, and we assume that r_t follows a simple time series model such as a stationary ARMA(p,q) model. For example, consider the monthly log returns of Intel stock. As shown before, the Ljung–Box statistics show that the returns have no serial correlations and a simple one-sample test confirms that the mean of r_t is significantly different from zero. More specifically, the t-ratio of testing $H_0 : \mu = 0$ versus $H_a : \mu \neq 0$ is 2.38 with p value 0.018. Thus, for the Intel log returns, we have $r_t = \mu_t + a_t$ with $\mu_t = \mu$, a constant.

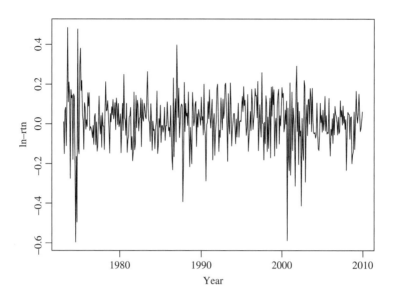

Figure 4.1. Time plot of the monthly log returns of Intel stock from January 1973 to December 2009.

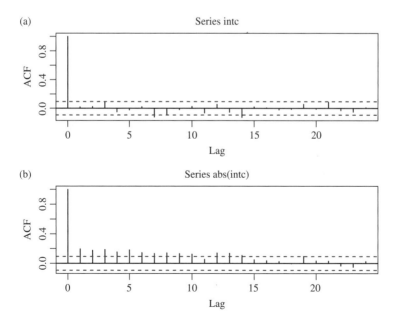

Figure 4.2. Sample ACF of the monthly log returns of Intel stock from January 1973 to December 2009: (a) ACF of the log returns and (b) ACF of the absolute log returns.

In general, we assume that r_t follows an ARMA(p,q) model so that $r_t = \mu_t + a_t$, where μ_t is given by

$$\mu_t = \phi_0 + \sum_{i=1}^{p} \phi_i r_{t-i} - \sum_{j=1}^{q} \theta_j a_{t-j}. \qquad (4.2)$$

If some explanatory variables are available, then we entertain the model $r_t = \mu_t + a_t$, where

$$\mu_t = \phi_0 + \sum_{i=1}^{k} \beta_i x_{i,t-1} + \sum_{i=1}^{p} \phi_i y_{t-i} - \sum_{j=1}^{q} \theta_j a_{t-j}, \qquad (4.3)$$

where $y_{t-i} = r_{t-i} - \phi_0 - \sum_{i=1}^{k} \beta_i x_{i,t-i-1}$ denotes the adjusted return series after removing the effect of explanatory variables, and $x_{i,t-j}$ are explanatory variables available at time $t-j$.

R Demonstration

```
> da=read.table("m-intcsp7309.txt",header=T)
> head(da)
        date       intc         sp
1 19730131   0.010050 -0.017111
...
> intc=log(da$intc+1)
> rtn=ts(intc,frequency=12,start=c(1973,1))
> plot(rtn,type='l',xlab='year',ylab='ln-rtn') % time plot
> t.test(intc)   % testing the mean of returns
        One Sample t-test
data:   intc
t = 2.3788, df = 443, p-value = 0.01779
alternative hypothesis: true mean is not equal to 0

> Box.test(intc,lag=12,type='Ljung')
        Box-Ljung test

data:   intc
X-squared = 18.6761, df = 12, p-value = 0.09665

> par(mfcol=c(2,1))
> acf(intc,lag=24) % ACF plots
> acf(abs(intc),lag=24)
> Box.test(abs(intc),lag=12,type='Ljung')
        Box-Ljung test

data:   abs(intc)
X-squared = 124.9064, df = 12, p-value < 2.2e-16
```

Model (4.3) illustrates a possible application of the regression model with time series errors of Chapter 2 in finance. The order (p, q) of an ARMA model may depend on the frequency of the return series. For example, daily returns of a market index often show some minor serial correlations, but monthly returns of the index may not contain any significant serial correlation. The explanatory variables $x_{i,t-j}$ in Equation (4.3) are flexible. For example, a dummy variable can be used for the Mondays to study the effect of weekend on daily stock returns. In the capital asset pricing model (CAPM), the mean equation of r_t can be written as $r_t = \phi_0 + \beta r_{m,t} + a_t$, where $r_{m,t}$ denotes the market return.

Combining Equations (4.1) and (4.3), we have

$$\sigma_t^2 = \text{Var}(r_t | F_{t-1}) = \text{Var}(a_t | F_{t-1}), \tag{4.4}$$

where the positive square root σ_t is the volatility. The conditional heteroscedastic models of this chapter are concerned with the evolution of σ_t^2. The manner under which σ_t^2 evolves over time distinguishes one volatility model from another.

Conditional heteroscedastic models can be classified into two general categories. Those in the first category use an exact function to govern the evolution of σ_t^2, whereas those in the second category use a stochastic equation to describe σ_t^2. The GARCH model belongs to the first category, whereas the SV model is in the second category.

Throughout the book, a_t is referred to as the *shock* or *innovation* of an asset return at time t. The model for μ_t in Equation (4.3) is referred to as the *mean* equation for r_t and the model for σ_t^2 is the *volatility* equation for r_t. Therefore, modeling conditional heteroscedasticity amounts to augmenting a dynamic equation, which governs the time evolution of the conditional variance of the asset return to a time series model.

4.3 MODEL BUILDING

Building a volatility model for an asset return series consists of four steps:

1. Specify a mean equation by testing for serial dependence in the data and, if necessary, building an econometric model (e.g., an ARMA model) for the return series to remove any linear dependence.
2. Use the residuals of the mean equation to test for ARCH effects.
3. Specify a volatility model if ARCH effects are statistically significant, and perform a joint estimation of the mean and volatility equations.
4. Check the fitted model carefully and refine it if necessary.

In what follows, we describe each step of the modeling procedure in detail and introduce various volatility models.

Specifying the Mean Equation. For most asset return series, the serial cor-relations are weak, if any. Thus, building a mean equation amounts to removing the sample mean from the data if the sample mean is significantly different from zero. For some daily return series, a simple AR model might be needed. In some cases, the mean equation may employ some explanatory variables such as an indicator variable for weekend or January effects. As discussed before, for the monthly log returns of Intel stock, the mean equation consists a constant only.

4.4 TESTING FOR ARCH EFFECT

For ease in notation, let $a_t = r_t - \mu_t$ be the residuals of the mean equation. The squared series a_t^2 is then used to check for conditional heteroscedasticity, which is also known as the *ARCH* effects. Two tests are available. The first test is to apply the usual Ljung–Box statistics $Q(m)$ to the $\{a_t^2\}$ series; see McLeod and Li (1983). The null hypothesis of the test statistic is that the first m lags of ACF of the a_t^2 series are zero. The second test for conditional heteroscedasticity is the Lagrange multiplier test of Engle (1982). This test is equivalent to the usual F statistic for testing $\alpha_i = 0$ $(i = 1, \ldots, m)$ in the linear regression

$$a_t^2 = \alpha_0 + \alpha_1 a_{t-1}^2 + \cdots + \alpha_m a_{t-m}^2 + e_t, \quad t = m+1, \ldots, T,$$

where e_t denotes the error term, m is a prespecified positive integer, and T is the sam-ple size. Specifically, the null hypothesis is $H_0 : \alpha_1 = \cdots = \alpha_m = 0$ and the alternative hypothesis is $H_a : \alpha_i \neq 0$ for some i between 1 and m. Let $SSR_0 = \sum_{t=m+1}^{T} (a_t^2 - \overline{\omega})^2$, where $\overline{\omega} = (1/T) \sum_{t=1}^{T} a_t^2$ is the sample mean of a_t^2, and $SSR_1 = \sum_{t=m+1}^{T} \hat{e}_t^2$, where $\hat{e}_t$ is the least squares residual of the prior linear regression. Then we have

$$F = \frac{(SSR_0 - SSR_1)/m}{SSR_1/(T - 2m - 1)},$$

which follows an F distribution with degrees of freedom m and $T - 2m - 1$ under H_0. When T is sufficiently large, one can use mF as the test statistic, which is asymptotically a chi-squared distribution with m degrees of freedom under the null hypothesis. The decision rule is to reject the null hypothesis if $mF > \chi_m^2(\alpha)$, where $\chi_m^2(\alpha)$ is the upper $100(1 - \alpha)$th percentile of χ_m^2, or the p value of mF is less than α, type I error.

To demonstrate, we consider the monthly log stock returns of Intel Corporation from 1973 to 2009. As the mean equation is simply a constant plus innovations, we use $y_t = r_t - \overline{r}$ to test for the ARCH effect, where $\overline{r}$ is the sample mean of r_t and y_t is an estimate of a_t. The Ljung–Box statistics of y_t^2 shows strong ARCH effects with $Q(12) = 92.94$, the p value of which is close to zero. Applying the Lagrange multiplier test of Engle with $m = 12$, we have $F = 4.978$ with p value 9.74×10^{-8}. Again, the test confirms strong ARCH effects in the monthly log returns of Intel stock.

R Demonstration

```
> y=intc-mean(intc)
> Box.test(y^2,lag=12,type='Ljung')
        Box-Ljung test

data:  y^2
X-squared = 92.9389, df = 12, p-value = 1.332e-14

> source("archTest.R")   % R script available on the book web site.
> archTest(y,12)    % output edited.
Coefficients:
            Estimate Std. Error t value Pr(>|t|)
(Intercept) 0.005977   0.002249   2.658 0.008162 **
x1          0.093817   0.048147   1.949 0.052013 .
x2          0.153085   0.048102   3.183 0.001569 **
x3          0.146087   0.048614   3.005 0.002815 **
 ....
x12         0.161945   0.045965   3.523 0.000473 ***
---
Residual standard error: 0.03365 on 419 degrees of freedom
Multiple R-squared: 0.1248,    Adjusted R-squared: 0.0997
F-statistic: 4.978 on 12 and 419 DF, p-value: 9.742e-08  <== F-ratio
```

The ARCH effect also occurs in other financial time series. Figure 4.3a shows the time plot of the log returns for the daily exchange rate between US Dollar and Euro

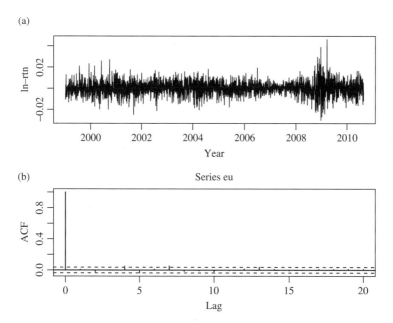

Figure 4.3. Daily log returns of the exchange rate between Dollar and Euro from January 4, 1999, to August 20, 2010: (a) time plot of the returns and (b) sample ACF of the returns.

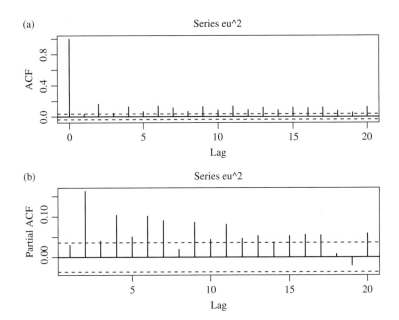

Figure 4.4. (a) Sample autocorrelations and (b) partial autocorrelations of the squared series of daily log returns of the exchange rate between Dollar and Euro from January 4, 1999, to August 20, 2010.

from January 4, 1999, to August 20, 2010. As expected, the variability of the exchange rate was higher during the second half of 2008 and the first half of 2009. Figure 4.3b gives the sample ACF of the series. These ACFs suggest that there are no strong serial correlations in the log return series. The Ljung–Box statistics of the log returns give $Q(20) = 30.59$ with p value 0.061. The null hypothesis of zero correlations is not rejected at the 5% level, but is rejected at the 10% level. The t-ratio for testing zero expected return is 0.20 with p value 0.84. Therefore, the mean equation for the daily log returns of Dollar/Euro exchange rate is $r_t = a_t$ if one uses the 5% type I error.

Figure 4.4 shows the sample ACF and PACF of the r_t^2 series. Clearly, the squared log return series has some serial correlations. Both plots confirm that there are serial correlations in the squared series. Consequently, the sample ACF and PACF of r_t^2 show that the daily log returns of the Dollar/Euro exchange rate have significant ARCH effects. For further confirmation, the Ljung–Box statistics of r_t^2 show $Q(20) = 661.45$ with p value close to zero, and the Lagrange multiplier test gives an F-ratio of 14.74 with p value also close to zero. These tests confirm that there are strong ARCH effects in the exchange rate series.

R Demonstration

```
> fx=read.table("d-useu9910.txt",header=T)
> fxeu=log(fx$rate)
```

```
> eu=diff(fxeu)
> Box.test(eu,lag=20,type='Ljung')
        Box-Ljung test
data:   eu
X-squared = 30.585, df = 20, p-value = 0.06091

> t.test(eu)
        One Sample t-test
data:   eu
t = 0.2022, df = 2928, p-value = 0.8398
alternative hypothesis: true mean is not equal to 0

> Box.test(eu^2,lag=20,type='Ljung')
        Box-Ljung test
data:   eu^2
X-squared = 661.4545, df = 20, p-value < 2.2e-16

> source("archTest.R")
> archTest(eu,20)
Coefficients:
             Estimate  Std. Error t value Pr(>|t|)
(Intercept) 1.281e-05  2.535e-06    5.054 4.60e-07 ***
x1          -3.022e-02 1.858e-02   -1.626 0.103966
x2           9.441e-02 1.859e-02    5.080 4.02e-07 ***
....
x20          5.844e-02 1.859e-02    3.144 0.001683 **
---

Residual standard error: 8.483e-05 on 2888 degrees of freedom
Multiple R-squared: 0.09265,    Adjusted R-squared: 0.08636
F-statistic: 14.74 on 20 and 2888 DF,   p-value: < 2.2e-16
```

4.5 THE ARCH MODEL

The first model that provides a systematic framework for volatility modeling is the ARCH model of Engle (1982). The basic idea of ARCH models is that (i) the shock a_t of an asset return is serially uncorrelated, but dependent, and (ii) the dependence of a_t can be described by a simple quadratic function of its lagged values. Specifically, an ARCH(m) model assumes that

$$a_t = \sigma_t \epsilon_t, \quad \sigma_t^2 = \alpha_0 + \alpha_1 a_{t-1}^2 + \cdots + \alpha_m a_{t-m}^2, \tag{4.5}$$

where $\{\epsilon_t\}$ is a sequence of independent and identically distributed (iid) random variables with mean zero and variance 1, $\alpha_0 > 0$, and $\alpha_i \geq 0$ for $i > 0$. The coefficients α_i must satisfy some regularity conditions to ensure that the unconditional variance of a_t is finite. In practice, ϵ_t is often assumed to follow the standard normal or a standardized Student t distribution or a generalized error distribution (GED). In some applications, a skew distribution is used for ϵ_t.

From the structure of the model, it is seen that large past squared shocks $\{a_{t-i}^2\}_{i=1}^m$ imply a large conditional variance σ_t^2 for the innovation a_t. Consequently, a_t tends to assume a large value (in modulus). This means that, under the ARCH framework, large shocks tend to be followed by another large shock. Here, I use the word *tend* because a large variance does not necessarily produce a large realization. It only says that the probability of obtaining a large variate is greater than that of a smaller variance. This feature is similar to the volatility clusterings observed in asset returns.

Remark. Some authors use h_t to denote the conditional variance in Equation (4.5). In this case, the shock becomes $a_t = \sqrt{h_t}\epsilon_t$.   □

4.5.1 Properties of ARCH Models

To understand the ARCH models, it pays to carefully study the ARCH(1) model

$$a_t = \sigma_t\epsilon_t, \quad \sigma_t^2 = \alpha_0 + \alpha_1 a_{t-1}^2,$$

where $\alpha_0 > 0$ and $\alpha_1 \geq 0$. First, the unconditional mean of a_t remains zero because

$$E(a_t) = E[E(a_t|F_{t-1})] = E[\sigma_t E(\epsilon_t)] = 0.$$

Second, the unconditional variance of a_t can be obtained as

$$\text{Var}(a_t) = E(a_t^2) = E[E(a_t^2|F_{t-1})]$$
$$= E(\alpha_0 + \alpha_1 a_{t-1}^2) = \alpha_0 + \alpha_1 E(a_{t-1}^2).$$

Because a_t is a stationary process with $E(a_t) = 0$, $\text{Var}(a_t) = \text{Var}(a_{t-1}) = E(a_{t-1}^2)$. Therefore, we have $\text{Var}(a_t) = \alpha_0 + \alpha_1 \text{Var}(a_t)$ and $\text{Var}(a_t) = \alpha_0/(1 - \alpha_1)$. Because the variance of a_t must be positive, we require $0 \leq \alpha_1 < 1$. Third, in some applications, we need higher order moments of a_t to exist and, hence, α_1 must also satisfy some additional constraints. For instance, to study its tail behavior, we require that the fourth moment of a_t is finite. Under the normality assumption of ϵ_t in Equation (4.5), we have

$$E(a_t^4|F_{t-1}) = 3[E(a_t^2|F_{t-1})]^2 = 3(\alpha_0 + \alpha_1 a_{t-1}^2)^2.$$

Therefore,

$$E(a_t^4) = E[E(a_t^4|F_{t-1})] = 3E(\alpha_0 + \alpha_1 a_{t-1}^2)^2$$
$$= 3E\left(\alpha_0^2 + 2\alpha_0\alpha_1 a_{t-1}^2 + \alpha_1^2 a_{t-1}^4\right).$$

If a_t is fourth-order stationary with $m_4 = E(a_t^4)$, then we have

$$m_4 = 3[\alpha_0^2 + 2\alpha_0\alpha_1 \text{Var}(a_t) + \alpha_1^2 m_4]$$
$$= 3\alpha_0^2\left(1 + 2\frac{\alpha_1}{1 - \alpha_1}\right) + 3\alpha_1^2 m_4.$$

Consequently,

$$m_4 = \frac{3\alpha_0^2(1 + \alpha_1)}{(1 - \alpha_1)(1 - 3\alpha_1^2)}.$$

This result has two important implications: (i) as the fourth moment of a_t is positive, we see that α_1 must also satisfy the condition $1 - 3\alpha_1^2 > 0$; that is, $0 \le \alpha_1^2 < \frac{1}{3}$; and (ii) the unconditional kurtosis of a_t is

$$\frac{E(a_t^4)}{[\text{Var}(a_t)]^2} = 3\frac{\alpha_0^2(1 + \alpha_1)}{(1 - \alpha_1)(1 - 3\alpha_1^2)} \times \frac{(1 - \alpha_1)^2}{\alpha_0^2} = 3\frac{1 - \alpha_1^2}{1 - 3\alpha_1^2} > 3.$$

Thus, the excess kurtosis of a_t is positive and the tail distribution of a_t is heavier than that of a normal distribution. In other words, the shock a_t of a conditional Gaussian ARCH(1) model is more likely than a Gaussian white noise series to produce "outliers." This is in agreement with the empirical finding that "outliers" appear more often in asset returns than that implied by an iid sequence of normal random variates.

These properties continue to hold for general ARCH models, but the formulas become more complicated for higher order ARCH models. The condition $\alpha_i \ge 0$ in Equation (4.5) can be relaxed. It is a condition to ensure that the conditional variance σ_t^2 is positive for all t. In fact, a natural way to achieve positiveness of the conditional variance is to rewrite an ARCH(m) model as

$$a_t = \sigma_t \epsilon_t, \quad \sigma_t^2 = \alpha_0 + A'_{m,t-1}\Omega A_{m,t-1}, \tag{4.6}$$

where $A_{m,t-1} = (a_{t-1}, \ldots, a_{t-m})'$ and Ω is an $m \times m$ nonnegative definite matrix. The ARCH(m) model in Equation (4.5) requires Ω to be diagonal. Thus, Engle's model uses a parsimonious approach to approximate a quadratic function. For further discussion, see Tsay (2010, Chapter 3).

4.5.2   Advantages and Weaknesses of ARCH Models

As discussed in the previous subsection, ARCH models have several advantages in analyzing asset returns. The key advantages include the following:

1. The model can produce volatility clusters.
2. The shocks a_t of the model have heavy tails.

The models also have some weaknesses:

1. The model assumes that positive and negative shocks have the same effects on volatility because it depends on the square of the previous shocks. In practice, it is well known that price of a financial asset responds differently to positive and negative shocks.

2. The ARCH model is rather restrictive. For instance, α_1^2 of an ARCH(1) model must be in the interval $[0, \frac{1}{3}]$ if the series has a finite fourth moment. The constraint becomes complicated for higher order ARCH models. In practice, it limits the ability of ARCH models with Gaussian innovations to capture excess kurtosis.

3. The ARCH model does not provide any new insight for understanding the source of variations of a financial time series. It merely provides a mechanical way to describe the behavior of the conditional variance. It gives no indication about what causes such behavior to occur.

4. ARCH models are likely to overpredict the volatility because they respond slowly to large isolated shocks to the return series.

4.5.3 Building an ARCH Model

Among volatility models, specifying an ARCH model is relatively easy. Details are given below.

__Order Determination.__ If an ARCH effect is found to be significant, one can use the PACF of a_t^2 to determine the ARCH order. Using PACF of a_t^2 to select the ARCH order can be justified as follows. From the model in Equation (4.5), we have

$$\sigma_t^2 = \alpha_0 + \alpha_1 a_{t-1}^2 + \cdots + \alpha_m a_{t-m}^2.$$

For a given sample, a_t^2 is an unbiased estimate of σ_t^2. Therefore, we expect that a_t^2 is linearly related to $a_{t-1}^2, \ldots, a_{t-m}^2$ in a manner similar to that of an autoregressive model of order m. Note that a single a_t^2 is generally not an efficient estimate of σ_t^2, but it can serve as an approximation that could be informative in specifying the order m.

Alternatively, define $\eta_t = a_t^2 - \sigma_t^2$. It can be shown that $\{\eta_t\}$ is an uncorrelated series with mean 0. The ARCH model then becomes

$$a_t^2 = \alpha_0 + \alpha_1 a_{t-1}^2 + \cdots + \alpha_m a_{t-m}^2 + \eta_t,$$

which is in the form of an AR(m) model for a_t^2, except that $\{\eta_t\}$ is not an iid series. From Chapter 2, PACF of a_t^2 is a useful tool to determine the order m. Because $\{\eta_t\}$ are not identically distributed, the least squares estimates of the prior model are consistent, but not efficient. The PACF of a_t^2 may not be effective when the sample size is small.

Consider the PACF of the squared log returns of Dollar/Euro exchange rate in Figure 4.4b. There are significant PACFs at higher order lags, indicating that a high order ARCH model is needed for the series. In this situation, one would employ the more parsimonious GARCH model of the next section, instead of a pure ARCH model.

Estimation. Several likelihood functions are commonly used in ARCH estimation, depending on the distributional assumption of ϵ_t. Under the normality assumption, the likelihood function of an ARCH(m) model is

$$f(a_1, \ldots, a_T | \boldsymbol{\alpha}) = f(a_T | F_{T-1}) f(a_{T-1} | F_{T-2}) \cdots f(a_{m+1} | F_m) f(a_1, \ldots, a_m | \boldsymbol{\alpha})$$

$$= \prod_{t=m+1}^{T} \frac{1}{\sqrt{2\pi\sigma_t^2}} \exp\left(-\frac{a_t^2}{2\sigma_t^2}\right) \times f(a_1, \ldots, a_m | \boldsymbol{\alpha}),$$

where $\boldsymbol{\alpha} = (\alpha_0, \alpha_1, \ldots, \alpha_m)'$ and $f(a_1, \ldots, a_m | \boldsymbol{\alpha})$ is the joint probability density function of $a_1, \ldots, a_m$. As the exact form of $f(a_1, \ldots, a_m | \boldsymbol{\alpha})$ is complicated, it is commonly dropped from the prior likelihood function, especially when the sample size is sufficiently large. This results in using the conditional likelihood function

$$f(a_{m+1}, \ldots, a_T | \boldsymbol{\alpha}, a_1, \ldots, a_m) = \prod_{t=m+1}^{T} \frac{1}{\sqrt{2\pi\sigma_t^2}} \exp\left(-\frac{a_t^2}{2\sigma_t^2}\right),$$

where σ_t^2 can be evaluated recursively. We refer to estimates obtained by maximizing the prior likelihood function as the conditional maximum likelihood estimates (MLEs) under normality.

Maximizing the conditional likelihood function is equivalent to maximizing its logarithm, which is easier to handle. The conditional log likelihood function is

$$\ell(a_{m+1}, \ldots, a_T | \boldsymbol{\alpha}, a_1, \ldots, a_m) = \sum_{t=m+1}^{T} \left[-\frac{1}{2} \ln(2\pi) - \frac{1}{2} \ln(\sigma_t^2) - \frac{1}{2} \frac{a_t^2}{\sigma_t^2} \right].$$

As the first term $\ln(2\pi)$ does not involve any parameters, the log likelihood function becomes

$$\ell(a_{m+1}, \ldots, a_T | \boldsymbol{\alpha}, a_1, \ldots, a_m) = -\sum_{t=m+1}^{T} \left[\frac{1}{2} \ln(\sigma_t^2) + \frac{1}{2} \frac{a_t^2}{\sigma_t^2} \right],$$

where $\sigma_t^2 = \alpha_0 + \alpha_1 a_{t-1}^2 + \cdots + \alpha_m a_{t-m}^2$ can be evaluated recursively.

In some applications, it is more appropriate to assume that ϵ_t follows a heavy-tailed distribution such as a standardized Student t distribution. Let x_v be a Student t distribution with v degrees of freedom. Then $\text{Var}(x_v) = v/(v-2)$ for $v > 2$, and we use $\epsilon_t = x_v/\sqrt{v/(v-2)}$. The probability density function of ϵ_t is

$$f(\epsilon_t | v) = \frac{\Gamma((v+1)/2)}{\Gamma(v/2)\sqrt{(v-2)\pi}} \left(1 + \frac{\epsilon_t^2}{v-2}\right)^{-(v+1)/2}, \qquad v > 2, \qquad (4.7)$$

where $\Gamma(x)$ is the usual gamma function (i.e., $\Gamma(x) = \int_0^\infty y^{x-1} e^{-y} dy$). Using $a_t = \sigma_t \epsilon_t$, we obtain the conditional likelihood function of a_t as

$$f(a_{m+1},\ldots,a_T|\boldsymbol{\alpha},A_m) = \prod_{t=m+1}^{T} \frac{\Gamma((v+1)/2)}{\Gamma(v/2)\sqrt{(v-2)\pi}} \frac{1}{\sigma_t} \left(1 + \frac{a_t^2}{(v-2)\sigma_t^2}\right)^{-(v+1)/2},$$

where $v > 2$ and $A_m = (a_1, a_2, \ldots, a_m)$. We refer to the estimates that maximize the prior likelihood function as the conditional MLEs under t distribution. The degrees of freedom of the t distribution can be specified *a priori* or estimated jointly with other parameters. A value between 3 and 6 is often used if it is prespecified.

If the degrees of freedom v of the Student t distribution is prespecified, then the conditional log likelihood function is

$$\ell(a_{m+1},\ldots,a_T|\boldsymbol{\alpha},A_m) = -\sum_{t=m+1}^{T} \left[\frac{v+1}{2}\ln\left(1 + \frac{a_t^2}{(v-2)\sigma_t^2}\right) + \frac{1}{2}\ln(\sigma_t^2)\right]. \quad (4.8)$$

If one wishes to estimate v jointly with other parameters, then the log likelihood function becomes

$$\ell(a_{m+1},\ldots,a_T|\boldsymbol{\alpha}, v, A_m)$$
$$= (T-m)[\ln(\Gamma((v+1)/2)) - \ln(\Gamma(v/2)) - 0.5\ln((v-2)\pi)]$$
$$+ \ell(a_{m+1},\ldots,a_T|\boldsymbol{\alpha}, A_m),$$

where the second term is given in Equation (4.8).

Besides fat tails, empirical distributions of asset returns may also be skew. To handle this additional characteristic of asset returns, the Student t distribution has been modified to become a skew Student t distribution. There are multiple versions of skew Student t distribution, but we shall adopt the approach of Fernández and Steel (1998), which can introduce skewness into any continuous unimodal and symmetric (with respect to 0) univariate distribution. Specifically, for the innovation ϵ_t of an ARCH process, Lambert and Laurent (2001) apply the Fernández and Steel's method to the standardized Student t distribution in Equation (4.7) to obtain a standardized skew Student t distribution. The resulting probability density function is

$$g(\epsilon_t|\xi, v) = \begin{cases} \dfrac{2}{\xi + \frac{1}{\xi}}\varrho f[\xi(\varrho\epsilon_t + \varpi)|v] & \text{if } \epsilon_t < -\varpi/\varrho \\[4mm] \dfrac{2}{\xi + \frac{1}{\xi}}\varrho f[(\varrho\epsilon_t + \varpi)/\xi|v] & \text{if } \epsilon_t \geq -\varpi/\varrho, \end{cases} \quad (4.9)$$

where $f(\cdot)$ is the pdf of the standardized Student t distribution in Equation (4.7), ξ is the skewness parameter, $v > 2$ is the degrees of freedom, and the parameters ϱ and ϖ are given below

$$\varpi = \frac{\Gamma((v-1)/2)\sqrt{v-2}}{\sqrt{\pi}\Gamma(v/2)}(\xi - \frac{1}{\xi}), \quad \varrho^2 = (\xi^2 + \frac{1}{\xi^2} - 1) - \varpi^2.$$

In Equation (4.9), ξ^2 is equal to the ratio of probability masses above and below the mode of the distribution and, hence, it is a measure of the skewness.

To illustrate, Figure 4.5 shows some density functions of Student t and skew Student t distributions. For the Student t distribution, the densities shown are those of degrees of freedom 5, 10, and 30, respectively. The solid line denotes 5 degrees of freedom. As expected, as the degrees of freedom increase the tails become shorter and the peak becomes lower. For the skew Student t distribution, the densities shown all have 5 degrees of freedom, but their skew parameters are 0.75, 1, and 1.5, respectively. We see that these parameters produce left-skew, symmetric, and right-skew density, respectively.

Finally, ϵ_t may assume a GED with probability density function

$$f(x) = \frac{v \exp(-\frac{1}{2}|x/\lambda|^v)}{\lambda 2^{(1+1/v)}\Gamma(1/v)}, \qquad -\infty < x < \infty, \quad 0 < v \le \infty, \qquad (4.10)$$

where $\Gamma(\cdot)$ is the gamma function and

$$\lambda = [2^{(-2/v)}\Gamma(1/v)/\Gamma(3/v)]^{1/2}.$$

This distribution reduces to a Gaussian distribution if $v = 2$ and it has heavy tails when $v < 2$. The conditional log likelihood function $\ell(a_{m+1}, \ldots, a_T | \boldsymbol{\alpha}, \boldsymbol{A}_m)$ can be easily obtained.

Figure 4.6 shows some density functions for GED and skew GED random variables. They are zero mean and unit standard deviation. For GED, the shape

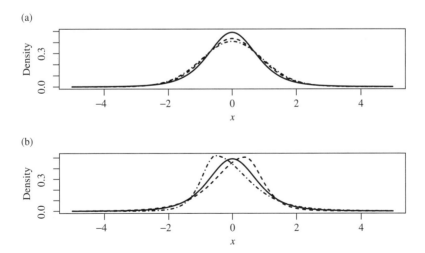

Figure 4.5. Density functions for (a) Student t and (b) skew Student t distributions. For Student t, the degrees of freedom are 5 (solid), 10 (dashed), and 30 (dot-dashed), respectively. For skew Student t, the degrees of freedom are 5, but the skew parameters are 0.75 (dashed), 1.0 (solid), and 1.5 (dot-dashed), respectively.

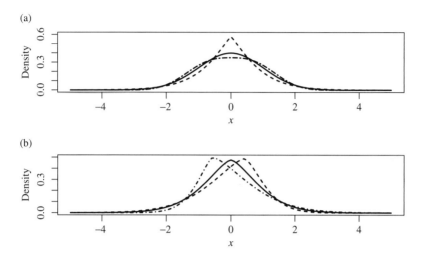

Figure 4.6. Density functions for (a) generalized error distributions (GED) and (b) skew GED. For GED, the shape parameters are 2 (solid), 1.2 (dashed), and 2.8, respectively. For skew GED, the shape parameter is 1.5, but the skew parameters are 1 (solid), 0.75 (dashed), and 1.5, respectively.

parameter used are 2, 1.2, and 2.8, respectively, so that they represent normal, heavier, and shorter tails. For skew GED, the shape parameter is 1.5 so that they all have heavy tails. The skew parameters are 1, 0.75, and 1.5, respectively, so that the densities are symmetric, left skew, and right skew, respectively.

Model Checking. For a properly specified ARCH model, the standardized residuals

$$\tilde{a}_t = \frac{a_t}{\sigma_t}$$

form a sequence of iid random variables. Therefore, one can check the adequacy of a fitted ARCH model by examining the series $\{\tilde{a}_t\}$. In particular, the Ljung–Box statistics of $\tilde{a}_t$ can be used to check the adequacy of the mean equation and that of $\tilde{a}_t^2$ can be used to test the validity of the volatility equation. The skewness, kurtosis, and quantile-to-quantile plot (i.e., QQ plot) of $\{\tilde{a}_t\}$ can be used to check the validity of the distribution assumption. The fGarch package provides many plots for a fitted volatility model.

Forecasting. Forecasts of the ARCH model in Equation (4.5) can be obtained recursively as those of an AR model. Consider an ARCH(m) model. At the forecast origin h, the 1-step ahead forecast of σ_{h+1}^2 is

$$\sigma_h^2(1) = \alpha_0 + \alpha_1 a_h^2 + \cdots + \alpha_m a_{h+1-m}^2.$$

The 2-step ahead forecast is

$$\sigma_h^2(2) = \alpha_0 + \alpha_1 \sigma_h^2(1) + \alpha_2 a_h^2 + \cdots + \alpha_m a_{h+2-m}^2,$$

and the ℓ-step ahead forecast for $\sigma_{h+\ell}^2$ is

$$\sigma_h^2(\ell) = \alpha_0 + \sum_{i=1}^{m} \alpha_i \sigma_h^2(\ell - i), \tag{4.11}$$

where $\sigma_h^2(\ell - i) = a_{h+\ell-i}^2$ if $\ell - i \leq 0$.

4.5.4 Some Examples

In this subsection, we illustrate ARCH modeling by considering two examples.

Example 4.1. We continue to demonstrate the volatility modeling by using the monthly log returns of Intel stock from 1973 to 2009. The ARCH tests of Section 4.4 showed that the log return series has significant ARCH effects. Figure 4.7 shows the sample ACF and PACF of the squared series of mean-adjusted returns. From the PACF plot, we see significant correlations at lags 1, 2, 3, and 11. To keep the model

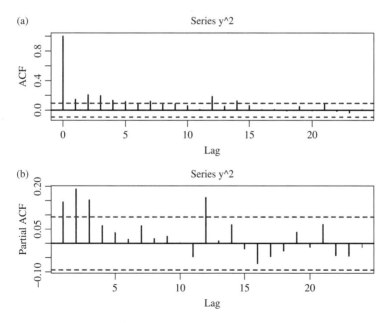

Figure 4.7. Sample ACF and PACF of the squared series of mean-adjusted monthly log returns of Intel stock from January 1973 to December 2009: (a) ACF and (b) PACF

simple, we entertain an ARCH(3) model for the volatility. Consequently, we specify
the model

$$r_t = \mu + a_t, \quad a_t = \sigma_t \epsilon_t, \quad \sigma_t^2 = \alpha_0 + \alpha_1 a_{t-1}^2 + \alpha_2 a_{t-2}^2 + \alpha_3 a_{t-3}^2$$

for the monthly log returns of Intel stock. Assuming that ϵ_t arc iid standard normal,
we obtain the fitted model

$$r_t = 0.0126 + a_t, \quad \sigma_t^2 = 0.0104 + 0.2329 a_{t-1}^2 + 0.0751 a_{t-2}^2 + 0.0520 a_{t-3}^2,$$

where the standard errors of the parameters are 0.0055, 0.0012, 0.1115, 0.0473, and
0.0451, respectively; see the R output below. While the estimates meet the general
requirement of an ARCH(3) model, the estimates of α_2 and α_3 appear to be statistically
insignificant at the 5% level. Therefore, the model can be simplified. The R command
for estimating GARCH model is `garchFit` of the `fGarch` package.   □

R Demonstration. Output edited and % marks explanation.

```
> library(fGarch) % Load package
> da=read.table("m-intcsp7309.txt",header=T)
> head(da)
        date        intc          sp
1 19730131   0.010050  -0.017111
  ....
6 19730629   0.133333  -0.006575
> intc=log(da$intc+1)
> m1=garchFit(~1+garch(3,0),data=intc,trace=F) % Fit an ARCH(3) model
    % Use subcommand "trace = F" to reduce the output.
> summary(m1)
Title:  GARCH Modelling

Mean and Variance Equation:  data ~ 1+garch(3,0) [data=intc]
Conditional Distribution:  norm

Coefficient(s):
          Estimate  Std. Error  t value  Pr(>|t|)
mu        0.012567  0.005515    2.279    0.0227 *
omega     0.010421  0.001238    8.418    <2e-16 ***
alpha1    0.232889  0.111541    2.088    0.0368 *
alpha2    0.075069  0.047305    1.587    0.1125
alpha3    0.051994  0.045139    1.152    0.2494
---
> m2=garchFit(~1+garch(1,0),data=intc,trace=F)
> summary(m2)
Title: GARCH Modelling
Call: garchFit(formula=~1+garch(1,0),data=intc,trace=F)

Mean and Variance Equation:  data ~ 1+garch(1,0) [data=intc]
Conditional Distribution:  norm

Coefficient(s):
```

```
          Estimate   Std. Error   t value Pr(>|t|)
mu        0.013130   0.005318     2.469   0.01355 *
omega     0.011046   0.001196     9.238 < 2e-16 ***
alpha1    0.374976   0.112620     3.330   0.00087 ***
---
Log Likelihood: 299.9247     normalized: 0.675506
```

```
Standardized Residuals Tests:
                             Statistic p-Value
Jarque-Bera Test   R    Chi^2 144.3783  0
Shapiro-Wilk Test  R    W     0.9678164 2.669091e-08
Ljung-Box Test     R    Q(10) 12.12248  0.2769429
Ljung-Box Test     R    Q(20) 24.33412  0.2281016
Ljung-Box Test     R^2  Q(10) 16.57807  0.08423723
Ljung-Box Test     R^2  Q(20) 38.81395  0.007031558
LM Arch Test       R    TR^2  27.32897  0.006926822
```

```
Information Criterion Statistics:
      AIC        BIC        SIC       HQIC
-1.337499 -1.309824 -1.337589 -1.326585
  %% further model checking
> resi=residuals(m2,standardize=T)
> tdx=c(1:444)/12+1973
> par(mfcol=c(3,10)
> plot(tdx,resi,xlab='year',ylab='stand-resi',type='l')
> acf(resi,lag=20)
> pacf(resi^2,lag=20)
 % Use fGarch built-in plots
> plot(m2)
Make a plot selection (or 0 to exit):
 1:    Time Series
 2:    Conditional SD
 3:    Series with 2 Conditional SD Superimposed
 4:    ACF of Observations
 5:    ACF of Squared Observations
 6:    Cross Correlation
 7:    Residuals
 8:    Conditional SDs
 9:    Standardized Residuals
10:    ACF of Standardized Residuals
11:    ACF of Squared Standardized Residuals
12:    Cross Correlation between r^2 and r
13:    QQ-Plot of Standardized Residuals

Selection: 0
```

Dropping the two insignificant parameters, we obtain the model

$$r_t = 0.0131 + a_t, \quad \sigma_t^2 = 0.0110 + 0.3750a_{t-1}^2, \tag{4.12}$$

where the standard errors of the parameters are 0.0053, 0.0012, and 0.1126, respectively. All the estimates are statistically significant. Figure 4.8 shows the standardized residuals $\{\tilde{a}_t\}$, the sample ACF of $\{\tilde{a}_t\}$, and the sample PACF of $\{\tilde{a}_t^2\}$. The

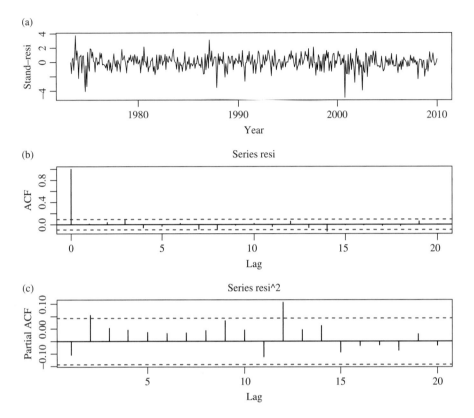

Figure 4.8. Model checking statistics of the Gaussian ARCH(1) model in Equation (4.12) for the monthly log returns of Intel stock from January 1973 to December 2009: (a) time plot of the standardized residuals, (b) sample ACF of the standardized residuals, and (c) sample PACF of the squared series of standardized residuals.

ACF plot indicates that the standardized residuals have no serial correlations, but the PACF plot suggests that certain serial dependence at higher lags remains in the squared standardized residuals. Indeed, the Ljung–Box statistics of standardized residuals give $Q(10) = 12.12$ with p value 0.28 and $Q(20) = 24.33$ with p value 0.23. On the other hand, the Q-statistics of $\{\tilde{a}_t^2\}$ give $Q(10) = 16.58$ with p value 0.08 and $Q(20) = 38.81$ with p value 0.007. Consequently, the ARCH(1) model in Equation (4.12) is adequate for describing the conditional heteroscedasticity of the data at the 5% significance level if one focuses only on the lower order models.

The ARCH(1) model in Equation (4.12) has some interesting properties. First, the expected monthly log return for Intel stock is about 1.31%, which is remarkable, especially because the data span includes the period after the Internet bubble and the recent financial crisis. Second, $\hat{\alpha}_1^2 = 0.375^2 < \frac{1}{3}$ so that the unconditional fourth moment of the monthly log returns of Intel stock exists. Third, the unconditional standard deviation of r_t is $\sqrt{0.0110/(1 - 0.375)} \approx 0.1327$, which is close to the sample standard error 0.1269 of the monthly log returns. Finally, the ARCH(1) model can be used to predict the monthly volatility of Intel stock returns.

t Innovation. For comparison, we also fit an ARCH(1) model with Student t innovations to the series. The resulting model is

$$r_t = 0.0172 + a_t, \quad \sigma_t^2 = 0.0118 + 0.2775a_{t-1}^2, \tag{4.13}$$

where the standard errors of the parameters are 0.0052, 0.0016, and 0.1072, respectively. The estimated degrees of freedom is 5.97 with standard error 1.53. All estimates are significant at the 5% level. The unconditional standard deviation of a_t is $\sqrt{0.0118/(1 - 0.27755)} \approx 0.1278$, which is close to that obtained under normality. The Ljung–Box statistics of the standardized residuals give $Q(10) = 12.86$ with p value 0.23, confirming that the mean equation is adequate. However, The Ljung–Box statistics for the squared standardized residuals show $Q(10) = 19.96$ with p value 0.0296. The volatility equation is inadequate at the 5% level.

Comparing models (4.12) and (4.13), we see that (i) using a heavy-tailed distribution for ϵ_t reduces the ARCH coefficient, and (ii) the difference between the two models is small for this particular instance. Finally, a more appropriate conditional heteroscedastic model for the monthly log returns of Intel stock is a GARCH(1,1) model, which is discussed in the next section.

R Demonstration. With Student-t innovations.

```
> m3=garchFit(~1+garch(1,0),data=intc,trace=F,cond.dist="std")
> summary(m3)
Title:  GARCH Modelling
Call:garchFit(formula=~1+garch(1,0),data=intc,cond.dist="std",trace=F)

Mean and Variance Equation:  data ~ 1 + garch(1, 0) [data = intc]
Conditional Distribution:  std

Coefficient(s):
        Estimate  Std. Error  t value Pr(>|t|)
mu       0.017202    0.005195    3.311 0.000929 ***
omega    0.011816    0.001560    7.574 3.62e-14 ***
alpha1   0.277476    0.107183    2.589 0.009631 **
shape    5.970266    1.529524    3.903 9.49e-05 ***
---

Standardized Residuals Tests:
                             Statistic p-Value
  Jarque-Bera Test    R    Chi^2  157.7799  0
  Shapiro-Wilk Test   R    W      0.9663975 1.488202e-08
  Ljung-Box Test      R    Q(10)  12.85940  0.2316396
  Ljung-Box Test      R    Q(20)  25.374    0.1874956
  Ljung-Box Test      R^2  Q(10)  19.96092  0.02962445
  Ljung-Box Test      R^2  Q(20)  44.06739  0.001473970
  LM Arch Test        R    TR^2   29.76071  0.003033508
```

Remark. In fGarch package, the command `garchFit` allows for several conditional distributions, including Student t and skew Student t distributions. They are specified by cond.dist = "std" or "sstd", respectively.   □

Example 4.2. Consider the log returns of daily exchange rate between US Dollar and Euro from January 4, 1999, to August 20, 2010. As shown in Section 4.4, the mean equation of the log returns is $r_t = a_t$ and there exist strong ARCH effects in the data. Thus, the series is a good example for pure conditional heteroscedastic models. If ARCH models are entertained, the sample PACF of r_t^2 shown in Figure 4.4b suggests an ARCH(11) model. Using the conditional Gaussian likelihood function, we obtain the fitted model $r_t = 0.00013 + \sigma_t \epsilon_t$ and

$$\sigma_t^2 = 1.89 \times 10^{-5} + 0.017 a_{t-1}^2 + 0.045 a_{t-2}^2 + 0.027 a_{t-3}^2 + \cdots + 0.039 a_{t-11}^2.$$

Details of the estimates and their standard errors are given in the R output below. Several coefficient estimates are not significant at the 5% level. Also given are some model checking statistics, including Ljung–Box statistics for standardized residuals and their squared series. On the basis of the diagnostic statistics, we see that, except for the normality assumption, the fitted model is adequate in modeling the daily log returns of the Dollar/Euro exchange rate. A simpler model for the series can be obtained if one entertains the GARCH models.   □

R Demonstration

```
> mm1=garchFit(~1+garch(11,0),data=eu,trace=F)
> summary(mm1)
Title: GARCH Modelling
Call: garchFit(formula=~1+garch(11,0),data=eu,trace=F)

Mean and Variance Equation: data ~ 1+garch(11,0) [data = eu]
Conditional Distribution:  norm

Coefficient(s):
           Estimate  Std. Error  t value  Pr(>|t|)
mu        1.265e-04   1.110e-04    1.140  0.254426
omega     1.890e-05   1.727e-06   10.944  < 2e-16  ***
alpha1    1.661e-02   1.575e-02    1.055  0.291568
alpha2    4.456e-02   2.085e-02    2.137  0.032592  *
alpha3    2.721e-02   1.700e-02    1.601  0.109353
alpha4    8.037e-02   2.363e-02    3.402  0.000669  ***
alpha5    5.011e-02   2.127e-02    2.355  0.018500  *
alpha6    9.219e-02   2.274e-02    4.053  5.05e-05  ***
alpha7    7.528e-02   2.406e-02    3.129  0.001755  **
alpha8    6.954e-02   2.455e-02    2.832  0.004622  **
alpha9    3.347e-02   2.022e-02    1.656  0.097822  .
alpha10   2.782e-02   1.820e-02    1.528  0.126412
```

```
alpha11 3.877e-02    1.906e-02     2.035 0.041896 *
---
Standardised Residuals Tests:
                               Statistic p-Value
Jarque-Bera Test    R     Chi^2   360.802   0
Shapiro-Wilk Test   R     W       0.9891754 3.90746e-14
Ljung-Box Test      R     Q(10)   15.77626  0.1062187
Ljung-Box Test      R     Q(20)   24.77444  0.2101971
Ljung-Box Test      R^2   Q(10)   4.801266  0.904052
Ljung-Box Test      R^2   Q(20)   27.56081  0.1202105
LM Arch Test        R     TR^2    11.96818  0.4482389
```

4.6 THE GARCH MODEL

Although the ARCH model is simple, it often requires many parameters to adequately describe the volatility process of an asset return. For instance, consider the daily log returns of the Dollar/Euro exchange rate of Example 4.2. An ARCH(11) is needed for the series. To keep the model simple, some alternative model must be sought. Boller-slev (1986) proposes a useful extension known as the *generalized ARCH (GARCH) model*. For a log return series r_t, let $a_t = r_t - \mu_t$ be the innovation at time t. Then, a_t follows a GARCH(m, s) model if

$$a_t = \sigma_t \epsilon_t, \qquad \sigma_t^2 = \alpha_0 + \sum_{i=1}^{m} \alpha_i a_{t-i}^2 + \sum_{j=1}^{s} \beta_j \sigma_{t-j}^2, \qquad (4.14)$$

where again $\{\epsilon_t\}$ is a sequence of iid random variables with mean 0 and variance 1.0, $\alpha_0 > 0$, $\alpha_i \geq 0$, $\beta_j \geq 0$, and $\sum_{i=1}^{\max(m,s)} (\alpha_i + \beta_i) < 1$. Here, it is understood that $\alpha_i = 0$ for $i > m$ and $\beta_j = 0$ for $j > s$. The latter constraint on $\alpha_i + \beta_i$ implies that the unconditional variance of a_t is finite, whereas its conditional variance σ_t^2 evolves over time. As before, ϵ_t is often assumed to follow a standard normal or standardized Student t distribution or GED. Equation (4.14) reduces to a pure ARCH(m) model if $s = 0$. The α_i and β_j are referred to as ARCH and GARCH parameters, respectively.

To understand properties of GARCH models, it is informative to use the following representation. Let $\eta_t = a_t^2 - \sigma_t^2$ so that $\sigma_t^2 = a_t^2 - \eta_t$. By plugging $\sigma_{t-i}^2 = a_{t-i}^2 - \eta_{t-i}$ $(i = 0, \cdots, s)$ into Equation (4.14), we can rewrite the GARCH model as

$$a_t^2 = \alpha_0 + \sum_{i=1}^{\max(m,s)} (\alpha_i + \beta_i) a_{t-i}^2 + \eta_t - \sum_{j=1}^{s} \beta_j \eta_{t-j}. \qquad (4.15)$$

It is easy to check that $\{\eta_t\}$ is a martingale difference series (i.e., $E(\eta_t) = 0$ and $\mathrm{cov}(\eta_t, \eta_{t-j}) = 0$ for $j \geq 1$). However, $\{\eta_t\}$ in general is not an iid sequence. Equation (4.15) is an ARMA form for the squared series a_t^2. Thus, a GARCH model can be regarded as an application of the ARMA idea to the squared series a_t^2. Using the unconditional mean of an ARMA model, we have

$$E(a_t^2) = \frac{\alpha_0}{1 - \sum_{i=1}^{\max(m,s)}(\alpha_i + \beta_i)},$$

provided the denominator of the prior fraction is positive.

The strengths and weaknesses of GARCH models can be easily seen by focusing on the simplest GARCH(1,1) model with

$$\sigma_t^2 = \alpha_0 + \alpha_1 a_{t-1}^2 + \beta_1 \sigma_{t-1}^2, \quad 0 \le \alpha_1, \beta_1 \le 1, (\alpha_1 + \beta_1) < 1. \tag{4.16}$$

First, a large a_{t-1}^2 or σ_{t-1}^2 gives rise to a large σ_t^2. This means that a large a_{t-1}^2 tends to be followed by another large a_t^2, generating, again, the well-known behavior of volatility clustering in financial time series. Second, it can be shown that if $1 - 2\alpha_1^2 - (\alpha_1 + \beta_1)^2 > 0$, then

$$\frac{E(a_t^4)}{[E(a_t^2)]^2} = \frac{3[1 - (\alpha_1 + \beta_1)^2]}{1 - (\alpha_1 + \beta_1)^2 - 2\alpha_1^2} > 3.$$

Consequently, similar to ARCH models, the tail distribution of a GARCH(1,1) process is heavier than that of a normal distribution. Third, the model provides a simple parametric function that can be used to describe the volatility evolution.

Forecasts of a GARCH model can be obtained using methods similar to those of an ARMA model. Consider the GARCH(1,1) model in Equation (4.16) and assume that the forecast origin is h. For 1-step ahead forecast, we have

$$\sigma_{h+1}^2 = \alpha_0 + \alpha_1 a_h^2 + \beta_1 \sigma_h^2,$$

where a_h and σ_h^2 are known at the time index h. Therefore, the 1-step ahead forecast is

$$\sigma_h^2(1) = \alpha_0 + \alpha_1 a_h^2 + \beta_1 \sigma_h^2.$$

For multistep ahead forecasts, we use $a_t^2 = \sigma_t^2 \epsilon_t^2$ and rewrite the volatility equation in Equation (4.16) as

$$\sigma_{t+1}^2 = \alpha_0 + (\alpha_1 + \beta_1)\sigma_t^2 + \alpha_1 \sigma_t^2(\epsilon_t^2 - 1).$$

When $t = h + 1$, the equation becomes

$$\sigma_{h+2}^2 = \alpha_0 + (\alpha_1 + \beta_1)\sigma_{h+1}^2 + \alpha_1 \sigma_{h+1}^2(\epsilon_{h+1}^2 - 1).$$

As $E(\epsilon_{h+1}^2 - 1|F_h) = 0$, the 2-step ahead volatility forecast at the forecast origin h satisfies the equation

$$\sigma_h^2(2) = \alpha_0 + (\alpha_1 + \beta_1)\sigma_h^2(1).$$

In general, we have

$$\sigma_h^2(\ell) = \alpha_0 + (\alpha_1 + \beta_1)\sigma_h^2(\ell - 1), \quad \ell > 1. \tag{4.17}$$

This result is exactly the same as that of an ARMA(1,1) model with AR polynomial $1 - (\alpha_1 + \beta_1)B$. By repeated substitutions in Equation (4.17), we obtain that the ℓ-step ahead forecast can be written as

$$\sigma_h^2(\ell) = \frac{\alpha_0[1 - (\alpha_1 + \beta_1)^{\ell-1}]}{1 - \alpha_1 - \beta_1} + (\alpha_1 + \beta_1)^{\ell-1}\sigma_h^2(1).$$

Therefore,

$$\sigma_h^2(\ell) \to \frac{\alpha_0}{1 - \alpha_1 - \beta_1}, \quad \text{as} \quad \ell \to \infty$$

provided that $\alpha_1 + \beta_1 < 1$. Consequently, the multistep ahead volatility forecasts of a GARCH(1,1) model converge to the unconditional variance of a_t as the forecast horizon increases to infinity provided that Var(a_t) exists.

The literature on GARCH models is enormous; see Bollerslev et al. (1992, 1994), and references therein. The model encounters the same weaknesses as the ARCH model. For instance, it responds equally to positive and negative shocks. In addition, recent empirical studies of high frequency financial time series indicate that the tail behavior of GARCH models remains too short even with standardized Student t innovations. For further information about kurtosis of GARCH models, see Tsay (2010, Chapter 3).

4.6.1 An Illustrative Example

The modeling procedure of ARCH models can also be used to build a GARCH model. However, there is little study about specifying the order of an GARCH model for a financial time series. Only lower order GARCH models are used in most applications, say GARCH(1,1), GARCH(2,1), and GARCH(1,2) models. In many situations, a GARCH(1,1) model appears to be adequate. For estimation, the conditional maximum likelihood method continues to apply provided that the starting values of the volatility $\{\sigma_t^2\}$ are assumed to be known. Consider, for instance, a GARCH(1,1) model. If σ_1^2 is treated as fixed, then σ_t^2 can be computed recursively for a GARCH(1,1) model. In some applications, the sample variance of a_t serves as a good starting value of σ_1^2. The fitted model can be checked by using the standardized residual $\tilde{a}_t = a_t/\sigma_t$ and its squared process.

In this section, we focus on the monthly log returns of Intel stock from January 1973 to December 2009 with 444 observations; see Figure 4.1. Our goals are (i) to demonstrate empirical analysis of GARCH processes, (ii) to compare different GARCH models, and (iii) to show the prediction of a GARCH model.

Example 4.3. Model checking for the ARCH(1) model with Gaussian innovations of Example 4.1 shows that the model needs some refinement, for example, the Ljung–Box statistics of the squared standardized residuals give $Q(20) = 38.81$ with p value 0.007. Here, we entertain a GARCH(1,1) model for the monthly log returns of Intel stock. We employ different innovations to provide a better understanding of the return series. Again, let r_t be the monthly log return and, for simplicity, standard errors of the estimates are given in the R output.

Using Gaussian innovations, we obtain the model

$$r_t = 0.0113 + a_t, \quad a_t = \sigma_t \epsilon_t, \quad \epsilon \sim N(0,1)$$
$$\sigma_t^2 = 0.00092 + 0.086a_{t-1}^2 + 0.853\sigma_{t-1}^2, \tag{4.18}$$

where all estimates are significant at the 5% level. Except for the normality tests, model checking statistics indicate that this Gaussian GARCH(1,1) model is adequate for r_t. AIC for the model is -1.3889. Let $\tilde{a}_t = \hat{a}_t/\sigma_t$ be the standardized residuals of the model. Figure 4.9b shows the time plot of $\tilde{a}_t$. Except for one or two possible outliers, the standardized residuals look reasonable. Figure 4.10 shows the ACF and PACF of $\{\tilde{a}_t\}$ and $\{\tilde{a}_t^2\}$. Except for a marginal correlation at lag 12 of $\tilde{a}_t^2$, these ACF and PACF confirm that the fitted model is adequate in describing the conditional mean and variance of the log return series. Figure 4.9a gives the fitted volatility series of

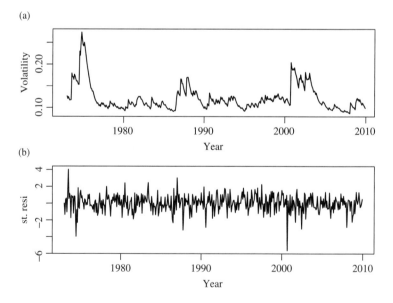

Figure 4.9. Time plots for a fitted GARCH(1,1) model with Gaussian innovations to the monthly log returns of Intel stock from January 1973 to December 2009: (a) volatility series and (b) standardized residuals.

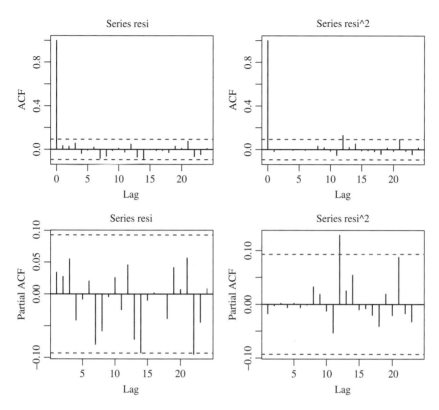

Figure 4.10. Sample ACF and PACF of the standardized residuals and their squared series of a Gaussian GARCH(1,1) model fitted to the monthly log returns of Intel stock from January 1973 to December 2009.

the model. As expected, the volatility was high during the oil crisis of 1973–1974 and the internet bubble around 2000. Finally, Figure 4.11 shows the time plot of the log returns with 95% pointwise predictive intervals. The intervals are calculated by $\hat{\mu} \pm \hat{\sigma}_t$, where $\hat{\mu} = 0.0113$ is the constant term of the mean equation. With some extreme exceptions, all returns are within the 95% predictive intervals. The implied unconditional variance for r_t is $0.000919/(1 - 0.0864 - 9.8526) = 0.0151$, which is slightly smaller than the sample variance 0.0161 of the data.

Using Student t innovations, we obtain the model

$$r_t = 0.0165 + a_t, \quad a_t = \sigma_t \epsilon_t, \quad \epsilon_t \sim t^*_{6.77}$$

$$\sigma_t^2 = 0.00116 + 0.1059 a_{t-1}^2 + 0.8171 \sigma_{t-1}^2, \tag{4.19}$$

where, again, all estimates are significantly different from zero at the 5% level, and t^*_d denotes a standardized Student t distribution with d degrees of freedom. Model checking statistics show that this fitted model is adequate for the log return series.

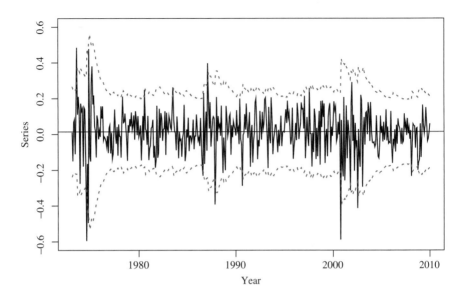

Figure 4.11. Time plot of the monthly log returns of Intel stock January 1973 to December 2009. The two dashed lines indicate pointwise 95% predictive intervals based on the Gaussian GARCH(1,1) model of Equation (4.18).

The AIC of the model in Equation (4.19) is -1.4470, and the implied unconditional variance of r_t is $0.0011576/(1 - 0.0159 - 0.8171) = 0.01503$.

The sample skewness of the log returns is -0.5526, which has a t-ratio of -4.75 so that the monthly log returns of Intel stock are negatively skew. To model this skewness, we employ a skew Student t distribution for the innovations ϵ_t. The resulting model is

$$r_t = 0.0133 + a_t, \quad a_t = \sigma_t \epsilon_t, \quad \epsilon_t \sim t^*_{0.87,7.23}$$
$$\sigma_t^2 = 0.00116 + 0.1049 a_{t-1}^2 + 0.8178 \sigma_{t-1}^2, \tag{4.20}$$

where $t^*_{\xi,d}$ denotes a standardized skew Student t distribution with skew parameter ξ and degrees of freedom d, and all estimates are significant at the 5% level. Model checking statistics also fail to indicate any inadequacy of the fitted model in Equation (4.20). AIC of the model is -1.4509. Note that the estimate of the skew parameter is 0.8717 with standard error 0.0629. The hypothesis of interest here is $H_0 : \xi = 1$ versus the alternative $H_a : \xi \neq 1$. In this particular case, the t-ratio is $t = (0.8717 - 1)/0.0629 = -2.04$ with a two-sided p value 0.041. Consequently, the null hypothesis of no skewness is rejected at the 5% level. Figure 4.12 shows the QQ plot of the standardized residuals of model (4.20) with skew Student t distribution with 7.23 degrees of freedom and skew parameter 0.8717. The plot looks reasonable.

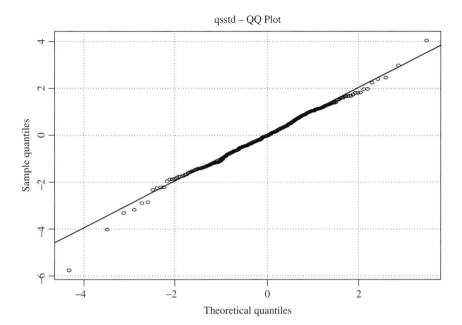

Figure 4.12. Quantile-to-quantile plot for the standardized residuals of the GARCH(1,1) model in Equation (4.20).

R Demonstration. Output edited.

```
> library(fGarch)
> m4=garchFit(~1+garch(1,1),data=intc,trace=F)
> summary(m4)
Title: GARCH Modelling
Call: garchFit(formula=~1+garch(1,1),data=intc,trace=F)

Mean and Variance Equation: data ~ 1 + garch(1, 1) [data = intc]
Conditional Distribution:  norm

Coefficient(s):
         Estimate  Std. Error  t value  Pr(>|t|)
mu      0.0112657   0.0053931    2.089   0.03672 *
omega   0.0009190   0.0003888    2.364   0.01808 *
alpha1  0.0864383   0.0265439    3.256   0.00113 **
beta1   0.8525855   0.0394322   21.622  < 2e-16 ***
---
Standardised Residuals Tests:
                              Statistic p-Value
 Jarque-Bera Test   R    Chi^2  174.904    0
 Shapiro-Wilk Test  R    W      0.9709618 1.030377e-07
 Ljung-Box Test     R    Q(10)  8.016844  0.6271916
 Ljung-Box Test     R    Q(20)  16.41549  0.6905368
 Ljung-Box Test     R^2  Q(10)  0.8746345 0.9999072
```

```
Ljung-Box Test     R^2   Q(20)   12.55994   0.8954573
LM Arch Test       R     TR^2    10.51401   0.5709617

Information Criterion Statistics:
     AIC         BIC         SIC         HQIC
-1.388877   -1.351978   -1.389037   -1.374326

> v1=volatility(m4)   % Obtain volatility
> resi=residuals(m4,standardize=T) % Standardized residuals
> vol=ts(v1,frequency=12,start=c(1973,1))
> res=ts(resi,frequency=12,start=c(1973,1))
> par(mfcol=c(2,1))   % Show volatility and residuals
> plot(vol,xlab='year',ylab='volatility',type='l')
> plot(res,xlab='year',ylab='st. resi',type='l')
> par(mfcol=c(2,2)) % Obtain ACF & PACF
> acf(resi,lag=24)
> pacf(resi,lag=24)
> acf(resi^2,lag=24)
> pacf(resi^2,lag=24)
> % Obtain plot of predictive intervals
> par(mfcol=c(1,1))
> upp=0.0113+2*v1
> low=0.0113-2*v1
> tdx=c(1:444)/12+1973
> plot(tdx,intc,xlab='year',ylab='series',type='l',ylim=c(-0.6,0.6))
> lines(tdx,upp,lty=2,col='red')
> lines(tdx,low,lty=2,col='red')
> abline(h=c(0.0113))
% Student-t innovations
> m5=garchFit(~1+garch(1,1),data=intc,trace=F,cond.dist="std")
> summary(m5)
Title:  GARCH Modelling
Call: garchFit(formula=~1+garch(1,1),data=intc,cond.dist="std",trace=F)

Mean and Variance Equation: data ~ 1+garch(1,1) [data = intc]
Conditional Distribution:  std

Coefficient(s):
       Estimate    Std. Error   t value  Pr(>|t|)
mu      0.0165075   0.0051031    3.235 0.001217 **
omega   0.0011576   0.0005782    2.002 0.045286 *
alpha1  0.1059030   0.0372047    2.846 0.004420 **
beta1   0.8171313   0.0580141   14.085 < 2e-16 ***
shape   6.7723503   1.8572380    3.646 0.000266 ***
---

Standardised Residuals Tests:
                                Statistic p-Value
Ljung-Box Test     R     Q(10)   7.877778  0.6407741
Ljung-Box Test     R     Q(20)  16.50475   0.6848581
Ljung-Box Test     R^2   Q(10)   1.066054  0.9997694
Ljung-Box Test     R^2   Q(20)  12.61496   0.8932865
LM Arch Test       R     TR^2   10.80739   0.5454935

Information Criterion Statistics:
```

```
        AIC         BIC         SIC        HQIC
-1.446966  -1.400841  -1.447215  -1.428776

> v2=volatility(m5)
> m6=garchFit(~1+garch(1,1),data=intc,trace=F,cond.dist='sstd')
> summary(m6)
Title:  GARCH Modelling
Call: garchFit(formula=~1+garch(1,1),data=intc,cond.dist="sstd",trace=F)

Mean and Variance Equation:  data ~ 1+garch(1,1) [data = intc]
Conditional Distribution:  sstd

Coefficient(s):
         Estimate  Std. Error  t value  Pr(>|t|)
mu       0.0133343  0.0053430    2.496  0.012572 *
omega    0.0011621  0.0005587    2.080  0.037519 *
alpha1   0.1049289  0.0358860    2.924  0.003456 **
beta1    0.8177875  0.0559863   14.607  < 2e-16 ***
skew     0.8717220  0.0629129   13.856  < 2e-16 ***
shape    7.2344224  2.1018041    3.442  0.000577 ***
---
Standardised Residuals Tests:
                               Statistic  p-Value
 Ljung-Box Test    R    Q(10)   7.882126  0.6403496
 Ljung-Box Test    R    Q(20)  16.57740   0.6802193
 Ljung-Box Test    R^2  Q(10)   1.078429  0.999757
 Ljung-Box Test    R^2  Q(20)  13.03792   0.8757513
 LM Arch Test      R    TR^2   11.18826   0.5128574

Information Criterion Statistics:
        AIC         BIC         SIC        HQIC
-1.450899  -1.395550  -1.451257  -1.429071

> v3=volatility(m6)
> par(mfcol=c(3,1))
> plot(tdx,v1,xlab='year',ylab='volatility',type='l',ylim=c(0.06,0.3))
> title(main='(a) Gaussian')
> plot(tdx,v2,xlab='year',ylab='volatility',type='l',ylim=c(0.06,0.3))
> title(main='(b) Student-t')
> plot(tdx,v3,xlab='year',ylab='volatility',type='l',ylim=c(0.06,0.3))
> title(main='(c) Skew Student-t')

> cor(cbind(v1,v2,v3))
           v1         v2         v3
v1  1.0000000  0.9936777  0.9944357
v2  0.9936777  1.0000000  0.9998430
v3  0.9944357  0.9998430  1.0000000
>
> library(fBasics)
> basicStats(intc)
                intc
nobs       444.000000
Minimum     -0.595420
Maximum      0.485508
```

```
Mean           0.014327
Variance       0.016106
Stdev          0.126910
Skewness      -0.552618
Kurtosis       3.124026
> tt=-0.5526/sqrt(6/444) % Testing skewness of the data
> tt
[1] -4.753645
> tt=(0.8717-1)/0.0629 % Testing skewness of the model.
> tt
[1] -2.039746
> pv=2*pnorm(tt)  % Compute p-value
> pv
[1] 0.04137567
> plot(m6)
Make a plot selection (or 0 to exit):
  1:    Time Series
  2:    Conditional SD
  3:    Series with 2 Conditional SD Superimposed
  4:    ACF of Observations
  5:    ACF of Squared Observations
  6:    Cross Correlation
  7:    Residuals
  8:    Conditional SDs
  9:    Standardized Residuals
 10:    ACF of Standardized Residuals
 11:    ACF of Squared Standardized Residuals
 12:    Cross Correlation between r^2 and r
 13:    QQ-Plot of Standardized Residuals

Selection: 13
```

Discussion and Comparison. We have applied three GARCH(1,1) models to
the monthly log returns of Intel stock from January 1973 to December 2009. All
three models fit the data well. If AIC is used in model selection, one selects the
one with skew Student t innovations as the best model for the data. This selection
is also supported by the preliminary analysis that shows significant skewness in the
returns. On the other hand, if BIC is used, then one selects the model with Student
t innovations as the best one. This is not surprising because BIC, in this particular
case, puts a heavier penalty for each parameter used and the p value for testing no
skewness is 0.041, which is only slightly smaller than 0.05. In other words, under
BIC, the penalty is heavier than the contribution of the skew parameter. This example
illustrates that different criteria may select different models in volatility modeling.

Figure 4.13 provides time plots of volatility of the three models entertained. The
plots are in the same scale so that a direct comparison is possible. From the plots, the
three estimated volatility series are essentially the same; it is hard to see any major
difference between the three volatility series. As a matter of fact, the correlation
coefficients between the three volatility series are all close to one; see the attached R
output. Thus, the three entertained models are close to each other.

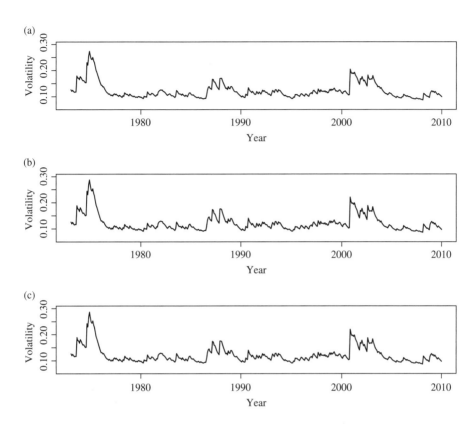

Figure 4.13. Time plots of volatility of a GARCH(1,1) model for the monthly log returns of Intel stock with different innovations: (a) Gaussian, (b) Student t, and (c) Skew Student t.

Table 4.1 shows some volatility forecasts of the three GARCH(1,1) models for the monthly log returns of Intel stock. The forecast origin is December 2009. As expected, the forecasts of volatility are also close to each other, supporting further that the difference between the three innovations is small.

TABLE 4.1. Volatility forecasts for monthly log returns of Intel stock[a]

	Forecast Horizon							
Model	1	2	3	4	5	6	9	12
$N(0,1)$	0.0975	0.0993	0.1009	0.1023	0.1037	0.105	0.108	0.111
t	0.0951	0.0975	0.0997	0.1016	0.1034	0.105	0.109	0.112
Skew t	0.0954	0.0979	0.1000	0.1019	0.1037	0.105	0.109	0.112

[a]The forecast origin is December 2009 and the models used are GARCH(1,1) models with Gaussian, Student t, and skew Student t innovations, respectively.

4.6.2   Forecasting Evaluation

As the volatility of an asset return is not directly observable, comparing the forecasting performance of different volatility models is a challenge to data analysts. In the literature, some researchers use out-of-sample forecasts and compare the volatility forecasts $\sigma_h^2(\ell)$ with the shock $a_{h+\ell}^2$ in the forecasting sample to assess the forecasting performance of a volatility model. This approach often finds a low correlation coefficient between $a_{h+\ell}^2$ and $\sigma_h^2(\ell)$, that is, low R^2. However, such a finding is not surprising because $a_{h+\ell}^2$ alone is not an adequate measure of the volatility at time index $h + \ell$. Consider the 1-step ahead forecasts. From a statistical point of view, $E(a_{h+1}^2|F_h) = \sigma_{h+1}^2$ so that a_{h+1}^2 is a consistent estimate of σ_{h+1}^2. But it is not an accurate estimate of σ_{h+1}^2 because a single observation of a random variable with a known mean value cannot provide an accurate estimate of its variance. Consequently, such an approach to evaluate forecasting performance of volatility models is strictly speaking not proper. For more information concerning forecasting evaluation of GARCH models, readers are referred to Andersen and Bollerslev (1998).

4.6.3   A Two-Pass Estimation Method

On the basis of Equation (4.15), a two-pass estimation method can be used to estimate GARCH models. First, ignoring any ARCH effects, one estimates the mean equation of a return series using the methods discussed in Chapter 2 (e.g., maximum likelihood method). Denote the residual series by a_t. Second, treating $\{a_t^2\}$ as an observed time series, one applies the maximum likelihood method to estimate parameters of Equation (4.15). Denote the AR and MA coefficient estimates by $\hat{\phi}_i$ and $\hat{\theta}_i$. The GARCH estimates are obtained as $\hat{\beta}_i = \hat{\theta}_i$ and $\hat{\alpha}_i = \hat{\phi}_i - \hat{\theta}_i$. Obviously, such estimates are approximations to the true parameters and their statistical properties have not been rigorously investigated. However, limited experience shows that this simple approach often provides good approximations, especially when the sample size is moderate or large. For instance, consider the monthly log returns of the Intel stock of Example 4.3. Using the conditional MLE method, we obtain the model

$$r_t = 0.0143 + a_t, \quad (1 - 0.9119B)(a_t^2 - 0.0161) = (1 - 0.7915B)\eta_t, \qquad (4.21)$$

where all estimates are significantly different from zero at the 5% level and the variance of η_t is 0.00122. From the estimates, we have $\hat{\beta}_1 = 0.7915$, $\hat{\alpha}_1 = 0.9119 - 0.7915 = 0.1204$, and $\hat{\phi}_0 = (1 - 0.9119) \times (0.0161) = 0.00142$. These approximate estimates are very close to those in Equation (4.19) or (4.20). The fitted volatility of this two-pass procedure is also close to those of GARCH(1,1) models. For instance, the correlation of fitted volatilities between models in Equations (4.20) and (4.21) is 0.9976.

R Demonstration

```
> yt=intc-mean(intc)
> m1=arima(yt^2,order=c(1,0,1))
> m1
```

```
Call: arima(x = yt^2, order = c(1,0,1))
Coefficients:
          ar1       ma1    intercept
       0.9119   -0.7915      0.0161
s.e.   0.0430    0.0635      0.0039

sigma^2 estimated as 0.001223: log likelihood=858.64,aic=-1709.28
> mean(intc)
[1] 0.0143273
> fit=yt^2-m1$residuals
> v3=volatility(m6)   % m6 is GARCH(1,1) with skew-t innovations.
> cor(v3,sqrt(fit))
[1] 0.9976242
```

4.7 THE INTEGRATED GARCH MODEL

If the AR polynomial of the GARCH representation in Equation (4.15) has a unit root, then we have an IGARCH (integrated generalized autoregressive conditional heteroscedastic) model. Thus, IGARCH models are unit-root GARCH models. Similar to ARIMA models, a key feature of IGARCH models is that the impact of past squared shocks $\eta_{t-i} = a_{t-i}^2 - \sigma_{t-i}^2$ for $i > 0$ on a_t^2 is persistent.

An IGARCH(1,1) model can be written as

$$a_t = \sigma_t \epsilon_t, \quad \sigma_t^2 = \alpha_0 + \beta_1 \sigma_{t-1}^2 + (1 - \beta_1)a_{t-1}^2,$$

where $\{\epsilon_t\}$ is defined as before and $1 > \beta_1 > 0$. For the monthly log returns of Intel stock, an estimated IGARCH(1,1) model is

$$r_t = 0.0097 + a_t, \quad a_t = \sigma_t \epsilon_t, \quad \epsilon_t \sim N(0,1),$$
$$\sigma_t^2 = 0.000348 + 0.1278a_{t-1}^2 + 0.8722\sigma_{t-1}^2, \qquad (4.22)$$

where standard error of the estimate in the mean equation is 0.0053, whereas those of the volatility equation are 0.00018 and 0.0336, respectively. Model checking statistics show that the fitted IGARCH(1,1) model is adequate in describing the mean and volatility of Intel stock returns.

The parameter estimates of the IGARCH(1,1) model are not far away from those of the GARCH(1,1) model in Equation (4.18), but there is a major difference between the two models. The unconditional variance of a_t, hence that of r_t, is not defined under the above IGARCH(1,1) model. This seems hard to justify for a log return series. From a theoretical point of view, the IGARCH phenomenon might be caused by occasional level shifts in volatility. The actual cause of persistence in volatility deserves a careful investigation.

When $\alpha_1 + \beta_1 = 1$, repeated substitutions in Equation (4.17) give

$$\sigma_h^2(\ell) = \sigma_h^2(1) + (\ell - 1)\alpha_0, \quad \ell \geq 1, \qquad (4.23)$$

where h is the forecast origin. Consequently, the effect of $\sigma_h^2(1)$ on future volatilities is also persistent, and the volatility forecasts form a straight line with slope α_0. Nelson (1990) studied some probability properties of the volatility process σ_t^2 under an IGARCH model. The process σ_t^2 is a martingale for which some nice results are available in the literature. Under certain conditions, the volatility process is strictly stationary, but not weakly stationary because it does not have the first two moments.

The case of $\alpha_0 = 0$ is of particular interest in studying the IGARCH(1,1) model. In this case, the volatility forecasts are simply $\sigma_h^2(1)$ for all forecast horizons; see Equation (4.23). This special IGARCH(1,1) model is the volatility model used in RiskMetrics, which is an approach for calculating value at risk; see Tsay (2010, Chapter 7). The model is also an exponential smoothing model for the $\{a_t^2\}$ series. To see this, rewrite the model as

$$\sigma_t^2 = (1 - \beta_1)a_{t-1}^2 + \beta_1 \sigma_{t-1}^2$$
$$= (1 - \beta_1)a_{t-1}^2 + \beta_1[(1 - \beta)a_{t-2}^2 + \beta_1 \sigma_{t-2}^2]$$
$$= (1 - \beta_1)a_{t-1}^2 + (1 - \beta_1)\beta_1 a_{t-2}^2 + \beta_1^2 \sigma_{t-2}^2.$$

By repeated substitutions, we have

$$\sigma_t^2 = (1 - \beta_1)[a_{t-1}^2 + \beta_1 a_{t-2}^2 + \beta_1^2 a_{t-3}^2 + \cdots],$$

which is the well-known exponential smoothing formation with β_1 being the discounting factor. Exponential smoothing methods can thus be used to estimate such an IGARCH(1,1) model.

R Demonstration

```
> source("Igarch.R")
> mm=Igarch(intc)
[1] -261.3556
  0:     -261.35565: 0.0143273 0.00161062 0.100000
  3:     -299.82581: 0.00754877 0.00211435 0.325506
  .....
 21:     -308.24866: 0.00962287 0.000349070 0.128039

Coefficient(s):
          Estimate    Std. Error   t value   Pr(>|t|)
mu     0.009666313  0.005338563   1.81066   0.0701937 .
omega  0.000347561  0.000181948   1.91022   0.0561052 .
alpha  0.127818122  0.033601301   3.80396   0.0001424 ***
> names(mm)
[1] "par"           "volatility"
```

Remark. We use a simple R script to estimate the IGARCH(1,1) model with Gaussian innovations. The script is available on the web page of the book.   □

4.8   THE GARCH-M MODEL

In finance, the return of an asset may depend on its volatility. To model such a phenomenon, one may consider the GARCH-M model, where "M" stands for GARCH *in the mean*. A simple GARCH(1,1)-M model can be written as

$$r_t = \mu + c\sigma_t^2 + a_t, \quad a_t = \sigma_t\epsilon_t,$$
$$\sigma_t^2 = \alpha_0 + \alpha_1 a_{t-1}^2 + \beta_1 \sigma_{t-1}^2, \tag{4.24}$$

where μ and c are constants. The parameter c is called the risk premium parameter. A positive c indicates that the return is positively related to its past volatility. Other specifications of risk premium have also been used in the literature, including $r_t = \mu + c\sigma_t + a_t$ and $r_t = \mu + c\ln(\sigma_t^2) + a_t$.

The formulation of the GARCH-M model in Equation (4.24) implies that there are serial correlations in the return series r_t. These serial correlations are introduced by those in the volatility process $\{\sigma_t^2\}$. The existence of risk premium is, therefore, another reason that some historical stock returns have serial correlations.

For illustration, we consider a GARCH(1,1)-M model with Gaussian innovations for the monthly log returns of Intel stock from January 1973 to December 2009. For numerical stability, we use the percentage returns in this section. In other words, r_t now denotes the log returns in percentages. The fitted model is

$$r_t = 1.025 + 0.00081\sigma_t^2 + a_t, \quad a_t = \sigma_t\epsilon_t, \quad \epsilon_t \sim N(0,1) \tag{4.25}$$
$$\sigma_t^2 = 9.376 + 0.087a_{t-1}^2 + 0.851\sigma_{t-1}^2, \tag{4.26}$$

where the standard errors for the two parameters in Equation (4.25) are 1.347 and 0.0092, respectively, and those for the parameters in Equation (4.26) are 3.977, 0.027, and 0.040, respectively. The estimated risk premium for the monthly log returns of Intel stock is small and insignificant. Here, the estimation is carried out in R with a simple script available on the web page of the book.

As a second demonstration, Figure 4.14 shows the monthly excess returns of the S&P 500 index from 1926 to 1991. See Tsay (2010, Chapter 3) for various volatility models fitted to this series. Again, the returns are in percentages. A simple GARCH(1,1) model for the series is

$$r_t = 0.745 + a_t, \quad a_t = \sigma_t\epsilon_t, \quad \epsilon_t \sim N(0,1),$$
$$\sigma_t^2 = 0.806 + 0.122a_{t-1}^2 + 0.854\sigma_{t-1}^2,$$

where the standard error of the parameter in the mean equation is 0.154 and those of the parameters in the volatility equation are 0.283, 0.022, and 0.022, respectively. Clearly, all estimates are significant at the 5% level. Model checking statistics show that, except for normality, the model fits the data well.

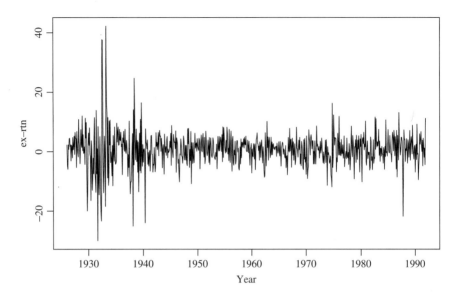

Figure 4.14. Time plot of monthly excess returns, in percentages, of the S&P 500 index from January 1926 to December 1991.

We also apply a GARCH(1,1)-M model to the returns and obtain

$$r_t = 0.6561 + 0.0041\sigma_t^2 + a_t, \quad a_t = \sigma_t\epsilon_t, \quad \epsilon_t \sim N(0,1),$$
$$\sigma_t^2 = 0.7936 + 0.1222a_{t-1}^2 + 0.8547\sigma_{t-1}^2, \tag{4.27}$$

where the standard errors of the parameters in the mean equation are 0.2458 and 0.0089, respectively, and those of the parameters in Equation (4.27) are 0.2830, 0.0221, and 0.0218, respectively. Again, the risk premium is not significant at the usual 5% level. Consequently, the model reduces to the usual GARCH(1,1) model for the data.

R Demonstration

```
> y=intc*100    % Intel stock returns in percentages
> source("garchM.R")   % Compile the script
> garchM(y)
[1] "initial estimates:"
 (Intercept)        v1            ar0            ar1            ma1
 0.472685490   0.0059504  14.1506718   0.1203339   0.7915216

Estimation results of GARCH(1,1)-M model:
estimates:  1.02491 0.000806 9.37589 0.086950  0.850682
std.errors: 1.34736 0.009245 3.97716 0.026644  0.040035
t-ratio:    0.76068 0.087235 2.35744 3.263355  21.2486
```

```
% Monthly excess returns, in percentages, of the S&P 500 index
> sp5=scan(file='sp500.txt')
> sp5=sp5*100
> m2=garchFit( 1+garch(1,1),data=sp5,trace=F)
> summary(m2)
Title:  GARCH Modelling
Mean and Variance Equation: data ~ 1+garch(1,1) [data=sp5]
Conditional Distribution:  norm
Coefficients:
        Estimate  Std. Error  t value Pr(>|t|)
mu       0.74497     0.15377    4.845 1.27e-06 ***
omega    0.80615     0.28333    2.845  0.00444 **
alpha1   0.12198     0.02202    5.540 3.02e-08 ***
beta1    0.85436     0.02175   39.276 < 2e-16 ***
---

Standardised Residuals Tests:
                             Statistic p-Value
 Jarque-Bera Test   R    Chi^2  80.32111  0
 Shapiro-Wilk Test  R    W      0.9850518 3.14157e-07
 Ljung-Box Test     R    Q(10)  11.22050  0.340599
 Ljung-Box Test     R    Q(20)  24.29896  0.2295768
 Ljung-Box Test     R^2  Q(10)  9.920157  0.4475259
 Ljung-Box Test     R^2  Q(20)  16.75081  0.6690903

> garchM(sp5)
Estimation results of GARCH(1,1)-M model:
estimates:  0.65613 0.004096 0.793615 0.12219  0.85465
std.errors: 0.24579 0.008938 0.283022 0.02209  0.02176
t-ratio:    2.66943 0.458335 2.804075 5.53226 39.2788
```

4.9   THE EXPONENTIAL GARCH MODEL

To overcome some weaknesses of the GARCH model in handling financial time series, Nelson (1991) proposes the EGARCH model. In particular, to allow for asymmetric effects between positive and negative asset returns, he considered the weighted innovation

$$g(\epsilon_t) = \theta\epsilon_t + \gamma[|\epsilon_t| - E(|\epsilon_t|)], \tag{4.28}$$

where θ and γ are real constants. Both ϵ_t and $|\epsilon_t| - E(|\epsilon_t|)$ are zero-mean iid sequences with continuous distributions. Therefore, $E[g(\epsilon_t)] = 0$. The asymmetry of $g(\epsilon_t)$ can be easily seen by rewriting it as

$$g(\epsilon_t) = \begin{cases} (\theta + \gamma)\epsilon_t - \gamma E(|\epsilon_t|) & \text{if } \epsilon_t \geq 0, \\ (\theta - \gamma)\epsilon_t - \gamma E(|\epsilon_t|) & \text{if } \epsilon_t < 0. \end{cases}$$

Remark. For the standard Gaussian random variable ϵ_t, $E(|\epsilon_t|) = \sqrt{2/\pi}$. For the standardized Student t distribution in Equation (4.7), we have

$$E(|\epsilon_t|) = \frac{2\sqrt{v-2}\,\Gamma((v+1)/2)}{(v-1)\Gamma(v/2)\sqrt{\pi}}.$$

$\square$

An EGARCH(m, s) model can be written as

$$a_t = \sigma_t \epsilon_t, \quad \ln(\sigma_t^2) = \alpha_0 + \frac{1 + \beta_1 B + \cdots + \beta_{s-1}B^{s-1}}{1 - \alpha_1 B - \cdots - \alpha_m B^m} g(\epsilon_{t-1}), \tag{4.29}$$

where α_0 is a constant, B is the back-shift (or lag) operator such that $Bg(\epsilon_t) = g(\epsilon_{t-1})$, and $1 + \beta_1 B + \cdots + \beta_{s-1}B^{s-1}$ and $1 - \alpha_1 B - \cdots - \alpha_m B^m$ are polynomials with zeros outside the unit circle and have no common factors. By outside the unit circle, we mean that absolute values of the zeros are greater than 1. Again, Equation (4.29) uses the usual ARMA parameterization to describe the evolution of the conditional variance of a_t. On the basis of this representation, some properties of the EGARCH model can be obtained in a similar manner as those of the GARCH model. For instance, the unconditional mean of $\ln(\sigma_t^2)$ is α_0. However, the model differs from the GARCH model in several ways. First, it uses logged conditional variance to relax the positiveness constraint of model coefficients. Second, the use of $g(\epsilon_t)$ enables the model to respond asymmetrically to positive and negative lagged values of a_t. Some additional properties of the EGARCH model can be found in Nelson (1991).

To better understand the EGARCH model, let us consider the simple model with order (1,1):

$$a_t = \sigma_t \epsilon_t, \quad (1 - \alpha B)\ln(\sigma_t^2) = (1 - \alpha)\alpha_0 + g(\epsilon_{t-1}), \tag{4.30}$$

where the ϵ_t are iid standard normal and the subscript of α_1 is omitted. In this case, $E(|\epsilon_t|) = \sqrt{2/\pi}$ and the model for $\ln(\sigma_t^2)$ becomes

$$(1 - \alpha B)\ln(\sigma_t^2) = \begin{cases} \alpha_* + (\gamma + \theta)\epsilon_{t-1} & \text{if } \epsilon_{t-1} \geq 0, \\ \alpha_* + (\gamma - \theta)(-\epsilon_{t-1}) & \text{if } \epsilon_{t-1} < 0, \end{cases} \tag{4.31}$$

where $\alpha_* = (1 - \alpha)\alpha_0 - \sqrt{2/\pi}\gamma$. This is a nonlinear function similar to that of the threshold autoregressive (TAR) model of Tong (1978, 1990). It suffices to say that for this simple EGARCH model the conditional variance evolves in a nonlinear manner depending on the sign of a_{t-1}. Specifically, we have

$$\sigma_t^2 = \sigma_{t-1}^{2\alpha} \exp(\alpha_*) \begin{cases} \exp\left[(\gamma + \theta)\dfrac{a_{t-1}}{\sigma_{t-1}}\right] & \text{if } a_{t-1} \geq 0, \\ \exp\left[(\gamma - \theta)\dfrac{|a_{t-1}|}{\sigma_{t-1}}\right] & \text{if } a_{t-1} < 0. \end{cases}$$

The coefficients $(\gamma + \theta)$ and $(\gamma - \theta)$ show the asymmetry in response to positive and negative a_{t-1}. The model is, therefore, nonlinear if $\theta \neq 0$. Since negative shocks tend to have larger impacts, we expect θ to be negative. For higher order EGARCH models, the nonlinearity becomes much more complicated. Cao and Tsay (1992) used nonlinear models, including EGARCH models, to obtain multistep ahead volatility forecasts.

4.9.1 An Illustrative Example

Nelson (1991) applies an EGARCH model to the daily excess returns of the value-weighted market index from the Center for Research in Security Prices from July 1962 to December 1987. The excess returns are obtained by removing monthly Treasury bill returns from the value-weighted index returns, assuming that the Treasury bill return was constant for each calendar day within a given month. There are 6408 observations. Denote the excess return by r_t. The model used is as follows:

$$r_t = \phi_0 + \phi_1 r_{t-1} + c\sigma_t^2 + a_t, \tag{4.32}$$

$$\ln(\sigma_t^2) = \alpha_0 + \ln(1 + wN_t) + \frac{1 + \beta B}{1 - \alpha_1 B - \alpha_2 B^2} g(\epsilon_{t-1}),$$

where σ_t^2 is the conditional variance of a_t given F_{t-1}, N_t is the number of nontrading days between trading days $t - 1$ and t, α_0 and w are real parameters, $g(\epsilon_t)$ is defined in Equation (4.28), and ϵ_t follows a GED in Equation (4.10). Similar to a GARCH-M model, the parameter c in Equation (4.32) is the risk premium parameter. Table 4.2 gives the parameter estimates and their standard errors of the model. The mean equation of model (4.32) has two features that are of interest. First, it uses an AR(1) model to take care of possible serial correlation in the excess returns. Second, it uses the volatility σ_t^2 as a regressor to account for risk premium. The estimated risk premium is negative, but statistically insignificant.

TABLE 4.2. Estimated AR(1)–EGARCH(2,2) Model for the Daily Excess Returns of the Value-Weighted CRSP Market Index: July 1962 to December 1987.

Par.	α_0	w	γ	α_1	α_2	β
Est.	-10.06	0.183	0.156	1.929	-0.929	-0.978
Err.	0.346	0.028	0.013	0.015	0.015	0.006

Par.	θ	ϕ_0	ϕ_1	c	v
Est.	-0.118	$3.5 \cdot 10^{-4}$	0.205	-3.361	1.576
Err.	0.009	$9.9 \cdot 10^{-5}$	0.012	2.026	0.032

4.9.2   An Alternative Model Form

For ease in estimation, one can use an alternative form for the EGARCH(m, s) model:

$$\ln(\sigma_t^2) = \alpha_0 + \sum_{i=1}^{m} \alpha_i \frac{|a_{t-i}| + \gamma_i a_{t-i}}{\sigma_{t-i}} + \sum_{j=1}^{s} \beta_j \ln(\sigma_{t-j}^2). \qquad (4.33)$$

Here, a positive a_{t-i} contributes $\alpha_i(1 + \gamma_i)|\epsilon_{t-i}|$ to the log volatility, whereas a negative a_{t-i} gives $\alpha_i(1 - \gamma_i)|\epsilon_{t-i}|$, where $\epsilon_{t-i} = a_{t-i}/\sigma_{t-i}$. The γ_i parameter thus signifies the leverage effect of a_{t-i}. Again, we expect γ_i to be negative in real applications. This model form has been used by some packages, for example, S-Plus. We use an R script `Egarch` to estimate an EGARCH(1,1) model.

4.9.3   Second Example

As another illustration, we consider the monthly log returns of IBM stock from January 1967 to December 2009 for 516 observations. Figure 4.15a shows the time plot of the series. Sample ACF and the Ljung–Box statistics indicate that there are no significant serial correlations in the data so that we proceed to volatility modeling. For example, $Q(12) = 7.40$ with p value 0.83 for the log return series.

Using the alternative parameterization in Equation (4.33) and noting that $\epsilon_t = a_t/\sigma_t$, we rewrote Equation (4.30) as

$$\ln(\sigma_t^2) = \alpha_0 + \alpha_1(|\epsilon_{t-1}| + \gamma_1\epsilon_{t-1}) + \beta_1 \ln(\sigma_{t-1}^2).$$

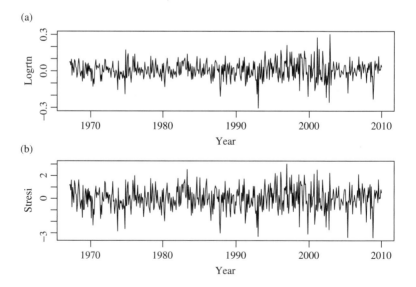

Figure 4.15. Time plots of monthly log returns of IBM stock from January 1967 to December 2009: (a) log returns and (b) standardized residuals of an EGARCH(1,1) model.

Entertaining this EGARCH(1,1) model, we obtain, for IBM monthly log returns,

$$r_t = 0.0067 + a_t, \quad a_t = \sigma_t \epsilon_t, \quad \epsilon_t \sim N(0,1),$$
$$\ln(\sigma_t^2) = -0.5975 + 0.213(|\epsilon_{t-1}| - 0.4355\epsilon_{t-1}) + 0.9196 \ln(\sigma_{t-1}^2), \quad (4.34)$$

where all parameter estimates are statistically significant at the 5% level. See the R demonstration for details. For model checking, the Ljung–Box statistics give $Q(10) = 5.28(0.87)$ and $Q(20) = 20.97(0.40)$ for the standardized residuals $\tilde{a}_t = a_t/\sigma_t$ and $Q(10) = 5.01(0.89)$ and $Q(20) = 14.26(0.82)$ for the squared series $\tilde{a}_t^2$, where the number in parentheses denotes p value. The model fits the data reasonably well.

From the estimated equation (Eq. 4.34) and using $\sqrt{2/\pi} \approx 0.7979$, we obtain the volatility equation as

$$\ln(\sigma_t^2) = -0.5975 + 0.9196 \ln(\sigma_{t-1}^2) + \begin{cases} 0.1203\epsilon_{t-1} & \text{if } \epsilon_{t-1} \geq 0, \\ -0.3058\epsilon_{t-1} & \text{if } \epsilon_{t-1} < 0. \end{cases}$$

Taking antilog transformation, we have

$$\sigma_t^2 = \sigma_{t-1}^{2 \times 0.9196} e^{-0.5975} \times \begin{cases} e^{0.1203\epsilon_{t-1}} & \text{if } \epsilon_{t-1} \geq 0, \\ e^{-0.3058\epsilon_{t-1}} & \text{if } \epsilon_{t-1} < 0. \end{cases}$$

This equation highlights the asymmetric responses in volatility to the past positive and negative shocks under an EGARCH model. For example, for a standardized shock with magnitude 2 (i.e., two standard deviations), we have

$$\frac{\sigma_t^2(\epsilon_{t-1} = -2)}{\sigma_t^2(\epsilon_{t-1} = 2)} = \frac{\exp[-0.3058 \times (-2)]}{\exp(0.1203 \times 2)} = e^{0.371} = 1.449.$$

Therefore, the impact of a negative shock of size two standard deviations is about 44.9% higher than that of a positive shock of the same size. This example clearly demonstrates the asymmetric feature of EGARCH models. In general, the bigger the shock, the larger the difference in volatility impact.

R Demonstration. Output edited.

```
> source("Egarch.R") % Compile R script
> da=read.table("m-ibmsp6709.txt",header=T) % Load data
> dim(da)   % Check sample size of the data
[1] 516    3
> ibm=log(da$ibm+1) % Take log transformation
> Box.test(ibm,lag=12,type='Ljung') % Check serial correlations
        Box-Ljung test
data:  ibm
X-squared = 7.4042, df = 12, p-value = 0.8298
```

```
> m1=Egarch(ibm) % Model fitting
Estimation results of EGARCH(1,1) model:
estimates:   0.00671172 -0.5975393 0.21298   -0.4355012 0.919648
std.errors: 0.00287457   0.2371713 0.058016   0.1721217 0.039416
t-ratio:     2.334859    -2.519442  3.671078 -2.530194 23.33172
> names(m1)
[1] "residuals"  "volatility"
> stresi=m1$residuals/m1$volatility % Obtain standardized residuals
> tdx=c(1:516)/12+1967 % Compute time index
> par(mfcol=c(2,1)) % Plotting
> plot(tdx,ibm,xlab='year',ylab='logrtn',type='l')
> plot(tdx,stresi,xlab='year',ylab='stresi',type='l')
> Box.test(stresi,lag=10,type='Ljung')    % Model checking
        Box-Ljung test
data:   stresi
X-squared = 5.2807, df = 10, p-value = 0.8717
> Box.test(stresi,lag=20,type='Ljung')
        Box-Ljung test
data:   stresi
X-squared = 20.971, df = 20, p-value = 0.3988
> Box.test(stresi^2,lag=10,type='Ljung')
        Box-Ljung test
data:   stresi^2
X-squared = 5.0127, df = 10, p-value = 0.8903
> Box.test(stresi^2,lag=20,type='Ljung')
        Box-Ljung test
data:   stresi^2
X-squared = 14.2643, df = 20, p-value = 0.8168
```

4.9.4 Forecasting Using an EGARCH Model

We use the EGARCH(1,1) model to illustrate multistep ahead forecasts of EGARCH models, assuming that the model parameters are known and the innovations are standard Gaussian. For such a model, we have

$$\ln(\sigma_t^2) = (1 - \alpha_1)\alpha_0 + \alpha_1 \ln(\sigma_{t-1}^2) + g(\epsilon_{t-1}),$$

$$g(\epsilon_{t-1}) = \theta\epsilon_{t-1} + \gamma(|\epsilon_{t-1}| - \sqrt{2/\pi}).$$

Taking exponential, the model becomes

$$\sigma_t^2 = \sigma_{t-1}^{2\alpha_1} \exp[(1 - \alpha_1)\alpha_0] \exp[g(\epsilon_{t-1})],$$

$$g(\epsilon_{t-1}) = \theta\epsilon_{t-1} + \gamma(|\epsilon_{t-1}| - \sqrt{2/\pi}). \tag{4.35}$$

Let h be the forecast origin. For the 1-step ahead forecast, we have

$$\sigma_{h+1}^2 = \sigma_h^{2\alpha_1} \exp[(1 - \alpha_1)\alpha_0] \exp[g(\epsilon_h)],$$

where all of the quantities on the right-hand side are known. Thus, the 1-step ahead volatility forecast at the forecast origin h is simply $\hat{\sigma}_h^2(1) = \sigma_{h+1}^2$ given earlier. For the 2-step ahead forecast, Equation (4.35) gives

$$\sigma_{h+2}^2 = \sigma_{h+1}^{2\alpha_1} \exp[(1 - \alpha_1)\alpha_0] \exp[g(\epsilon_{h+1})].$$

Taking conditional expectation at time h, we have

$$\hat{\sigma}_h^2(2) = \hat{\sigma}_h^{2\alpha_1}(1) \exp[(1 - \alpha_1)\alpha_0] E_h\{\exp[g(\epsilon_{h+1})]\},$$

where E_h denotes a conditional expectation taken at the time origin h. The prior expectation can be obtained as follows:

$$
\begin{aligned}
E\{\exp[g(\epsilon)]\} &= \int_{-\infty}^{\infty} \exp[\theta\epsilon + \gamma(|\epsilon| - \sqrt{2/\pi})]f(\epsilon)d\epsilon \\
&= \exp\left(-\gamma\sqrt{2/\pi}\right)\left[\int_0^{\infty} e^{(\theta+\gamma)\epsilon}\frac{1}{\sqrt{2\pi}}e^{-\epsilon^2/2}d\epsilon\right.\\
&\quad \left. + \int_{-\infty}^0 e^{(\theta-\gamma)\epsilon}\frac{1}{\sqrt{2\pi}}e^{-\epsilon^2/2}d\epsilon\right] \\
&= \exp\left(-\gamma\sqrt{2/\pi}\right)\left[e^{(\theta+\gamma)^2/2}\Phi(\theta+\gamma) + e^{(\theta-\gamma)^2/2}\Phi(\gamma-\theta)\right],
\end{aligned}
$$

where $f(\epsilon)$ and $\Phi(x)$ are the probability density function and CDF (cumulative distribution function) of the standard normal distribution, respectively. Consequently, the 2-step ahead volatility forecast is

$$
\begin{aligned}
\hat{\sigma}_h^2(2) = \hat{\sigma}_h^{2\alpha_1}(1) \exp&\left[(1 - \alpha_1)\alpha_0 - \gamma\sqrt{2/\pi}\right] \\
&\times \left\{\exp[(\theta+\gamma)^2/2]\Phi(\theta+\gamma) + \exp[(\theta-\gamma)^2/2]\Phi(\gamma-\theta)\right\}.
\end{aligned}
$$

Repeating the previous procedure, we obtain a recursive formula for a j-step ahead forecast:

$$
\begin{aligned}
\hat{\sigma}_h^2(j) = \widehat{\sigma_h^{2\alpha_1}}(j-1) \exp&(\omega) \\
&\times \left\{\exp[(\theta+\gamma)^2/2]\Phi(\theta+\gamma) + \exp[(\theta-\gamma)^2/2]\Phi(\gamma-\theta)\right\},
\end{aligned}
$$

where $\omega = (1 - \alpha_1)\alpha_0 - \gamma\sqrt{2/\pi}$. The values of $\Phi(\theta+\gamma)$ and $\Phi(\gamma-\theta)$ can be obtained from most statistical packages.

4.10   THE THRESHOLD GARCH MODEL

Another volatility model commonly used to handle leverage effects is the threshold generalized autoregressive conditional heteroscedastic (or TGARCH) model; see Glosten et al. (1993) and Zakoian (1994). A TGARCH(m, s) model assumes the form

$$\sigma_t^2 = \alpha_0 + \sum_{i=1}^{m}(\alpha_i + \gamma_i N_{t-i})a_{t-i}^2 + \sum_{j=1}^{s}\beta_j \sigma_{t-j}^2, \qquad (4.36)$$

where N_{t-i} is an indicator for *negative* a_{t-i}, that is,

$$N_{t-i} = \begin{cases} 1 & \text{if } a_{t-i} < 0, \\ 0 & \text{if } a_{t-i} \ge 0, \end{cases}$$

and α_i, γ_i, and β_j are nonnegative parameters satisfying conditions similar to those of GARCH models. From the model, it is seen that a positive a_{t-i} contributes $\alpha_i a_{t-i}^2$ to σ_t^2, whereas a negative a_{t-i} has a larger impact $(\alpha_i + \gamma_i)a_{t-i}^2$ with $\gamma_i > 0$. The model uses zero as its *threshold* to separate the impacts of past shocks. Other threshold values can also be used; see Tsay (2010, Chapter 4) for the general concept of threshold models. Model (4.36) is also called the GJR model because Glosten et al. (1993) proposed essentially the same model.

For illustration, consider the daily log returns, in percentages, of the exchange rate between US Dollar and Euro from January 4, 1999, to August 20, 2010; see Figure 4.3. Using a simple R script available on the book web page, we obtain the fitted TGARCH(1,1) as

$$r_t = 0.0122 + a_t, \quad a_t = \sigma_t \epsilon_t, \quad \epsilon_t \sim N(0, 1),$$

$$\sigma_t^2 = 0.00128 + (0.0223 + 0.0125N_{t-1})a_{t-1}^2 + 0.9687\sigma_{t-1}^2, \qquad (4.37)$$

where the standard error of the parameter for the mean equation is 0.0107 and the standard errors of the parameters in the volatility equation are 0.00061, 0.0052, 0.0071, and 0.0044, respectively. To check the fitted model, we have $Q(10) = 13.38(0.20)$ and $Q(20) = 22.87(0.30)$ for the standardized residual $\tilde{a}_t$ and $Q(10) = 12.89(0.23)$ and $Q(20) = 27.23(0.13)$ for $\tilde{a}_t^2$. The model is adequate in modeling the first two conditional moments of the log returns of exchange rate. The coefficient of the mean equation is insignificant. For the volatility equation, all estimates are significant. To test the leverage effect, we consider the null hypothesis $H_0 : \gamma_1 \le 0$ versus the alternative hypothesis $H_a : \gamma_1 > 0$. The t-ratio of the test is 1.772 with p value 0.038 so that the leverage effect is significant at the 5% level.

Figure 4.16 provides time plots of the fitted volatility and the standardized residuals of the TGARCH(1,1) model. As expected, the volatility was higher during the recent financial crisis. The standardized residuals appear to be random, but their magnitudes indicate the possibility of heavy tails.

(a)

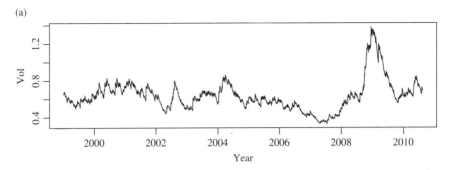

(b)

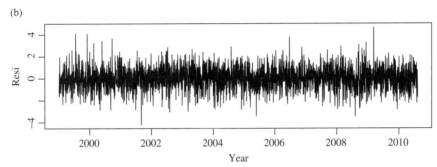

Figure 4.16. Time plots of a TGARCH(1,1) model for the daily log returns of Dollar/Euro exchange rate from January 4, 1999, to August 20, 2010: (a) fitted volatility and (b) standardized residuals.

R Demonstration

```
> da=read.table("d-useu9910.txt",header=T)
> fx=log(da$rate)
> eu=diff(fx)*100
> source('Tgarch11.R')
> m1=Tgarch11(eu)
[1] 2834.995
  0:     2834.9946: 0.00243980 0.0426591 0.100000 0.100000 0.800000
  ......
 48:     2731.8319: 0.0122412 0.00127505 0.0223472 0.0125162 0.968720
Coefficient(s):
          Estimate   Std. Error   t value   Pr(>|t|)
mu     0.012241549 0.010729840   1.14089    0.253916
omega  0.001275045 0.000618464   2.06163    0.039243  *
alpha  0.022347138 0.005249457   4.25704 2.0715e-05 ***
gam1   0.012516327 0.007062559   1.77221    0.076360  .
beta   0.968720367 0.004357861 222.29265 < 2.22e-16 ***
---
> names(m1)
[1] "residuals"  "volatility" "par"
> at=m1$residuals
```

```
> sigt=m1$volatility
> resi=at/sigt
> Box.test(resi,lag=10,type='Ljung')
        Box-Ljung test
data:   resi
X-squared = 13.3818, df = 10, p-value = 0.2031
> Box.test(resi,lag=20,type='Ljung')
        Box-Ljung test
data:   resi
X-squared = 22.8733, df = 20, p-value = 0.2951
> Box.test(resi^2,lag=10,type='Ljung')
        Box-Ljung test
data:   resi^2
X-squared = 12.8935, df = 10, p-value = 0.2297
> Box.test(resi^2,lag=20,type='Ljung')
        Box-Ljung test
data:   resi^2
X-squared = 27.2298, df = 20, p-value = 0.1289
```

4.11 ASYMMETRIC POWER ARCH MODELS

The TGARCH model belongs to the class of asymmetric power autoregressive conditional heteroscedastic (APARCH) models of Ding et al. (1993). A general APARCH(m, s) model can be written as

$$r_t = \mu_t + a_t, \quad a_t = \sigma_t \epsilon_t, \quad \epsilon_t \sim D(0,1)$$

$$\sigma_t^\delta = \omega + \sum_{i=1}^{m} \alpha_i (|a_{t-i}| + \gamma_i a_{t-i})^\delta + \sum_{j=1}^{s} \beta_j \sigma_{t-j}^\delta, \tag{4.38}$$

where μ_t is the conditional mean, $D(0,1)$ denotes a distribution with mean zero and variance 1, δ is a positive real number, and the coefficients ω, α_i, γ_i, and β_j satisfy some regularity conditions so that the volatility is positive. Similar to GARCH models, the APARCH(1,1) model is often used in practice. Three special cases of the APARCH models are of interest. When $\delta = 2$, the APARCH model reduces to a TGARCH model. When $\delta = 1$, the model uses volatility directly in the volatility equation. The case of $\delta = 0$ is taking as the limit of $\delta \to 0$ and in this case the model becomes the EGARCH model of Nelson (1991).

The power function in Equation (4.38) is a transformation used to improve the goodness of fit of the model. This appears to be a sensible approach if one is interested in prediction. On the other hand, except for some special values, it seems hard to find a good interpretation for the power parameter δ. To demonstrate, consider the log return series of the Dollar/Euro exchange rate. A fitted APARCH(1,1) model with Gaussian innovations is

$$r_t = 0.0128 + a_t, \quad a_t = \sigma_t \epsilon_t, \quad \epsilon_t \sim N(0,1)$$

$$\sigma_t^{1.67} = 0.0016 + 0.0313(a_{t-1} + 0.1135a_{t-1})^{1.67} + 0.9689\sigma_{t-1}^{1.67}.$$

Model checking statistics suggest that this model fits the data reasonably well. However, it is hard to find the meaning of $\hat{\delta} = 1.67$. On the other hand, the standard error of $\hat{\delta}$ is 0.406 so that $\hat{\delta}$ is not significantly different from 2. Thus, in this particular case, one can simply fix $\delta = 2$, which results in using a TGARCH(1,1) model. As expected, fixing $\delta = 2$ gives results very close to the TGARCH(1,1) model obtained before.

R Demonstration

```
> m1=garchFit(~1+aparch(1,1),data=eu,trace=F)
> summary(m1)
Title: GARCH Modelling
Call: garchFit(formula=~1+aparch(1,1),data=eu,trace=F)

Mean and Variance Equation:
 data ~ 1 + aparch(1, 1) [data = eu]
Conditional Distribution:  norm

Error Analysis:
        Estimate  Std. Error  t value Pr(>|t|)
mu      0.0127649  0.0107626    1.186   0.2356
omega   0.0015919  0.0007226    2.203   0.0276 *
alpha1  0.0313680  0.0053350    5.880 4.11e-09 ***
gamma1  0.1135337  0.0711911    1.595   0.1108
beta1   0.9689156  0.0038405  252.292 < 2e-16 ***
delta   1.6743115  0.4057131    4.127 3.68e-05 ***
---
Standardised Residuals Tests:
                          Statistic p-Value
 Jarque-Bera Test    R    Chi^2  50.20525  1.253342e-11
 Shapiro-Wilk Test   R    W      0.9956706 1.606077e-07
 Ljung-Box Test      R    Q(10)  13.37689  0.2033562
 Ljung-Box Test      R    Q(20)  22.84736  0.2963516
 Ljung-Box Test      R^2  Q(10)  13.1561   0.2150747
 Ljung-Box Test      R^2  Q(20)  27.44886  0.1231014
 LM Arch Test        R    TR^2   14.35738  0.2784714

Information Criterion Statistics:
     AIC       BIC       SIC      HQIC
1.869014 1.881269 1.869006 1.873428
>
> m2=garchFit(~1+aparch(1,1),data=eu,delta=2,include.delta=F,trace=F)
> summary(m2)
Title:  GARCH Modelling
Call:
garchFit(formula=~1+aparch(1,1),data=eu,delta=2,include.delta=F,trace=F)

Mean and Variance Equation:
 data ~ 1 + aparch(1, 1) [data = eu]
Conditional Distribution: norm

Error Analysis:
```

```
          Estimate  Std. Error   t value  Pr(>|t|)
mu        0.0122646  0.0107289    1.143    0.2530
omega     0.0012745  0.0005752    2.216    0.0267 *
alpha1    0.0282723  0.0038637    7.317  2.53e-13 ***
gamma1    0.1100242  0.0649051    1.695    0.0900 .
beta1     0.9687115  0.0039421  245.735  < 2e-16 ***
---
Log Likelihood: -2731.850      normalized:  -0.9326902

Standardised Residuals Tests:
                                Statistic  p-Value
  Jarque-Bera Test   R    Chi^2  49.97677  1.405021e-11
  Shapiro-Wilk Test  R    W      0.9956783 1.645761e-07
  Ljung-Box Test     R    Q(10)  13.38285  0.2030469
  Ljung-Box Test     R    Q(20)  22.87265  0.2950908
  Ljung-Box Test     R^2  Q(10)  12.89586  0.2295531
  Ljung-Box Test     R^2  Q(20)  27.24036  0.128636
  LM Arch Test       R    TR^2   14.29661  0.2821695

Information Criterion Statistics:
     AIC       BIC       SIC      HQIC
1.868795  1.879007  1.868789  1.872472
> plot(m2)
```

4.12 NONSYMMETRIC GARCH MODEL

Another GARCH family model that can capture asymmetric volatility responses to past positive and negative shocks is proposed by Engle and Ng (1993) and studied by Duan (1995). It assumes the form

$$r_t = \mu_t + a_t, \quad a_t = \sigma_t \epsilon_t, \quad \epsilon_t \sim D(0,1)$$
$$\sigma_t^2 = \beta_0 + \beta_1 \sigma_{t-1}^2 + \beta_2 (a_{t-1} - \theta \sigma_{t-1})^2, \tag{4.39}$$

where μ_t is the conditional mean, $D(0,1)$ denotes a distribution with mean zero and variance 1, β_i are nonnegative parameters with $\beta_0 > 0$, and θ is a leverage parameter. The model in Equation (4.39) is referred to as a nonsymmetric GARCH(1,1), or NGARCH(1,1), model. It reduces to a GARCH(1,1) model if $\theta = 0$.

To study properties of NGARCH(1,1) model, we rewrite Equation (4.39) as

$$\sigma_t^2 = \beta_0 + \beta_1 \sigma_{t-1}^2 + \beta_2 \sigma_{t-1}^2 (\epsilon_{t-1} - \theta)^2. \tag{4.40}$$

Taking expectation and using independence between ϵ_{t-1} and σ_{t-1}, we have

$$E(\sigma_t^2) = \beta_0 + \beta_1 E(\sigma_{t-1}^2) + \beta_2 E(\sigma_{t-1}^2) E(\epsilon_{t-1} - \theta)^2$$
$$= \beta_0 + \beta_1 E(\sigma_{t-1}^2) + \beta_2 E(\sigma_{t-1}^2)(1 + \theta^2).$$

If r_t is weakly stationary, $E(\sigma_t^2) = E(\sigma_{t-1}^2)$ and we have

$$E(\sigma_t^2) = \frac{\beta_0}{1 - \beta_1 - \beta_2(1 + \theta^2)},$$

which is the unconditional variance of r_t. Consequently, we require that $1 - \beta_1 - \beta_2(1 + \theta^2) > 0$ for the NGARCH(1,1) model. Multiplying Equation (4.40) by ϵ_{t-1} and taking expectation, we obtain

$$E(\epsilon_{t-1}\sigma_t^2) = -2\theta\beta_2 E(\sigma_{t-1}^2) = \frac{-2\theta\beta_0\beta_2}{1 - \beta_1 - \beta_2(1 + \theta^2)}.$$

This result says that if $\theta > 0$ and $\beta_2 > 0$, then ϵ_{t-1} is negatively related to σ_t^2. Therefore, θ is a leverage parameter and should be positive. Finally, it can be shown that, under certain conditions, the shock a_t of a NGARCH(1,1) model has heavy tails even if ϵ_t is Gaussian; see Duan (1995).

For demonstration, we apply an NGARCH(1,1) model to the log returns, in percentages, of the Dollar/Euro exchange rate from January 4, 1999, to August 20, 2010. Using a simple R script available on the book web page, we obtain the model

$$r_t = -0.0011 + a_t, \quad a_t = \sigma_t\epsilon_t, \quad \epsilon_t \sim N(0, 1)$$

$$\sigma_t^2 = 0.00237 + 0.9618\sigma_{t-1}^2 + 0.02119\sigma_{t-1}^2(\epsilon_{t-1} - 0.731)^2, \tag{4.41}$$

where all estimates, but the mean return -0.0011, are significant at the 5% level. The Ljung–Box statistics of the standardized residuals $\tilde{a}_t$ and their squared series fail to reject the model. For instance, we have $Q(10) = 14.78(0.14)$ and $12.94(0.23)$, respectively, for $\tilde{a}_t$ and $\tilde{a}_t^2$, where the number in parentheses denotes p value. Consider the leverage parameter θ. The estimate 0.731 has a t-ratio 2.92, so that the leverage effect is significant at the 5% level. Therefore, the NGARCH(1,1) model supports similar inference as the TGARCH(1,1) model for the exchange rate series.

Figure 4.17 provides time plots of the fitted volatility of NGARCH(1,1) model in Equation (4.41) and the squared residuals of the log returns. Compared with that of Figure 4.16, we see that the two models provide similar volatility estimates.

R Demonstration

```
> da=read.table("d-useu9910.txt",header=T)
> fx=log(da$rate)
> eu=diff(fx)*100
> source("Ngarch.R")
> m1=Ngarch(eu)

Estimation results of NGARCH(1,1) model:
estimates:   -0.001094 0.0023667 0.961805 0.021186 0.730962
std.errors:   0.010809 0.0005806 0.006046 0.003605 0.250155
```

```
t-ratio:       -0.101217 4.076674   159.0863 5.877186 2.922037
> res=m1$residuals
> vol=m1$volatility
> resi=res/vol
> Box.test(resi,lag=10,type='Ljung')
        Box-Ljung test
data:   resi
X-squared = 14.776, df = 10, p-value = 0.1404

> Box.test(resi^2,lag=10,type='Ljung')
        Box-Ljung test
data:   resi^2
X-squared = 12.9434, df = 10, p-value = 0.2269
```

Figure 4.18 compares the volatilities of NGARCH(1,1) and GARCH(1,1) models with Gaussian innovations for the daily log returns of Dollar/Euro exchange rate. The solid line denotes the volatility of the traditional GARCH(1,1) model. Some differences are seen from the plot, indicating the effect of negative returns on the volatility. The overall pattern of the volatility, however, remains unchanged.

4.13   THE STOCHASTIC VOLATILITY MODEL

An alternative approach to describe the volatility evolution of a financial time series is to introduce an innovation to the conditional variance equation of a_t; see Melino

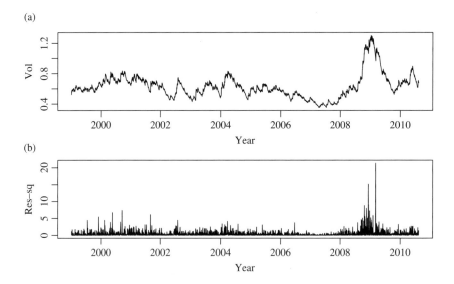

Figure 4.17. Time plots of an NGARCH(1,1) model for the daily log returns of Dollar/Euro exchange rate from January 4, 1999, to August 20, 2010: (a) fitted volatility and (b) squared residuals.

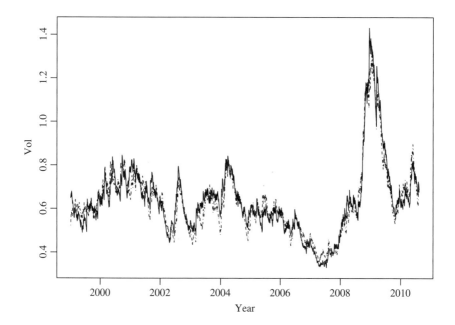

Figure 4.18. Time plots of NGARCH and GARCH volatility for daily log returns of Dollar/Euro exchange rate from January 4, 1999, to August 20, 2010. The solid and dashed lines are for GARCH(1,1) and NGARCH(1,1) models, respectively.

and Turnbull (1990), Taylor (1994), Harvey et al. (1994), and Jacquier et al. (1994). The resulting model is referred to as a *stochastic volatility (SV) model*. Similar to EGARCH models, to ensure positiveness of the conditional variance, SV models use $\ln(\sigma_t^2)$ instead of σ_t^2. An SV model is defined as

$$a_t = \sigma_t \epsilon_t, \qquad (1 - \alpha_1 B - \cdots - \alpha_m B^m) \ln(\sigma_t^2) = \alpha_0 + v_t, \qquad (4.42)$$

where the ϵ_t are iid $N(0,1)$, the v_t are iid $N(0,\sigma_v^2)$, $\{\epsilon_t\}$ and $\{v_t\}$ are independent, α_0 is a constant, and all zeros of the polynomial $1 - \sum_{i=1}^{m} \alpha_i B^i$ are greater than 1 in modulus. Adding the innovation v_t substantially increases the flexibility of the model in describing the evolution of σ_t^2, but it also increases the difficulty in parameter estimation. To estimate an SV model, we need a quasi-likelihood method via Kalman filtering or a Monte Carlo method. Jacquier et al. (1994) provide some comparison of estimation results between quasi-likelihood and Markov chain Monte Carlo (MCMC) methods. The difficulty in estimating an SV model is understandable because for each shock a_t the model uses two innovations ϵ_t and v_t. Readers are referred to Tsay (2010, Chapter 12) for using MCMC methods to estimate SV models. For more discussions on SV models, see Taylor (1994).

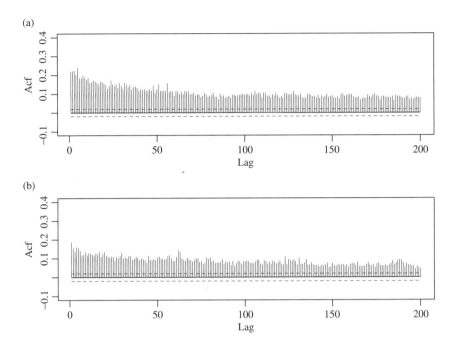

Figure 4.19. The sample ACF of daily absolute log returns for (a) the S&P 500 index and (b) IBM stock for the period from July 3, 1962, to December 31, 2003. The two horizontal lines denote the asymptotic 5% limits.

The appendices of Jacquier et al. (1994) provide some properties of the SV model when $m = 1$. For instance, with $m = 1$, we have

$$\ln(\sigma_t^2) \sim N\left(\frac{\alpha_0}{1 - \alpha_1}, \frac{\sigma_v^2}{1 - \alpha_1^2}\right) \equiv N(\mu_h, \sigma_h^2),$$

and $E(a_t^2) = \exp(\mu_h + \sigma_h^2/2)$, $E(a_t^4) = 3\exp(2\mu_h^2 + 2\sigma_h^2)$, and $\text{corr}(a_t^2, a_{t-i}^2) = [\exp(\sigma_h^2\alpha_1^i) - 1]/[3\exp(\sigma_h^2) - 1]$. Limited experience shows that SV models often provided improvements in model fitting, but their contributions to out-of-sample volatility forecasts received mixed results.

4.14 LONG-MEMORY STOCHASTIC VOLATILITY MODELS

More recently, the SV model is further extended to allow for long memory in volatility, using the idea of fractional difference. As stated in Chapter 2, a time series is a long-memory process if its autocorrelation function decays at a hyperbolic, instead of an exponential, rate as the lag increases. The extension to long-memory models

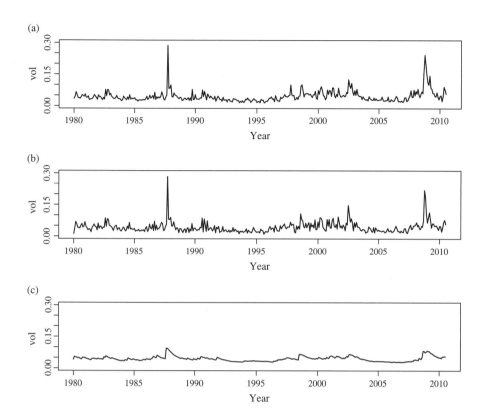

Figure 4.20. Time plots of estimated monthly volatility for the log returns of the S&P 500 index from January 1980 to August 2010: (a) daily log returns form a white noise series, (b) daily log returns follow an MA(1) model, and (c) based on monthly returns from February 1967 to August 2010 and a Gaussian GARCH(1, 1) model.

in volatility study is motivated by the fact that the autocorrelation function of the squared or absolute-valued series of an asset return often decays slowly, even though the return series has no serial correlation; see Ding et al. (1993). Figure 4.19 shows the sample ACF of the daily absolute returns for IBM stock and the S&P 500 index from July 3, 1962, to December 31, 2003. These sample ACFs are positive with moderate magnitude, but decay slowly.

A simple long-memory stochastic volatility (LMSV) model can be written as

$$a_t = \sigma_t \epsilon_t, \qquad \sigma_t = \sigma \exp(u_t/2), \qquad (1-B)^d u_t = \eta_t, \qquad (4.43)$$

where $\sigma > 0$, the ϵ_t are iid $N(0,1)$, the η_t are iid $N(0,\sigma_\eta^2)$ and independent of ϵ_t, and $0 < d < 0.5$. The feature of long memory stems from the fractional difference $(1-B)^d$, which implies that the ACF of u_t decays slowly at a hyperbolic, instead of an exponential, rate as the lag increases. For model lmsv, we have

$$\ln(a_t^2) = \ln(\sigma^2) + u_t + \ln(\epsilon_t^2)$$
$$= [\ln(\sigma^2) + E(\ln \epsilon_t^2)] + u_t + [\ln(\epsilon_t^2) - E(\ln \epsilon_t^2)]$$
$$\equiv \mu + u_t + e_t.$$

Thus, the $\ln(a_t^2)$ series is a Gaussian long-memory signal plus a non-Gaussian white noise; see Breidt et al. (1998). Estimation of the LMSV model is complicated, but the fractional difference parameter d can be estimated by using either a quasi-maximum likelihood method or a regression method. Using the log series of squared daily returns for companies in the S&P 500 index, Bollerslev and Jubinski (1999) and Ray and Tsay (2000) found that the median estimate of d is about 0.38. For applications, Ray and Tsay (2000) studied common long-memory components in daily stock volatilities of groups of companies classified by various characteristics. They found that companies in the same industrial or business sector tend to have more common long-memory components (e.g., big US national banks and financial institutions).

4.15 ALTERNATIVE APPROACHES

In this section, we discuss two alternative methods to volatility modeling.

4.15.1 Use of High Frequency Data

French et al. (1987) consider an alternative approach for volatility estimation that uses high frequency data to calculate volatility of low frequency returns. In recent years, this approach has attracted substantial interest owing to the availability of high frequency financial data; see Andersen et al. (2001a,b).

Suppose that we are interested in the monthly volatility of an asset for which daily returns are available. Let r_t^m be the monthly log return of the asset at month t. Assume that there are n trading days in month t and the daily log returns of the asset in the month are $\{r_{t,i}\}_{i=1}^n$. Using properties of log returns, we have

$$r_t^m = \sum_{i=1}^n r_{t,i}.$$

Assuming that the conditional variance and covariance exist, we have

$$\mathrm{Var}(r_t^m|F_{t-1}) = \sum_{i=1}^n \mathrm{Var}(r_{t,i}|F_{t-1}) + 2\sum_{i<j} \mathrm{Cov}[(r_{t,i}, r_{t,j})|F_{t-1}], \qquad (4.44)$$

where F_{t-1} denotes the information available at month $t-1$ (inclusive). The prior equation can be simplified if additional assumptions are made. For example, if we assume that $\{r_{t,i}\}$ is a white noise series, then

$$\text{Var}(r_t^m | F_{t-1}) = n\,\text{Var}(r_{t,1}),$$

where $\text{Var}(r_{t,1})$ can be estimated from the daily returns $\{r_{t,i}\}_{i=1}^n$ by

$$\hat{\sigma}^2 = \frac{\sum_{i=1}^n (r_{t,i} - \bar{r}_t)^2}{n-1},$$

where $\bar{r}_t$ is the sample mean of the daily log returns in month t (i.e., $\bar{r}_t = (\sum_{i=1}^n r_{t,i})/n$). The estimated monthly volatility is then

$$\hat{\sigma}_m^2 = \frac{n}{n-1} \sum_{i=1}^n (r_{t,i} - \bar{r}_t)^2. \tag{4.45}$$

If $\{r_{t,i}\}$ follows an MA(1) model, then

$$\text{Var}(r_t^m | F_{t-1}) = n\,\text{Var}(r_{t,1}) + 2(n-1)\text{Cov}(r_{t,1}, r_{t,2}),$$

which can be estimated by

$$\hat{\sigma}_m^2 = \frac{n}{n-1} \sum_{i=1}^n (r_{t,i} - \bar{r}_t)^2 + 2 \sum_{i=1}^{n-1} (r_{t,i} - \bar{r}_t)(r_{t,i+1} - \bar{r}_t). \tag{4.46}$$

The previous approach for volatility estimation is simple, but it encounters several difficulties in practice. First, the model for daily returns $\{r_{t,i}\}$ is unknown. This complicates the estimation of covariances in Equation (4.44). Second, there are roughly 21 trading days in a month, resulting in a small sample size. The accuracy of the estimates of variance and covariance in Equation (4.44) might be questionable. The accuracy depends on the dynamic structure of $\{r_{t,i}\}$ and their distribution. If the daily log returns have high excess kurtosis and serial correlations, then the sample estimates $\hat{\sigma}_m^2$ in Equations (4.45) and (4.46) may not even be consistent; see Bai et al. (2004). Further research is needed to make this approach valuable.

Example 4.4. Consider the volatility of monthly log returns of the S&P 500 index from January 1980 to August 2010. The daily data are downloaded from Yahoo Finance. We calculate the volatility by three methods. In the first method, we use daily log returns and Equation (4.45) (i.e., assuming that the daily log returns form a white noise series). The second method also uses daily returns but assumes an MA(1) model (i.e., using Eq. (4.46)). The third method applies a GARCH(1, 1) model with Gaussian innovations to the monthly log returns from February 1967 to August 2010. Here, the monthly returns are obtained using the first daily closing index of each month. We use a longer data span for monthly series to obtain a more accurate estimate of the volatility. The GARCH(1, 1) model used is

$$r_t^m = 0.00535 + a_t, \quad a_t = \sigma_t \epsilon_t, \quad \epsilon_t \sim N(0, 1),$$
$$\sigma_t^2 = 9.326 \times 10^{-5} + 0.1142a_{t-1}^2 + 0.8486\sigma_{t-1}^2.$$

Model checking statistics show that the model is adequate. For instance, the Ljung–Box statistics give $Q(20) = 17.56(0.61)$ and $Q(20) = 8.20(0.99)$, respectively, for the standardized residuals and their squared series. Figure 4.20 shows the time plots of the estimated monthly volatility of the three methods. They are in the same scale. Clearly, the estimated volatilities based on daily returns are much higher than those based on monthly returns and a GARCH(1, 1) model. All three methods show that high volatility occurred in October 1987 and in the recent financial crisis.

R Demonstration

```
> da=read.table("d-sp58010.txt", header=T)
> x=da[, c(1:3, 9)]
> dim(x)
[1] 7737     4

> source("vold2m.R") %% Compile the script
> m1=vold2m(x)
> names(m1)
[1] "volatility" "ndays"
> v1=m1$volatility
> cnt=m1$ndays
> cnt[1:5]
[1] 20 20 21 21 21

> m2=vold2m(x, ma=1) % Use MA(1) dependence
> names(m2)
[1] "volatility" "ndays"
> v2=m2$volatility

> da1=read.table("m-sp56710.txt", header=T)
> sp=log(da1[, 9])
> sp5=diff(sp)
> library(fGarch)
> m3=garchFit(~1+garch(1, 1), data=sp5, trace=F)
> summary(m3)
Title: GARCH Modelling
Call:
 garchFit(formula=~1+garch(1, 1), data=sp5, trace=F)

Mean and Variance Equation: data ~ 1+garch(1, 1)[data=sp5]
Conditional Distribution:   norm
```

```
Coefficient(s):
          Estimate  Std. Error   t value Pr(>|t|)
mu       5.347e-03   1.742e-03    3.069 0.002149 **
omega    9.326e-05   4.859e-05    1.919 0.054942 .
alpha1   1.142e-01   3.003e-02    3.804 0.000142 ***
beta1    8.486e-01   3.186e-02   26.634 < 2e-16 ***
---
> v3=volatility(m3)
> v3=v3[158:524]
> v1=ts(v1, frequency=12, start=c(1980, 1))
> v2=ts(v2, frequency=12, start=c(1980, 1))
> v3=ts(v3, frequency=12, start=c(1980, 1))
> max(v1, v2, v3)
[1] 0.2870294
> par(mfcol=c(3, 1))
> plot(v1, xlab='year', ylab='vol', type='l', ylim=c(0, .3))
> title(main='(a) No correlations')
> plot(v2, xlab='year', ylab='vol', type='l', ylim=c(0, .3))
> title(main='(b) Lag-1 correlation')
> plot(v3, xlab='year', ylab='vol', type='l', ylim=c(0, .3))
> title(main='(c) GARCH(1, 1)')
```

4.15.2 Use of Daily Open, High, Low, and Close Prices

For many assets, daily opening, high, low, and closing prices are available. Parkinson (1980), Garman and Klass (1980), Rogers and Satchell (1991), and Yang and Zhang (2000) showed that one can use such information to improve volatility estimation. Figure 4.21 shows a time plot of price versus time for the tth trading day, assuming that time is continuous. For an asset, define the following variables:

- C_t = the closing price of the tth trading day.
- O_t = the opening price of the tth trading day.
- f = fraction of the day (in interval [0,1]) that trading is closed.
- H_t = the highest price of the tth trading period.
- L_t = the lowest price of the tth trading period.
- F_{t-1} = public information available at time $t-1$.

The conventional variance (or volatility) is $\sigma_t^2 = E[(C_t - C_{t-1})^2|F_{t-1}]$. Garman and Klass (1980) considered several estimates of σ_t^2 assuming that the price follows a simple diffusion model without drift; see Tsay (2010, Chapter 6) for more information about stochastic diffusion models. The following are the estimators considered:

- $\hat{\sigma}_{0,t}^2 = (C_t - C_{t-1})^2.$

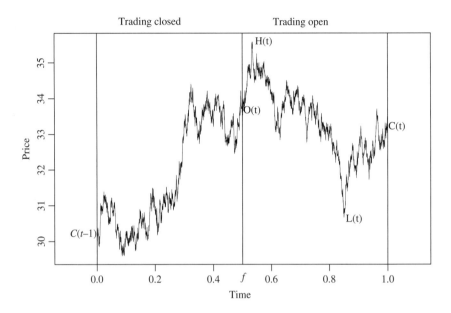

Figure 4.21. Time plot of price over time: scale for price is arbitrary.

- $$\hat\sigma_{1,t}^2 = \frac{(O_t - C_{t-1})^2}{2f} + \frac{(C_t - O_t)^2}{2(1-f)}, \quad 0 < f < 1.$$

- $$\hat\sigma_{2,t}^2 = \frac{(H_t - L_t)^2}{4\ln(2)} \approx 0.3607(H_t - L_t)^2.$$

- $$\hat\sigma_{3,t}^2 = 0.17\frac{(O_t - C_{t-1})^2}{f} + 0.83\frac{(H_t - L_t)^2}{(1-f)4\ln(2)}, \quad 0 < f < 1.$$

- $\hat\sigma_{5,t}^2 = 0.5(H_t - L_t)^2 - [2\ln(2) - 1](C_t - O_t)^2$, which is $\approx 0.5(H_t - L_t)^2 -$
  $0.386(C_t - O_t)^2.$

- $$\hat\sigma_{6,t}^2 = 0.12\frac{(O_t - C_{t-1})^2}{f} + 0.88\frac{\hat\sigma_{5,t}^2}{1-f}, \quad 0 < f < 1.$$

A more precise, but complicated, estimator $\hat\sigma_{4,t}^2$ was also considered. However, it is close to $\hat\sigma_{5,t}^2$. Defining the efficiency factor of a volatility estimator as

$$\text{Eff}(\hat\sigma_{i,t}^2) = \frac{\text{Var}(\hat\sigma_{0,t}^2)}{\text{Var}(\hat\sigma_{i,t}^2)},$$

Garman and Klass (1980) found that $\text{Eff}(\hat\sigma_{i,t}^2)$ is approximately 2, 5.2, 6.2, 7.4, and 8.4 for $i = 1, 2, 3, 5$, and 6, respectively, for the simple diffusion model entertained. Note that $\hat\sigma_{2,t}^2$ was derived by Parkinson (1980) with $f = 0$.

Turn to log returns. Define the following:

- $o_t = \ln(O_t) - \ln(C_{t-1})$, the normalized open.
- $u_t = \ln(H_t) - \ln(O_t)$, the normalized high.
- $d_t = \ln(L_t) - \ln(O_t)$, the normalized low.
- $c_t = \ln(C_t) - \ln(O_t)$, the normalized close.

Suppose that there are n days of data available and the volatility is constant over the period. Yang and Zhang (2000) recommend the estimate

$$\hat{\sigma}_{yz}^2 = \hat{\sigma}_o^2 + k\hat{\sigma}_c^2 + (1-k)\hat{\sigma}_{rs}^2 \qquad (4.47)$$

as a robust estimator of the volatility, where

$$\hat{\sigma}_o^2 = \frac{1}{n-1}\sum_{t=1}^{n}(o_t - \bar{o})^2 \quad \text{with} \quad \bar{o} = \frac{1}{n}\sum_{t=1}^{n} o_t,$$

$$\hat{\sigma}_c^2 = \frac{1}{n-1}\sum_{t=1}^{n}(c_t - \bar{c})^2 \quad \text{with} \quad \bar{c} = \frac{1}{n}\sum_{t=1}^{n} c_t,$$

$$\hat{\sigma}_{rs}^2 = \frac{1}{n}\sum_{t=1}^{n}[u_t(u_t - c_t) + d_t(d_t - c_t)],$$

$$k = \frac{0.34}{1.34 + (n+1)/(n-1)}.$$

The estimate $\hat{\sigma}_{rs}^2$ was proposed by Rogers and Satchell (1991), and the quantity k is chosen to minimize the variance of the estimator of $\hat{\sigma}_{yz}^2$, which is a linear combination of three estimates.

The quantity $H_t - L_t$ is called the *range* of the price in the tth day. This estimator has led to the use of range-based volatility estimates; see, for instance, Alizadeh et al. (2002). In practice, stock prices are only observed at discrete time points. As such, the observed daily high is likely lower than H_t and the observed daily low is likely higher than L_t. Consequently, the observed daily price range tends to underestimate the actual range and, hence, may lead to underestimation of volatility. This bias in volatility estimation depends on the trading frequency and tick size of the stocks. For intensively traded stocks, the bias should be negligible. For other stocks, further study is needed to better understand the performance of range-based volatility estimation.

Example 4.5. Again, consider the daily values of the S&P 500 index from January 3, 1980, to August 31, 2010, for 7737 trading days. Unlike Example 4.4, we focus on the daily volatility of the index in this example. Figure 4.22 provides a daily range plot of the index for June, July, and August 2010. In the plot, each vertical bar denotes a trading range of the index, a small left horizontal bar marks the open value of the

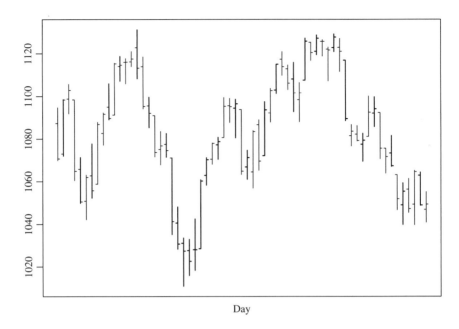

Figure 4.22. Daily range plot for S&P 500 index from June 1 to August 31, 2010. A vertical bar denotes the daily range whereas the left and right horizontal bars denote the open and close values of the index.

index, and a small right horizontal bar shows the closing value of the index. Thus, this range plot provides daily open, high, low, and close values of the index.

We consider three ways to estimate the daily volatility. The first approach applies the Yang–Zhang method in Equation (4.47) with window size $n = 63$, which is approximately the number of trading days in 3 months. The second approach also uses the Yang–Zhang method, but with window size 32. The results of these two approaches enable us to see the sensitivity of the volatility estimate with respect to the choice of window size n. The third approach is to fit an ARMA–GARCH model to the daily log returns. In this particular instance, the fitted model is

$$r_t = 0.00055 + 0.0145r_{t-1} + 0.0111r_{t-2} + 0.0221r_{t-3} + 0.034r_{t-4} + a_t,$$

$$a_t = \sigma_t \epsilon_t, \quad \epsilon_t \sim N(0, 1),$$

$$\sigma_t^2 = 1.248 \times 10^{-6} + 0.0756a_{t-1}^2 + 0.9158\sigma_{t-1}^2,$$

where only the lag-4 AR coefficient is significant at the 5% level for the mean equation, but all estimates in the volatility equation are highly significant. Model checking indicates that, except for normality, the fitted model is adequate. For instance, we have $Q(10) = 8.76(0.55)$ and $Q(20) = 23.15(0.28)$ for the standardized residual $\tilde{a}_t = a_t/\sigma_t$ and $Q(10) = 3.65(0.96)$ and $Q(20) = 8.52(0.99)$ for $\tilde{a}_t^2$, where the number

(a)

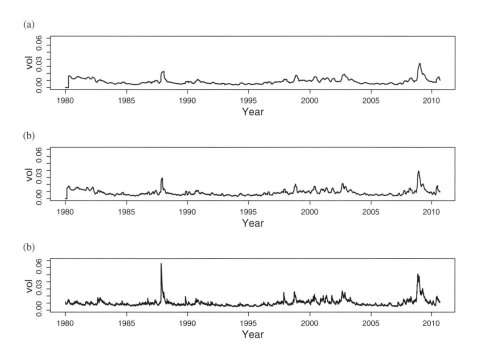

(b)

(b)

Figure 4.23. Daily volatility for the S&P 500 index from January 3, 1980, to August 31, 2010: (a) and (b) are based on the method of Yang and Zhang (2000) with window size 63 and 32, respectively, whereas part (c) is from a Gaussian ARMA(4,0)–GARCH(1,1) model.

in parentheses denotes p value. Figure 4.23 shows the time plots of the volatilities obtained by the three different approaches. The plots are in the same scale so that they can be used in direct comparison. From the plots, the three volatility series exhibit similar characteristics. It is clear that the GARCH volatility, estimated from daily log returns only, is more variable than those obtained by the Yang and Zhang method. It is also clear that a smaller window results in a more variable volatility series when the method of Yang and Zhang (2000) is used.   □

EXERCISES

To answer questions in this exercise, (a) use 5% significance level in tests, and (b) use 10 lags of serial correlations for return series.

1. Consider the daily returns of the exchange trade fund (ETF) SPDR S&P 500 of State Street Global Advisors from September 4, 2001, to September 30, 2011. The tick symbol is SPY and there are 2535 observations. The simple returns are available from CRSP and in the file d-spy-0111.txt. Transform the simple returns to log returns.

(a) Is the expected log return zero? Are there any serial correlations in the log returns? Is there ARCH effect in the log returns?

(b) Fit a Gaussian ARMA–GARCH model for the log return series. Perform model checking, obtain the QQ plot of the standardized residuals, and write down the fitted model. [Hint: Try GARCH(2,1).]

(c) Build an ARMA–GARCH model with Student t innovations for the log return series. Perform model checking and write down the fitted model.

2. Consider, again, the SPY log return series of Problem 1.

   (a) Fit an ARMA–APACRH model with Gaussian innovations to the data. Perform model checking and write down the model. You may ignore the lag-1 ARCH parameter as it is not statistically significant.]

   (b) Fit an ARMA–APARCH model with Student t innovations to the data. Write down the fitted model and perform 1- to 5-step ahead predictions of the series and its volatility.

3. Consider the monthly stock returns of the Coca-Cola Company (KO) from January 1961 to September 2011. The simple returns are available from CRSP and in the file m-ko-6111.txt. Transform the simple returns to log returns.

   (a) Is the expected monthly log return zero? Is there any serial correlation in the log returns? Is there any ARCH effect in the log returns?

   (b) Build a Gaussian GARCH model for the log returns. Perform model checking and write down the fitted model.

   (c) Build a GARCH model with Student t innovations for the log returns. Perform model checking, obtain the QQ plot of the standardized residuals, and write down the fitted model. Also, obtain 1- to 5-step ahead volatility predictions.

4. Consider again the monthly log returns of KO stock. Multiple the log returns by 100. That is, use percentage log returns.

   (a) Fit a TGARCH model to the series. Perform model checking and write down the fitted model. Is the level effect different from zero?

   (b) Fit an NGARCH model to the series. Perform model checking and write down the fitted model.

5. Consider the daily stock returns of Procter & Gamble from September 1, 2001, to September 30, 2011. The simple returns are available from CRSP and in the file d-pg-0111.txt. Transform the simple returns to log returns.

   (a) Is there any serial correlation in the log returns?

   (b) Fit an ARMA model to the log returns to remove serial correlations. Write down the fitted model.

   (c) Let r_t be the residuals of the ARMA model and $x_t = 100 \times r_t$. Is there ARCH effect in x_t?

   (d) Fit an EGARCH model to x_t. Perform model checking and write down the fitted model.

6. Use the `quantmod` package to obtain the daily prices of Apple stock from January 2, 2007, to November 30, 2011.

   (a) Consider the log prices. Use Yang and Zhang method with window sizes 63 and 32 to obtain the daily volatility of the stock.

   (b) Use an ARMA–GARCH model to obtain the daily volatility of the stock. Compare the three volatility series.

REFERENCES

Alizadeh S, Brandt M, Diebold FX. Range-based estimation of stochastic volatility models. J Finance 2002; 57: 1047–1092.

Andersen TG, Bollerslev T. Answering the skeptics: yes, standard volatility models do provide accurate forecasts. Int Econ Rev 1998; 39: 885–905.

Andersen T, Bollerslev T, Diebold FX, Ebens H. The distribution of realized stock return volatility. J Financ Econ 2001; 61: 43–76.

Andersen TG, Bollerslev T, Diebold FX, Labys P. The distribution of realized exchange rate volatility. J Amer Statist Asso 2001; 96: 42–55.

Bai X, Russell JR, Tiao GC. Effects of non-normality and dependence on the precision of variance estimates using high-frequency financial data. Revised working paper, Graduate School of Business, University of Chicago; 2004.

Bollerslev T. Generalized autoregressive conditional heteroskedasticity. J Econom 1986; 31: 307–327.

Bollerslev T, Chou RY, Kroner KF. ARCH modeling in finance. J Econom 1992; 52: 5–59.

Bollerslev T, Engle RF, Nelson DB. ARCH model. In: Engle RF, McFadden DC, editors. Handbook of Econometrics IV. Amsterdam: Elsevier Science; 1994. p 2959–3038.

Bollerslev T, Jubinski D. Equality trading volume and volatility: latent information arrivals and common long-run dependencies. J Bus Econ Stat 1999; 17: 9–21.

Breidt FJ, Crato N, de Lima P. On the detection and estimation of long memory in stochastic volatility. J Econom 1998; 83: 325–348.

Cao C, Tsay RS. Nonlinear time series analysis of stock volatilities. J Appl Econom 1992; 7: s165–s185.

Ding Z, Granger CWJ, Engle RF. A long memory property of stock returns and a new model. J Empir Finance 1993; 1: 83–106.

Duan J. The GARCH option pricing model. Math Finance 1995; 5: 13–32.

Engle RF. Autoregressive conditional heteroscedasticity with estimates of the variance of United Kingdom inflation. Econometrica 1982; 50: 987–1007.

Engle RE, Ng V. Measuring and testing the impact of news on volatility. J Finance 1993; 48: 1749–1778.

Ferbabdez C, Steel MFJ. On Bayesian modelling of fat tails and skewness. J Amer Statist Asso 1998; 93: 359–371.

French KR, Schwert GW, Stambaugh RF. Expected stock returns and volatility. J Financ Econ 1987; 19: 3–29.

Garman MB, Klass MJ. On the estimation of security price volatilities from historical data. J Bus 1980; 53: 67–78.

Glosten LR, Jagannathan R, Runkle DE. On the relation between the expected value and the volatility of nominal excess return on stocks. J Finance 1993; 48: 1779–1801.

Harvey AC, Ruiz E, Shephard N. Multivariate stochastic variance models. Rev Econ Stud 1994; 61: 247–264.

Jacquier E, Polson NG, Rossi P. Bayesian analysis of stochastic volatility models (with discussion). J Bus Econ Stat 1994; 12: 371–417.

Lambert P, Laurent S. Modelling financial time series using GARCH-type models and a skewed student density. Working paper. Universite de Liege.

McLeod AI, Li WK. Diagnostic checking ARMA time series models using squared-residual autocorrelations. J Time Ser Anal 1983; 4: 269–273.

Melino A, Turnbull SM. Pricing foreign currency options with stochastic volatility. J Econom 1990; 45: 239–265.

Nelson DB. Stationarity and persistence in the GARCH(1,1) model. Econom Theory 1990; 6: 318–334.

Nelson DB. Conditional heteroskedasticity in asset returns: a new approach. Econometrica 1991; 59: 347–370.

Parkinson M. The extreme value method for estimating the variance of the rate of return. J Bus 1980; 53: 61–65.

Ray BK, Tsay RS. Long-range dependence in daily stock volatilities. J Bus Econ Stat 2000; 18: 254–262.

Rogers LCG, Satchell SE. Estimating variance from high, low and closing prices. Ann Appl Probab 1991; 1: 504–512.

Taylor SJ. Modeling stochastic volatility: a review and comparative study. Math Finance 1994; 4: 183–204.

Tong H. On a threshold model. In: Chen CH, editor. Pattern Recognition and Signal Processing. Amsterdam: Sijhoff & Noordhoff; 1978.

Tong H. Non-Linear Time Series: A Dynamical System Approach. Oxford: Oxford University Press; 1990.

Tsay RS. Analysis of Financial Time Series. 3rd ed. Hoboken (NJ): John Wiley & Sons; 2010.

Yang D, Zhang Q. Drift-independent volatility estimation based on high, low, open, and close prices. J Bus 2000; 73: 477–491.

Zakoian JM. Threshold heteroscedastic models. J Econ Dyn Control 1994; 18: 931–955.

5

APPLICATIONS OF VOLATILITY MODELS

We consider some applications of the volatility models in this chapter. For simplicity, we focus on GARCH(1,1) models with Gaussian innovations. The applications, of course, can be extended to other GARCH models with different types of innovations. An obvious advantage of GARCH models over the constant volatility model is that the former allow for time-varying volatility and volatility cluster. Our first application then is to consider GARCH volatility forecasts and their financial applications. The volatility forecasts enable us to construct volatility term structure for an asset returns. We demonstrate that the volatility estimates from a GARCH model can be used in portfolio selection and in obtaining time-varying betas of an asset. We also show that a fitted GARCH model can be used in pricing options. More specifically, in this chapter, we use daily log returns to demonstrate the GARCH volatility term structure and the GARCH applications in option pricing and hedging.

Furthermore, we show that GARCH models can be used to improve the modeling and prediction of pure ARMA models. Using backtesting, we showed that incorporating a simple GARCH(1,1) model can produce more accurate forecasts of the change in weekly US crude oil price. Finally, applications of GARCH models in risk management such as calculating value at risk and expected shortfall are discussed in a later chapter.

An Introduction to Analysis of Financial Data with R, First Edition. Ruey S. Tsay.
© 2013 John Wiley & Sons, Inc. Published 2013 by John Wiley & Sons, Inc.

In the following text, we use the true parameters in our discussion. It is understood that these parameters are replaced by their maximum likelihood estimates in applications. In this way, we do not consider the parameter uncertainty in this chapter. In real applications, volatility is usually annualized. We use a simple procedure to obtain annualized volatility. Let σ_t be the volatility of daily log returns, then the annual volatility is $\sqrt{252}\sigma_t$. In general, if σ_t is the volatility obtained from an asset return series that has h observations per year then the annualized volatility is $\sqrt{h}\sigma_t$. An advantage of using annual volatility is that it simplifies the comparison of different volatility forecasts.

5.1   GARCH VOLATILITY TERM STRUCTURE

Again, we express a log return series as $r_t = \mu_t + a_t = \mu_t + \sigma_t\epsilon_t$, where ϵ_t forms a sequence of independent standard normal random variates, that is, $\epsilon_t \sim N(0,1)$, and $\mu_t = E(r_t|F_{t-1})$. The equation for a GARCH(1,1) model is

$$\sigma_t^2 = \alpha_0 + \alpha_1 a_{t-1}^2 + \beta_1 \sigma_{t-1}^2, \tag{5.1}$$

where $\alpha_1 + \beta_1 < 1$, $\alpha_0 > 0$, $0 \le \alpha_1, \beta_1 < 1$. As mentioned in Chapter 4, the unconditional (or long-term) variance of r_t is $\sigma^2 = \frac{\alpha_0}{1-\alpha_1-\beta_1}$. Using this result, Equation (5.1) can be rewritten as,

$$(\sigma_t^2 - \sigma^2) = \alpha_1(a_{t-1}^2 - \sigma^2) + \beta_1(\sigma_{t-1}^2 - \sigma^2). \tag{5.2}$$

The GARCH(1,1) model thus quantifies the reactions to deviations of squared unexpected returns from the long-term variance. Also, as shown in Chapter 4, the one-step ahead volatility prediction of a GARCH(1,1) model at the forecast origin t is

$$\sigma_t^2(1) = \alpha_0 + \alpha_1 a_t^2 + \beta_1 \sigma_t^2,$$

whereas the ℓ-step ahead volatility forecast is

$$\sigma_t^2(\ell) = \alpha_0 + (\alpha_1 + \beta_1)\sigma_t^2(\ell - 1), \quad \ell = 2,\dots.$$

Again, using $\sigma^2 = \alpha_0/(1 - \alpha_1 - \beta_1)$, we can rewrite the above equation as

$$[\sigma_t^2(\ell) - \sigma^2] = (\alpha_1 + \beta_1)[\sigma_t^2(\ell - 1) - \sigma^2].$$

By repeated substitution, we see that

$$[\sigma_t^2(\ell) - \sigma^2] = (\alpha_1 + \beta_1)^{\ell-1}[\sigma_t^2(1) - \sigma^2].$$

Thus, under the condition $\alpha_1 + \beta_1 < 1$, we have $\sigma_t^2(\ell) \to \sigma^2$ as $\ell \to \infty$. In other words, the volatility series of a GARCH(1,1) model with $\alpha_1 + \beta_1 < 1$ is mean reverting and the speed of mean reverting to the long-term variance can also be measured by the half-life $\ell = \log(0.5)/\log(\alpha_1 + \beta_1)$.

To demonstrate, we consider daily log returns of three US stocks from January 2, 2001 to December 31, 2010. The three stocks are Caterpillar (CAT), Cisco Systems (CSCO), and General Electric (GE). Table 5.1 summarizes the estimation results of fitting a GARCH(1,1) model with Gaussian innovations to three log return series. Figure 5.1 shows the time plot of CAT returns and the fitted volatility series. Model checking statistics show that, except for the normality assumption, the GARCH(1,1) model fits the data well. Similarly in Chapter 4, the fitted GARCH models are highly persistent, especially for the GE stock. The high persistence gives rise to large values

TABLE 5.1. Estimation results of Gaussian GARCH(1,1) models for daily log returns of three US stocks

Asset	$\hat{\mu} \times 10^3$	$\hat{\alpha}_0 \times 10^6$	$\hat{\alpha}_1$	$\hat{\beta}_1$	$\hat{\alpha}_1 + \hat{\beta}_1$	Half-life
			Parameters			
CAT	1.037	9.567	0.0531	0.9245	0.9776	30.60
CSCO	0.322	15.58	0.0823	0.8947	0.9770	29.79
GE	0.318	0.736	0.0507	0.9484	0.9991	769.82

The sample period is from January 2, 2001 to December 31, 2010 with 2515 observations.

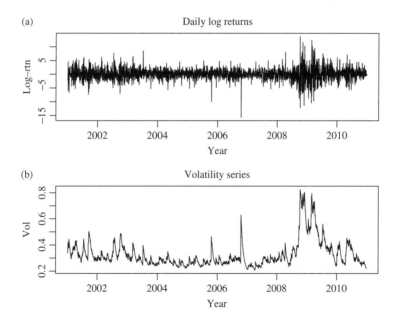

Figure 5.1. Time plots of (a) the daily log returns and (b) GARCH(1,1) volatility for Caterpillar stock from January 2, 2001 to December 31, 2010.

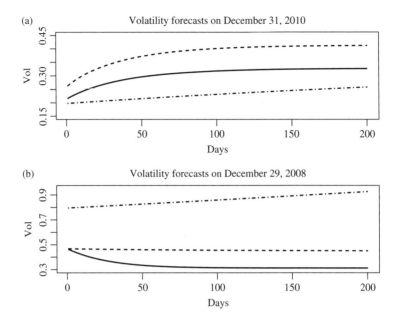

Figure 5.2. Out-of-sample volatility forecasts for daily log returns of CAT, CSCO, and GE stocks. The upper and lower plots are for forecast origin December 31, 2010 and December 29, 2008, respectively. The solid and dashed lines are for CAT and CSCO stock, respectively. The volatility is annualized.

of half-life. For both CAT and CSCO stocks, the half-life is approximately 30 trading days. The model for GE returns is essentially an IGARCH(1,1) model so that the half-life is high. It seems that, in this particular case, the financial crisis of 2007–2008 contributes significantly to the high persistence in stock volatility.

Figure 5.2 shows the time plots of volatility forecasts for the three stocks of Table 5.1 at two different forecast origins, namely, December 31, 2010 and December 29, 2008. These forecasts are out-of-sample point predictions and are annualized. The solid, dashed, and dot-dashed lines are for CAT, CSCO, and GE stock, respectively. For both forecast origins, the volatility forecasts of GE stock are not mean reverting. On the other hand, the volatility forecasts for CAT and CSCO stocks gradually approach their long-term standard deviations.

5.1.1 Term Structure

Turn to volatility term structure. The log return of an asset h periods from time t is

$$r_{t,h} = \sum_{i=1}^{h} r_{t+i}.$$

From which, we have $E(r_{t,h}|F_t) = \sum_{i=1}^{h} E(r_{t+i}|F_t)$ and

$$\text{Var}(r_{t,h}|F_t) = \sum_{i=1}^{h} \text{Var}(r_{t+i}|F_t) + 2 \sum_{i=1}^{h-1} \sum_{j=i+1}^{h} \text{Cov}[(r_{t+i}, r_{t+j})|F_t],$$

where, again, F_t denotes the information available at time t. Here the first term is the summation of conditional variances, and the second term consists of conditional autocovariances of r_t. For daily log returns, empirical evidence suggests that the autocovariances are typically close to zero. Indeed, the autocovariances are zero under the efficient market hypothesis. Therefore, as a reasonable approximation, we employ

$$\text{Var}(r_{t,h}|F_t) = \sum_{i=1}^{h} \text{Var}(r_{t+i}|F_t).$$

For GARCH models, the prior approximation shows that

$$\sigma_{t,h}^2 = \sum_{\ell=1}^{h} \sigma_t^2(\ell), \tag{5.3}$$

where $\sigma_{t,h}^2$ denotes the conditional variance of the h-period log return $r_{t,h}$ at the forecast origin t. Consequently, we can easily compute the h-period log return volatility from the volatility forecasts of a GARCH model.

For ease in comparing volatility forecasts for different frequencies of log returns, we annualize the volatility. For daily returns, the annualized h-period volatility is

$$\sigma_{t,h,a} = \sqrt{\frac{252}{h}} \sigma_{t,h},$$

where the subscript "a" is used to signify that the volatility is annualized. Volatility term structure is to study the behavior of $\sigma_{t,h,a}$ over a range of h values.

To illustrate this, we consider the daily log returns of CSCO stock. We use out-of-sample forecasts to compute $\sigma_t^2(\ell)$ for $h = 1, \ldots, 40$ and t from December 29, 2008 to December 31, 2010. For a given forecast origin t_0, we estimate the GARCH(1,1) model in Table 5.1 and use the fitted model to compute 1-step to 40-step ahead volatility forecasts. We then obtain the h-period volatility via Equation (5.3). Finally, the resulting volatility is annualized. Figure 5.3 shows the volatility term structure for CSCO stock from December 29, 2001 to December 31, 2010. The plot consists of annualized volatility for $h = 1, 5, 10, 15, 20, 25, 30, 35$, and 40. That is, the range of h is from 1 trading day to 2 months. The solid line of the plot is for $h = 1$. From the plot, it is seen that the volatility surface appears to be flat when volatility jumps. On the other hand, the volatility increases with h when the volatility is declining. Figure 5.4 shows the volatility term structure for the CAT stock in the sample period and obtained by the same procedure. Again, volatility surface shows some changes when the volatility is declining.

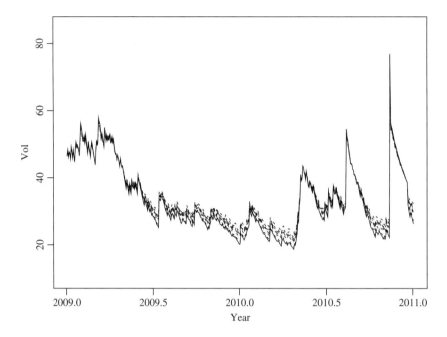

Figure 5.3. Volatility term structure for daily log returns of Cisco System stock from December 29, 2008 to December 31, 2010. The solid line is for 1-day volatility. The dashed lines are for $h = 5, 10, \ldots, 35$, and 40. The volatility is annualized, in percentages, and obtained from a GARCH(1,1) model.

5.2 OPTION PRICING AND HEDGING

An important application of volatility is to price financial derivatives. See, for instance, the well-known Black–Scholes option pricing formula, which assumes constant volatility. Hull and White (1987) studied the pricing of options when the volatility is stochastic. With the introduction of GARCH models to model systematically asset volatility, one would expect the use of GARCH volatility in option pricing and hedging (Engle and Rosenberg, 1995). Indeed, several authors have studied the use of GARCH models in option pricing. Duan (1995) derived the GARCH option pricing model. Amin and Ng (1993) and Hafner and Härdle (2000) found that the GARCH prices of short-maturity out-of-the-money equity options are closer to the observed market price than the Black–Scholes prices.

The basic model of the standard option pricing methods for an option on a single stock is that the price P_t of the stock follows the geometric Brownian motion

$$\frac{\mathrm{d}P_t}{P_t} = r\mathrm{d}t + \sigma\mathrm{d}W_t,$$

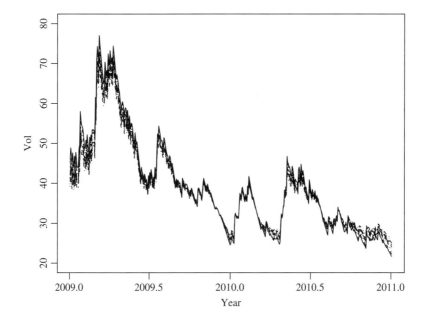

Figure 5.4. Volatility term structure for daily log returns of Caterpillar stock from December 29, 2008 to December 31, 2010. The solid line is for 1-day volatility. The dashed lines are for $h = 5, 10, \ldots, 35$, and 40. The volatility is annualized, in percentages, and obtained from a GARCH(1,1) model.

where r is the risk free interest rate, σ is the volatility, and $W(t)$ is the standard Wiener process. This is a model under the risk neural world. (Hull, 2011; Tsay, 2010, Chapter 6). Using Ito's lemma, the log price of the stock then follows the model

$$d \ln(P_t) = \left(r - \frac{\sigma^2}{2} \right) dt + \sigma \, dW_t.$$

In practice, assume that both r and σ are known and the only uncertainty involved in the model is the random noise $dW(t)$. Thus, simulation can be used to evaluate the stock price. The discrete time version of the model becomes

$$P_t = P_{t-1} \exp \left(r - 0.5\sigma^2 + \sigma \epsilon_t \right), \tag{5.4}$$

where ϵ_t is a sequence of independent standard normal random variates, that is, $\epsilon_t \sim N(0, 1)$.

Denote the current time as zero and the current stock price as P_0. Assume that the strike price of the option is K and the time to expiration is T. We can simulate the terminal stock price P_T by generating a sequence of independent standard normal random variates $\{\epsilon_1, \ldots, \epsilon_T\}$ and using Equation (5.4). In practice, P_T is a random quantity, but we can repeat the above generating procedure N times and treat the average of those P_T values as the expected value of P_T. That is, $E(P_T|F_t) = \sum_{i=1}^{N} P_T^{(i)}/N$, where the superscript (i) is used to denote the terminal price of the ith simulation.

Using the simulation results, the price of a European call option is then

$$C(P_0) = e^{-rT} E[\max(P_T - K, 0)] = e^{-rT} \sum_{i=1}^{T} \frac{\max\left(P_T^{(i)} - K, 0\right)}{N}.$$

The price of an Asian call option, which depends on the arithmetic average of the price path, is

$$C(P_0) = e^{-rT} \sum_{i=1}^{N} \max \frac{\left(\sum_{t=1}^{T} P_t^{(i)}/T - K, 0\right)}{N}.$$

For GARCH(1,1) models, one can extend the above simulation procedure by making use of the volatility equation. More specifically, we have

$$P_t = P_{t-1} \exp(r - 0.5\sigma_t^2 + \sigma_t \epsilon_t),$$

$$\sigma_t^2 = \alpha_0 + \alpha_1 \sigma_{t-1}^2 \epsilon_{t-1}^2 + \beta_1 \sigma_{t-1}^2,$$

where, again, ϵ_t is a sequence of independent standard normal random variates.

Finally, the corresponding GARCH option deltas and gammas can be calculated using finite difference approximations. For instance,

$$\delta = [C(P_0 + \Delta) - C(P_0 - \Delta)]/(2\Delta),$$

$$\gamma = [C(P_0 + \Delta) - 2C(P_0) + C(P_0 - \Delta)]/\Delta^2.$$

In practice, to improve the accuracy of simulation, one may need a large number of iterations in calculating δ and γ.

Strictly speaking, the above pricing formulas are based on perfect hedging, which may not be possible under the stochastic volatility framework. However, Duan (1995) showed that there is a local risk neural valuation relationship with the NGARCH volatility model

$$r_t = r - 0.5\sigma_t^2 + \lambda\sigma_t + \sigma_t \epsilon_t,$$

$$\sigma_t^2 = \alpha_0 + \alpha_1 \sigma_{t-1}^2(\epsilon_{t-1} - \theta)^2 + \beta_1 \sigma_{t-1}^2.$$

See Section 4.12. With estimated parameters, one can perform simulation using this NGARCH(1,1) model in pricing options.

5.3 TIME-VARYING CORRELATIONS AND BETAS

Another application of GARCH models is to obtain time-varying correlations between asset returns. It is obvious that a multivariate GARCH model can study the time evolution of variances and covariances simultaneously (Tsay, 2010, Chapter 10). However, in this section, we discuss a simple approach that uses directly the univariate GARCH models of the previous chapter to study time-varying correlations. The basic idea supporting the approach is as follows. Consider two asset returns x_t and y_t. Statistical theory states that

$$\text{Var}(x_t + y_t) = \text{Var}(x_t) + 2\text{Cov}(x_t, y_t) + \text{Var}(y_t),$$
$$\text{Var}(x_t - y_t) = \text{Var}(x_t) - 2\text{Cov}(x_t, y_t) + \text{Var}(y_t).$$

Therefore, we have

$$\text{Cov}(x_t, y_t) = \frac{\text{Var}(x_t + y_t) - \text{Var}(x_t - y_t)}{4}. \tag{5.5}$$

This identity continues to hold for conditional covariance. Therefore, time-varying covariances between asset returns x_t and y_t can be obtained by the volatilities of $x_t + y_t$ and $x_t - y_t$.

Let $\sigma_{x+y,t}$, $\sigma_{x-y,t}$, $\sigma_{x,t}$, and $\sigma_{y,t}$ be the volatility of $x_t + y_t$, $x_t - y_t$, x_t, and y_t, respectively. Using the identity in Equation (5.5), the time-varying correlation between the asset returns x_t and y_t can be obtained by

$$\rho_t = \frac{\sigma_{x+y,t}^2 - \sigma_{x-y,t}^2}{4\sigma_{x,t}\sigma_{y,t}}. \tag{5.6}$$

To demonstrate this, we consider the daily log returns of CAT and CSCO stocks from January 2, 2001 to December 31, 2010. The GARCH(1,1) models for the two daily asset returns are given in Table 5.1. Furthermore, applying GARCH(1,1) models with Gaussian innovations, we obtain the model

$$r_t = 0.00143 + a_t, \quad a_t = \sigma_t \epsilon_t, \quad \epsilon_t \sim N(0,1)$$
$$\sigma_t^2 = 2.188 \times 10^{-5} + 0.070 a_{t-1}^2 + 0.916 \sigma_{t-1}^2$$

for the sum of the two returns, and the model

$$r_t = -0.00084 + a_t, \quad a_t = \sigma_t \epsilon_t, \quad \epsilon_t \sim N(0,1)$$
$$\sigma_t^2 = 5.533 \times 10^{-6} + 0.0179 a_{t-1}^2 + 0.9726 \sigma_{t-1}^2$$

for the difference between the two returns. Except for the normality assumptions, model checking statistics fail to reject the adequacy of these two GARCH(1,1) models. Figure 5.5a shows the time plot of the time-varying correlations between CAT and

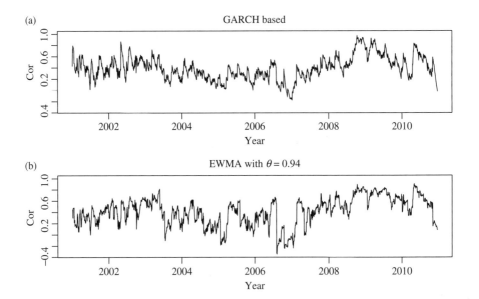

Figure 5.5. Time-varying correlations between daily log returns of Caterpillar and Cisco Systems stocks from January 2, 2001 to December 31, 2010: (a) based on univariate GARCH(1,1) models and (b) based on exponentially weighted moving average with $\theta = 0.94$.

CSCO stock returns. The correlations fluctuated substantially and were high in the later part of 2008.

For comparison, we also consider the time-varying correlations obtained by the exponentially weighted moving-average (EWMA) method. For a prespecified weight θ, where $0 < \theta < 1$, and a sequence of data $\{x_1, \ldots, x_n\}$, the EWMA of the sample is

$$\hat{x}_{n+1} = \frac{x_n + \theta x_{n-1} + \theta^2 x_{n-2} + \cdots + \theta^{n-1} x_1}{1 + \theta + \theta^2 + \cdots + \theta^{n-1}}.$$

This formula gives higher weights to observations closer to x_{n+1} and the weights decay exponentially. This simple and intuitive method has been widely used in the forecasting literature to obtain a point prediction for x_{n+1}. Using $1 + \theta + \theta^2 + \cdots + \theta^{n-1} = (1 - \theta^n)/(1 - \theta)$, we can rewrite the above formula as

$$\hat{x}_{n+1} = \frac{(1 - \theta) \sum_{i=0}^{n-1} \theta^i x_{n-i}}{1 - \theta^n}.$$

For a large n, we see that

$$\hat{x}_{n+1} = (1 - \theta) \sum_{i=0}^{\infty} \theta^i x_{n-i}.$$

In practice, this point prediction can be computed effectively because

$$\hat{x}_{n+1} = (1-\theta) \sum_{i=0}^{\infty} \theta^i x_{n-i}$$

$$= (1-\theta)x_n + (1-\theta) \sum_{i=1}^{\infty} \theta^i x_{n-i}$$

$$= (1-\theta)x_n + \theta(1-\theta) \sum_{i=0}^{\infty} \theta^i x_{n-1-i}$$

$$= (1-\theta)x_n + \theta\hat{x}_n.$$

In other words, with an initial value $\hat{x}_1$, the EWMA prediction $\hat{x}_{n+1}$ can be calculated recursively. The first term of the prior recursive formula denotes the contribution of the most recent observation to $\hat{x}_{n+1}$, whereas the second term shows the persistence in the prediction. A larger θ means higher persistence and less weight for the most recent data. A smaller θ implies higher weight for the recent data and less persistence. In practice, the range of θ is approximately between 0.75 and 0.98. If necessary, θ can be estimated by statistical methods. In fact, for volatility estimation, θ corresponds to the β_1 coefficient of an IGARCH(1,1) model when α_0 is fixed at zero.

For the estimation of time-varying covariances of (x_t, y_t), one can apply the EWMA method as follows:

- Variance estimation: use the sample variance as the initial estimate then apply the recursion. For instance, let $\sigma_{x,1}^2$ be the sample variance of x_t. Compute $\sigma_{x,t+1}^2 = (1-\theta)x_t^2 + \theta\sigma_{x,t}^2$ for $t = 1, 2, \ldots, T$.
- Covariance estimation: use the sample mean of the cross product $x_t y_t$ as the initial estimate then apply the recursion. That is, let $\sigma_{x,y,1}$ be the sample mean of $x_t y_t$. Compute $\sigma_{x,y,t+1} = (1-\theta)x_t y_t + \theta\sigma_{x,y,t}$ for $t = 1, 2, \ldots, T$, where T is the sample size.

The time-varying correlations are $\rho_t = \sigma_{x,y,t}/(\sigma_{x,t}\sigma_{y,t})$. Figure 5.5b shows the time plot of the time-varying correlations between CAT and CSCO stocks obtained by the EWMA method with $\theta = 0.94$. The correlations show a similar pattern as that of the GARCH approach.

In this demonstration, we use the sample variance as the starting value in applying the EWMA method. Other initial values can also be used. For instance, one can use the sample variance of the first 30 observations. As the weights decay exponentially, the effect of the initial value is small when the sample size is large.

R Demonstration. Output edited.

```
> da=read.table("d-c2c-0110.txt",header=T) % Load data
> csco=log(da$CSCO+1)
```

```
> cat=log(da$CAT+1)
> library(fGarch)
> m1=garchFit(~1+garch(1,1),data=csco,trace=F)
> summary(m1)
Title:   GARCH Modelling
Call: garchFit(formula=~1+garch(1,1),data=csco,trace=F)

Mean and Variance Equation:
 data ~ 1 + garch(1, 1)   [data = csco]
Conditional Distribution:  norm

Error Analysis:
        Estimate  Std. Error  t value Pr(>|t|)
mu      3.224e-04  4.067e-04    0.793    0.428
omega   1.558e-05  3.941e-06    3.954 7.69e-05 ***
alpha1  8.230e-02  1.784e-02    4.612 3.98e-06 ***
beta1   8.947e-01  2.142e-02   41.767 < 2e-16  ***
---
Standardized Residuals Tests:
                              Statistic p-Value
 Jarque-Bera Test   R    Chi^2 10943.10  0
 Shapiro-Wilk Test  R    W      0.9417184 0
 Ljung-Box Test     R    Q(10)  9.690227 0.4680781
 Ljung-Box Test     R    Q(20) 22.97434  0.2900562
 Ljung-Box Test     R^2  Q(10)  1.949594 0.996712
 Ljung-Box Test     R^2  Q(20)  4.167677 0.999935

> m2=garchFit(~1+garch(1,1),data=cat,trace=F)
> summary(m2)
Error Analysis:
        Estimate  Std. Error  t value Pr(>|t|)
mu      1.037e-03  3.713e-04    2.793  0.00522 **
omega   9.567e-06  3.385e-06    2.826  0.00471 **
alpha1  5.311e-02  1.134e-02    4.682 2.84e-06 ***
beta1   9.245e-01  1.787e-02   51.739 < 2e-16  ***
---
Standardized Residuals Tests:
                              Statistic p-Value
 Jarque-Bera Test   R    Chi^2 2397.073  0
 Shapiro-Wilk Test  R    W      0.9720346 0
 Ljung-Box Test     R    Q(10) 11.99378  0.2854729
 Ljung-Box Test     R    Q(20) 21.94146  0.3436954
 Ljung-Box Test     R^2  Q(10)  1.323397 0.9993882
 Ljung-Box Test     R^2  Q(20)  4.95693  0.9997407
 LM Arch Test       R    TR^2   2.674523 0.9974382

> vcsco=volatility(m1)
> vcat=volatility(m2)
> xp=csco+cat
```

```
> xm=csco-cat
> m3=garchFit(~1+garch(1,1),data=xp,trace=F)
> summary(m3)
Error Analysis:
         Estimate   Std. Error   t value  Pr(>|t|)
mu      1.430e-03   6.456e-04     2.215   0.02679 *
omega   2.188e-05   7.328e-06     2.986   0.00283 **
alpha1  7.002e-02   1.464e-02     4.782  1.74e-06 ***
beta1   9.158e-01   1.782e-02    51.404  < 2e-16 ***
---
Standardized Residuals Tests:
                                    Statistic p-Value
 Jarque-Bera Test     R     Chi^2   596.8557  0
 Shapiro-Wilk Test    R     W       0.9828594 0
 Ljung-Box Test       R     Q(10)   10.25332  0.4185566
 Ljung-Box Test       R     Q(20)   21.87410  0.347381
 Ljung-Box Test       R^2   Q(10)   6.559973  0.76623
 Ljung-Box Test       R^2   Q(20)   11.86000  0.9208104
 LM Arch Test         R     TR^2    9.634957  0.647951

> vxp=volatility(m3)
> m4=garchFit(~1+garch(1,1),data=xm,trace=F)
> summary(m4)
Error Analysis:
          Estimate   Std. Error   t value  Pr(>|t|)
mu       -8.312e-04   4.629e-04   -1.795    0.0726 .
omega     5.533e-06   1.189e-06    4.652  3.29e-06 ***
alpha1    1.789e-02   2.860e-03    6.257  3.93e-10 ***
beta1     9.726e-01   4.222e-03  230.360  < 2e-16 ***
---
Standardized Residuals Tests:
                                    Statistic p-Value
 Jarque-Bera Test     R     Chi^2   11268.66  0
 Shapiro-Wilk Test    R     W       0.9233508 0
 Ljung-Box Test       R     Q(10)   8.862264  0.5452234
 Ljung-Box Test       R     Q(20)   17.35761  0.6296453
 Ljung-Box Test       R^2   Q(10)   1.871286  0.9972348
 Ljung-Box Test       R^2   Q(20)   5.879934  0.9990502
 LM Arch Test         R     TR^2    3.000402  0.9955412

> vxm=volatility(m4)
> CoV=(vxp^2-vxm^2)/4
> COR=CoV/(vcat*vcsco)
> source("EWMAvol.R")
> M1=EWMAvol(rtn)
> tdx=c(1:2515)/252+2001
> par(mfcol=c(2,1))
> cr2=M1[,3]/sqrt(M1[,1]*M1[,2])
> range(cr2,COR)
```

```
[1] -0.3453691   0.9706971
> plot(tdx,COR,xlab='year',ylab='cor',ylim=c(-0.35,1),type='l')
> title(main='(a) GARCH based')
> plot(tdx,cr2,xlab='year',ylab='cor',ylim=c(-0.35,1),type='l')
> title(main='(b) EWMA with theta = 0.94')
```

5.3.1 Time-Varying Betas

One of the most common applications of a statistical model in finance is the capital asset pricing model (CAPM). The CAPM model can be written as

$$r_t = \alpha + \beta r_{m,t} + e_t, \quad t = 1, \ldots, T, \tag{5.7}$$

where $r_{m,t}$ denotes the return of the market and r_t is the asset return of interest. Here we use the returns, not the excess returns, in the study. (Black, 1972). This simple linear regression provides a method for estimating

- the stock sensitivity to the market risk, the β, factor;
- the mispricing of the stock relative to the market, α;
- the stock-specific return, e_t.

Roughly speaking, the stock-specific returns are considered to be diversifiable and, hence, are not important. In statistical terms, if a portfolio consists of many stocks then, by the law of large number, the average of stock-specific returns will tend to zero. The beta, on the other hand, provides a measure of how the stock responds to changes in the market. If β is not significantly different from zero then the market has no significant impact on the stock. If β is significantly greater than one then the stock responds aggressively to changes in the market. Thus, a stock with $\beta < 1$ is regarded as less risky than the market and stocks with $\beta > 1$ are classified as high risk investment. In practice, a positive α and small β are preferred if possible.

For the CAPM model in Equation (5.7), we have

$$\beta = \frac{\mathrm{Cov}(r_t, r_{m,t})}{\mathrm{Var}(r_{m,t})}.$$

The estimate of β thus depends on the sample and the choice of the market index $r_{m,t}$. As a matter of fact, it is commonly believed that β is likely time-varying. Using the method discussed in the previous section, the GARCH models can be used to model the time-varying β. Specifically, $\mathrm{Cov}(r_t, r_{m,t})$ can be obtained from the volatilities of $r_t + r_{m,t}$ and $r_t - r_{m,t}$ via Equation (5.5).

To demonstrate this, we consider the daily log returns of the CAT stock from January 2, 2001 to December 31, 2010. We use the daily log returns of the S&P 500 index as the market returns. The traditional CAPM model for the data is

$$r_t = 0.00068 + 1.146 r_{m,t} + e_t,$$

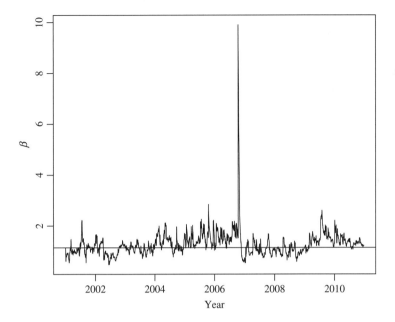

Figure 5.6. Time-varying betas for the daily log returns of Caterpillar stock from January 2, 2001 to December 31, 2010. The daily log returns of the S&P 500 index is used as the market returns. The horizontal line denotes the constant beta at 1.146.

where the standard error of e_t is 0.015 and the adjusted R^2 of the linear regression is 51.69%. Figure 5.6 shows the time-varying β for the daily log returns of CAT stock. The horizontal line denotes the constant β of the data with $\hat{\beta} = 1.146$. In this particular case, there exists a big outlying $\hat{\beta}_t$ at $t = 1460$. An examination of the returns shows that the CAT stock dropped more than 14.52% on October 20, 2006, which corresponds to $t = 1459$. Ignoring the big outlier, we see that the β varies within a certain range and there is no evidence to support a constant β.

R Demonstration. Output edited.

```
> da=read.table("d-sp500-0110.txt")
> sp5=da[,1]
> da=read.table("d-c2c-0110.txt",header=T)
> cat=log(da$CAT+1)
>
> xp=cat+sp5
> xm=cat-sp5
> m1=garchFit(~1+garch(1,1),data=xp,trace=F)
> summary(m1)
Title:   GARCH Modelling
Call:   garchFit(formula=~1+garch(1,1),data=xp,trace=F)
```

```
Mean and Variance Equation:
 data ~ 1 + garch(1, 1)  [data = xp]
Conditional Distribution:  norm
Error Analysis:
          Estimate   Std. Error   t value Pr(>|t|)
mu      1.393e-03   5.065e-04    2.750 0.005956 **
omega   1.402e-05   3.948e-06    3.551 0.000383 ***
alpha1  6.534e-02   9.868e-03    6.621 3.56e-11 ***
beta1   9.191e-01   1.254e-02   73.292 < 2e-16 ***
---
Standardized Residuals Tests:
                                  Statistic p-Value
 Jarque-Bera Test    R    Chi^2   539.8161  0
 Shapiro-Wilk Test   R    W       0.9854247 2.197766e-15
 Ljung-Box Test      R    Q(10)   8.708999  0.5599254
 Ljung-Box Test      R    Q(20)   22.16945  0.3313869
 Ljung-Box Test      R^2  Q(10)   8.114903  0.6176144
 Ljung-Box Test      R^2  Q(20)   10.86366  0.9496819
> m2=garchFit(~1+garch(1,1),data=xm,trace=F)
> summary(m2)
Error Analysis:
          Estimate   Std. Error   t value Pr(>|t|)
mu      8.527e-04   2.851e-04     2.991 0.002784 **
omega   3.562e-06   1.059e-06     3.363 0.000772 ***
alpha1  2.423e-02   5.045e-03     4.804 1.56e-06 ***
beta1   9.600e-01   8.561e-03   112.148 < 2e-16 ***
---
Standardized Residuals Tests:
                                  Statistic p-Value
 Jarque-Bera Test    R    Chi^2   11837.86  0
 Shapiro-Wilk Test   R    W       0.9329678 0
 Ljung-Box Test      R    Q(10)   21.67769  0.01683345
 Ljung-Box Test      R    Q(20)   27.82026  0.1137230
 Ljung-Box Test      R^2  Q(10)   1.764147  0.997849
 Ljung-Box Test      R^2  Q(20)   8.406545  0.988818

> m3=garchFit(~1+garch(1,1),data=sp5,trace=F)
> summary(m3)
Error Analysis:
          Estimate   Std. Error   t value Pr(>|t|)
mu      1.933e-04   1.762e-04     1.097   0.273
omega   1.233e-06   3.061e-07     4.028 5.63e-05 ***
alpha1  7.891e-02   9.468e-03     8.335 < 2e-16 ***
beta1   9.126e-01   9.848e-03    92.669 < 2e-16 ***
---
Standardized Residuals Tests:
                                  Statistic p-Value
 Jarque-Bera Test    R    Chi^2   199.5819  0
 Shapiro-Wilk Test   R    W       0.9890158 5.49156e-13
```

```
Ljung-Box Test      R     Q(10)   15.57277   0.1125331
Ljung-Box Test      R     Q(20)   28.50665   0.09793793
Ljung-Box Test      R^2   Q(10)   22.30558   0.01362134
Ljung-Box Test      R^2   Q(20)   24.89998   0.2052844

> vxp=volatility(m1)
> vxm=volatility(m2)
> vsp5=volatility(m3)
> beta=(vxp^2-vxm^2)/(4*vsp5^2)
> tdx=c(1:2515)/252+2001
> m4=lm(cat~sp5)
> summary(m4)
Call: lm(formula = cat ~ sp5)

Coefficients:
              Estimate Std. Error t value Pr(>|t|)
(Intercept) 0.0006688  0.0003041    2.199   0.0279 *
sp5         1.1457961  0.0220900   51.870   <2e-16 ***
---
Residual standard error: 0.01525 on 2513 degrees of freedom
Multiple R-squared: 0.5171,      Adjusted R-squared: 0.5169
> plot(tdx,beta,xlab='year',ylab='beta',type='l')
> abline(h=c(1.146))
> idx=c(1:2515)[beta==max(beta)]   #Locate the outlier
> idx
[1] 1460
```

5.4 MINIMUM VARIANCE PORTFOLIOS

Another application of GARCH models is to estimate time-varying covariances of asset returns for portfolio selection. We consider the mean–variance analysis of Markovitz (1959) to portfolio analysis. For simplicity, we focus on the minimum variance portfolios. Suppose that there are k risky assets in the portfolio and the standard error of the portfolio returns is taken as the risk measure. Let the k returns be $r_t = (r_{1t}, \ldots, r_{kt})'$ and the covariance matrix of r_t be V_t. Denote the weights of a portfolio by $w_t = (w_{1t}, \ldots, w_{kt})'$. Typically, weights are in percentages of investments. The return of the portfolio is then $w_t' r_t$, and the variance of the portfolio return is $w_t' V_t w_t$.

The idea of minimum variance portfolio is to choose the weights w_t that are the solution to the following simple optimization problem:

$$\min_{w} w' V_t w \quad \text{such that} \quad \sum_{i=1}^{k} w_i = 1.$$

If we allow for negative weights, that is, short selling is allowed then the solution to the prior optimization problem is easily available. The solution is

$$w_t = \frac{V_t^{-1}\mathbf{1}}{\mathbf{1}'V_t^{-1}\mathbf{1}}, \tag{5.8}$$

where $\mathbf{1}$ is the k-dimensional vector of ones. Here the denominator is simply the sum of all elements of V_t^{-1}, and the numerator is the vector consisting of the sums of each row of V_t^{-1}.

In practice, the weights would depend on the sample used to estimate V_t. To illustrate this, we consider the daily log returns of five stocks, namely, Boeing, CAT, IBM, Microsoft, and Procter and Gamble. We use sample covariance to estimate V_t. Table 5.2a gives the weights for the minimum variance portfolio for three sampling periods, whereas Table 5.2b shows the volatilities of each asset and the portfolio. As expected, the portfolio has smaller volatilities.

Turn to application of GARCH models. We use Equation (5.5) to estimate the covariance between two asset returns. These pairwise covariance estimates are then used to construct an estimate of V_t for all assets in the portfolio. A possible drawback of this approach to covariance matrix estimation is that the resulting covariance matrix V_t may not be positive definite. However, for small k and large T, we expect that V_t would be positive definite.

To demonstrate the application of GARCH models, we consider the daily log returns of Abbott Laboratories (ABT), International Business Machines (IBM), and Wal-Mart Stores (WMT) from January 2, 2001 to December 31, 2010. For a given t,

TABLE 5.2. Minimum Variance Portfolios for Five Stocks: (a) Portfolio Weights for Three Sampling Periods and (b) the Volatilities for the Assets and Portfolio

| | (a) Weights in (%) | | |
| | | Sample Period | |
Asset	$t \in [1 : 756]$	$[757 : 1512]$	$[1513 : 2515]$
Abbott Laboratories	6.73	9.09	2.05
Caterpillar	14.43	3.77	−8.05
Inter. Business Machines	11.14	28.99	34.34
Microsoft	4.45	15.79	0.42
Proter & Gamble	63.24	42.35	71.24
	(b) Volatilities (%)		
Abbott Laboratories	2.37	1.35	2.27
Caterpillar	2.08	1.64	2.61
Inter. Business Machines	2.22	0.99	1.67
Microsoft	2.45	1.11	2.17
Proter & Gamble	1.41	0.89	1.38
Minimum variance portfolio	1.22	0.70	1.30

The sample period is from January 2, 2001 to December 31, 2010.

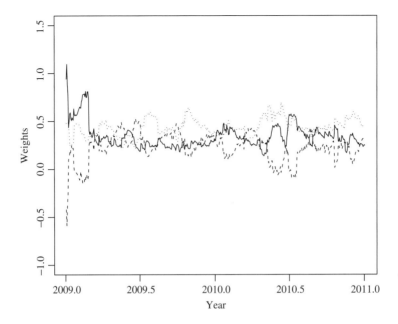

Figure 5.7. Weights of the minimum variance portfolio for three US stock returns from December 29, 2008 to December 31, 2010. Daily log returns are used. The solid, dashed, and dotted lines are for Abbott Laboratories, International Business Machines, and Wal-Mart Stores, respectively.

we estimate V_t by using GARCH(1,1) models for individual asset returns and their sums and differences. We then obtain the minimum variance portfolio and use the weights to compute the portfolio return for $t + 1$. The estimation process is repeated for time $t + 1$ to compute the portfolio return at $t + 2$. In other words, we estimate V_t, obtain the new weights, and rebalance the portfolio daily. This out-of-sample procedure starts with $t = 2011$, corresponding to December 29, 2008, and lasts until December 31, 2010. In this way, we consider the minimum variance portfolio for 2 years with 505 observations. The sample standard deviation of the portfolio returns in the out-of-sample period is 0.0106, whereas those of the individual assets are 0.0132, 0.0146, and 0.0117, respectively, for ABT, IBM, and WMT. Thus, as expected, the minimum variance portfolio reduces the risk. Figure 5.7 shows the time plots of portfolio weights. As expected, the weights evolve over time. Figure 5.8 shows the time plots of volatilities for each individual asset and the minimum variance portfolio. The plot confirms that the volatility of the portfolio is lower.

R Demonstration. Output edited. A simple R script GMVP.R is used.

```
> library(fGarch)
> rtn=cbind(ba,cat,ibm,msft,pg)
> V1=cov(rtn[1:756,])
> V1inv=solve(V1)
```

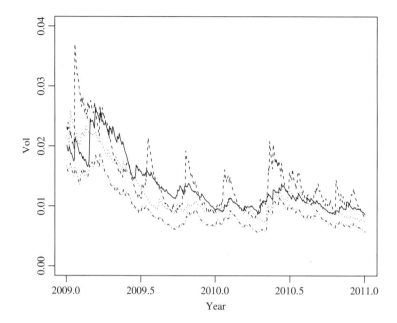

Figure 5.8. Volatility of log returns for stocks of Abbott Laboratories, International Business Machines, Wal-Mart Stores, and the minimum variance portfolio from December 29, 2008 to December 31, 2010. Daily log returns are used. The solid, dashed, dotted, and dot-dashed lines are for ABT, IBM, WMT, and portfolio, respectively.

```
> One=matrix(1,5,1)
> Wgt=V1inv%*%One
> D=sum(Wgt*One)
> Wgt=Wgt/D
> print(Wgt)
           [,1]
ba    0.06730200
cat   0.14432554
ibm   0.11141018
msft  0.04452799
pg    0.63243428
> print(1/sqrt(D))
[1] 0.01223689
> print(sqrt(diag(V1)))
        ba         cat         ibm        msft          pg
0.02374290  0.02076534  0.02222969  0.02452022  0.01409959

> rtn=cbind(abt,ibm,wmt)
> source("GMVP.R")
> M2=GMVP(rtn,start=2011)
> names(M2)
[1] "weights"   "minVariance"  "variances"   "returns"   "det"
> wgt=M2$weights
> range(wgt)
[1] -0.588097  1.095440
> prtn=M2$returns
```

```
> mean(prtn)
[1] 0.0001088868
> sqrt(var(prtn))
[1] 0.01026186
> Mean=apply(rtn[2012:2515,],2,mean)
> Mean
          abt           ibm           wmt
-8.356484e-05  1.180094e-03  1.122753e-05
> v1=sqrt(apply(rtn[2012:2515,],2,var))
> print(v1)
       abt         ibm          wmt
0.01315420 0.01460831 0.01171992
> minV=sqrt(M2$minVariance)
> Vol=sqrt(M2$variances)
> range(minV,Vol)
[1] 0.005651687 0.036997356
> tdx=c(1:505)/2515+2009
> plot(tdx,wgt[1,],xlab='year',ylab='weights',type='l',ylim=c(-.75,1.5))
> lines(tdx,wgt[2,],lty=2)
> lines(tdx,wgt[3,],lty=3)
> plot(tdx,Vol[,1],xlab='year',ylab='vol',type='l',ylim=c(0,0.04))
> lines(tdx,Vol[,2],lty=2)
> lines(tdx,Vol[,3],lty=3)
> lines(tdx,minV,lty=4)
```

5.5 PREDICTION

In this section, we apply the GARCH model to improve the modeling and forecasting of a time series. The increase in oil prices of Summer 2008 and Spring 2011 had substantial impacts on the global economy. Predicting oil price is, therefore, an interesting and important topic. Oil prices, however, are influenced by many factors and external shocks, and are not easy to analyze. In this application, we employ the weekly crude oil prices of the US from January 3, 1997 to September 24, 2010 with 717 observations. The prices are in dollars per barrel and are the spot price FOB (freight on board) weighted by estimated import volume. The data are downloaded from the US Energy Information Administration. Figure 5.9a shows the time plot of the oil prices. The increase in oil prices during Summer 2008 is clearly seen. The prices also exhibit an increasing trend so that the price series is nonstationary. Figure 5.9b shows the changes in the weekly price. This differenced series shows volatility clusters, but it has no obvious violation of being weakly stationary. We shall focus our analysis on the change series. Our analysis demonstrates that a pure ARMA model for the price change is not adequate because, among other reasons, ARMA models cannot handle volatility clusters. On the other hand, an ARMA-GARCH model can adequately handle the complexity of the data and produce improvement in out-of-sample prediction.

We treat our analysis as a case study for the application of GARCH models. As such, sufficient details are provided so that readers can gain insight into using GARCH models to improve time series analysis. Denote the price change series by C_t. Figure 5.10 gives the sample ACF and PACF of the C_t series. These correlations confirm the weak stationarity of C_t. They also show that there exists certain periodic behavior in the crude oil prices. The ACF and PACF have significant values at lags

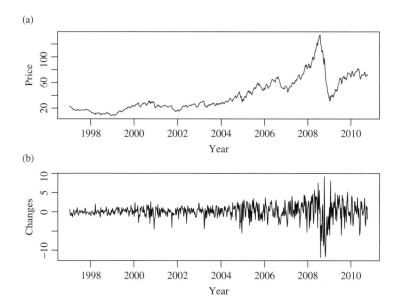

Figure 5.9. Weekly US crude oil prices, dollars per barrel, from January 3, 1997 to September 24, 2010: (a) the price series and (b) the changes in price. *Source*: The data are obtained from US Energy Information Administration.

5, 10, 20, and 25. We regard this periodic behavior as a weekly pattern because there are five working days in a week. Consequently, we employ a seasonal model for C_t.

Focusing on the PACF of C_t in Figure 5.10b, we see that, except for the seasonal lags, PACFs are significant mainly at lags 1 and 3. This implies that an AR(3) would be sufficient for the regular component of the change series. Turn to seasonal pattern. The ACF and PACF at the seasonal lags are not large, even though they are outside the asymptotic two standard-error limits. Therefore, it suffices to start with a lower order seasonal model. Consequently, we specify an ARMA$(3,0)(2,0)_5$ model for C_t, where the subscript 5 signifies the periodicity. The fitted model is

$$(1 - 0.319B + 0.069B^2 - 0.107B^3)(1 - 0.081B^5 - 0.118B^{10})C_t = a_t, \quad \sigma_a^2 = 3.63. \tag{5.9}$$

Figure 5.11 shows the model checking statistics for the model in Equation (5.9). The model appears to be reasonable except the higher volatilities from 2008 to 2010. Some of the standardized residuals have magnitudes around 6, which are rather high compared with standard normal distribution. Further improvement is needed and we turn to GARCH modeling.

As the fGarch package does not specifically handle seasonal mean equations, we decide to remove the weak seasonality from the change series. To this end, we fit

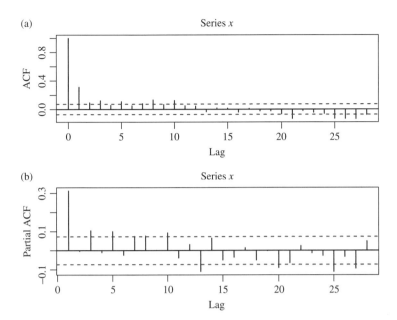

Figure 5.10. Sample ACF and PACF of the change series of weekly US crude oil prices from January 3, 1997 to September 24, 2010: (a) ACF and (b) PACF.

a pure seasonal model to C_t and obtain

$$(1 - 0.0983B^5 - 0.1152B^{10})C_t = b_t.$$

This model is not adequate, but it provides a simple filter that can be used to remove seasonality from C_t. In fact, for this particular instance, the residual series

$$C_t^* = C_t - 0.0983C_{t-5} - 0.1152C_{t-10}, \quad t = 11, 12, \ldots, 716.$$

has no significant serial correlations at the seasonal lags. Figure 5.12 shows the sample ACF and PACF of the C_t^* series. Except for lag 25, there are no significant correlations at the seasonal lags.

With the removal of seasonal component, we apply an AR(3)-GARCH(1,1) model to the adjusted series C_t^* of the price change. The fitted model is

$$C_t^* = 0.323C_{t-}^* - 0.092C_{t-2}^* + 0.040C_{t-3}^* + a_t, \quad a_t = \sigma_t \epsilon_t, \quad \epsilon_t \sim N(0, 1)$$
$$\sigma_t^2 = 0.0168 + 0.0904a_{t-1}^2 + 0.910\sigma_{t-1}^2. \tag{5.10}$$

Except for the AR(3) coefficient, all parameter estimates are statistically significant at the 5% level. Figure 5.13 shows the C_t^* series with estimated volatility and the

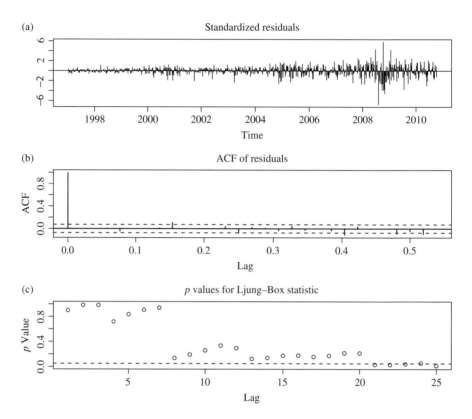

Figure 5.11. Model checking statistics of the seasonal model in Equation (5.9) for the change series of US weekly crude oil price: (a) standardized residuals, (b) ACF of the standardized residuals, and (c) p values of the Ljung-Box statistics.

QQ-plot for the standardized residuals. The nonnormality of standardized residuals is clearly seen.

To handle nonnormality, we entertain Student-t innovations and obtain the model

$$C_t^* = 0.325 C_{t-}^* - 0.065 C_{t-2}^* + 0.056 C_{t-3}^* + a_t, \quad a_t = \sigma_t \epsilon_t, \quad \epsilon_t \sim t_{6.76}$$
$$\sigma_t^2 = 0.0111 + 0.1197 a_{t-1}^2 + 0.8918 \sigma_{t-1}^2. \tag{5.11}$$

The AR(2) and AR(3) coefficients are insignificant. Figure 5.14a shows the QQ-plot of the standardized residuals. The Student-t distribution still encounters some difficulties as the plot shows some deviation from the straight line. Finally, we entertain skew Student-t innovations and obtain the model

$$C_t^* = 0.295 C_{t-}^* + a_t, \quad a_t = \sigma_t \epsilon_t, \quad \epsilon_t \sim t_{6.47, 0.862}$$
$$\sigma_t^2 = 0.0113 + 0.1216 a_{t-1}^2 + 0.8926 \sigma_{t-1}^2. \tag{5.12}$$

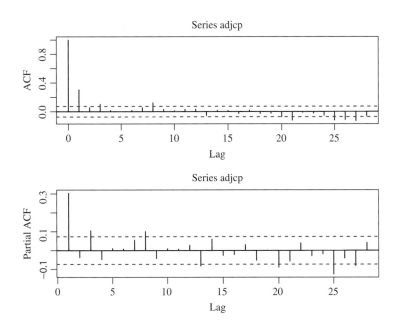

Figure 5.12. Sample ACF and PACF of seasonally adjusted data for the price changes of the US weekly crude oil price. The adjustment is to remove the seasonal component via a seasonal AR(2) model with periodicity 5.

All parameter estimates, but the constant of the volatility equation, are highly significant. Figure 5.14b shows the QQ-plot of the standardized residuals of the model in Equation (5.12). Except for a couple of outlying residuals, the plot exhibits a straight line. Thus, the innovational distribution seems reasonable. The Ljung–Box statistics of the standardized residuals and their squared series also fail to reject the fitted model. We have $Q(10) = 17.27(0.07)$ and $Q(20) = 24.93(0.20)$ for the standardized residuals, where the number in parentheses denotes p-value. For the squared residuals, we have $Q(10) = 4.37(0.93)$ and $Q(20) = 10.73(0.95)$. The model in Equation (5.12) seems adequate. Figure 5.15 shows the standardized residuals and time plots of C_t^* with fitted volatilities. The residual plot confirms the existence of a couple of outliers. On the basis of the fitted model, the t-ratio for the skewness is $t = (0.862 - 1)/0.048 = -2.875$, confirming that the innovational distribution is skewed to the left.

In terms of in-sample fitting, our analysis shows that the AR(1)-GARCH(1,1) model with skew Student-t innovations in Equation (5.12) fares well compared with the pure time series model Equation (5.9). For out-of-sample comparison, we use backtesting with the seasonally adjusted data. Here, for pure time series model, we use an AR(3) model. With starting point $t = 650$, we reserve 56 data points for forecasting comparison. The AR(3) model has mean square of forecast errors 2.368 and 2.558, respectively, for one-step and two-step ahead predictions. For the AR(1)-GARCH(1,1) model in Equation (5.12), the corresponding errors are 2.270 and 2.436,

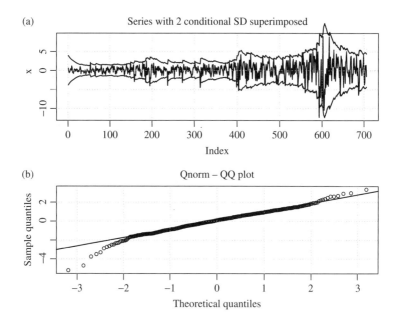

Figure 5.13. Time plots for an AR(3)-GARCH(1,1) model fitted to the seasonally adjusted series of changes in US weekly crude oil price: (a) adjusted series with volatility and (b) normal probability plot of standardized residuals.

respectively. Therefore, our analysis shows that using GARCH models also improved the out-of-sample prediction.

R Demonstration. Output edited.

```
> library(fGarch)
> da=read.table("w-petroprice.txt",header=T)
> price=ts(pet,frequency=52,start=c(1997,1))
> dp=ts(diff(price),frequency=52,start=c(1997,2))
> par(mfcol=c(2,1))
> plot(price,xlab='year',ylab='price')
> plot(dp,xlab='year',ylab='changes')
> cprice=diff(price)
> m2=arima(cprice,order=c(3,0,0),seasonal=list(order=c(2,0,0),period=5))
> m2=arima(cprice,order=c(3,0,0),seasonal=list(order=c(2,0,0),period=5),
     include.mean=F)
> m2
arima(x=cprice,order=c(3,0,0),seasonal=list(order=c(2,0,0),period=5),
     include.mean = F)
Coefficients:
          ar1       ar2       ar3      sar1      sar2
       0.3191   -0.0689    0.1075    0.0817    0.1181
s.e.   0.0372    0.0397    0.0375    0.0377    0.0376
```

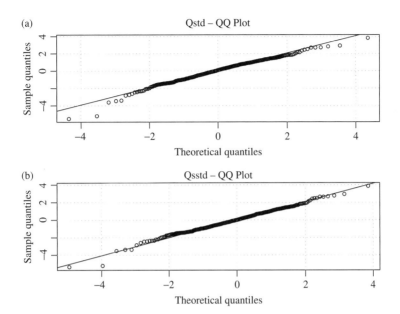

Figure 5.14. Quantile-to-quantile plots for the seasonally adjusted series of changes in the weekly US crude oil price: (a) an AR(3)-GARCH(1,1) model with Student-t innovations, and (b) an AR(1)-GARCH(1,1) model with skew Student-t innovations.

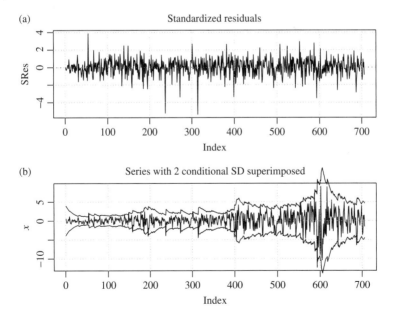

Figure 5.15. Model checking for an AR(1)-GARCH(1,1) model with skew Student-t innovations fitted to the seasonally adjusted series of changes in the US weekly crude oil price: (a) the standardized residuals; and (b) the adjusted price series and its volatilities.

```
sigma^2 estimated as 3.634:  log likelihood = -1478.02,  aic = 2968.04
> length(cprice)
[1] 716
> m2=arima(cprice,seasonal=list(order=c(2,0,0),period=5),include.mean=F)
> m2
arima(x=cprice,seasonal=list(order=c(2,0,0),period=5),include.mean=F)
Coefficients:
          sar1    sar2
        0.0983  0.1152
s.e.    0.0371  0.0372
> adjcp=cprice[11:716]-0.0983*cprice[6:711]-0.1152*cprice[1:706]
> acf(adjcp)
> pacf(adjcp)
>
> m3=garchFit(~arma(3,0)+garch(1,1),data=adjcp,trace=F,include.mean=F)
> summary(m3)
Title:  GARCH Modelling
Call:
 garchFit(formula=~arma(3,0)+garch(1,1),data=adjcp,include.mean=F,
    trace = F)
Error Analysis:
         Estimate  Std. Error  t value Pr(>|t|)
ar1      0.323491    0.040309    8.025 1.11e-15 ***
ar2     -0.092109    0.041690   -2.209   0.0271 *
ar3      0.040485    0.040117    1.009   0.3129
omega    0.016752    0.008171    2.050   0.0403 *
alpha1   0.090404    0.015674    5.768 8.04e-09 ***
beta1    0.910224    0.014837   61.348 < 2e-16 ***
---

Standardised Residuals Tests:
                                  Statistic p-Value
 Ljung-Box Test    R    Q(10)    15.30411  0.1213615
 Ljung-Box Test    R    Q(20)    22.75042  0.3012149
 Ljung-Box Test    R^2  Q(10)     3.155763 0.977517
 Ljung-Box Test    R^2  Q(20)     9.651636 0.9740579
> plot(m3)
> m4=garchFit(~arma(3,0)+garch(1,1),data=adjcp,trace=F,
    include.mean=F,cond.dist="std")
> summary(m4)
Call:
garchFit(formula=~arma(3,0)+garch(1,1),data=adjcp,
   cond.dist="std",include.mean = F, trace = F)
Conditional Distribution:  std
Error Analysis:
         Estimate  Std. Error  t value Pr(>|t|)
ar1      0.325394    0.038778    8.391 < 2e-16 ***
ar2     -0.065107    0.040791   -1.596   0.110
ar3      0.056019    0.039464    1.420   0.156
omega    0.011176    0.008849    1.263   0.207
alpha1   0.119702    0.023879    5.013 5.36e-07 ***
beta1    0.891777    0.019161   46.540 < 2e-16 ***
shape    6.761277    1.582183    4.273 1.93e-05 ***
---

Standardised Residuals Tests:
```

```
                                           Statistic p-Value
 Ljung-Box Test       R      Q(10)   14.96374   0.1333895
 Ljung-Box Test       R      Q(20)   22.95292   0.291112
 Ljung-Box Test       R^2    Q(10)    4.517846  0.920978
 Ljung-Box Test       R^2    Q(20)   10.55414   0.9569761
> plot(m4)
> m5=garchFit(~arma(1,0)+garch(1,1),data=adjcp,trace=F,
    include.mean=F,cond.dist="sstd")
> summary(m5)
Call: garchFit(formula=~arma(1,0)+garch(1,1),data=adjcp,
  cond.dist="sstd",include.mean = F, trace = F)
Conditional Distribution:  sstd
Error Analysis:
        Estimate   Std. Error   t value Pr(>|t|)
ar1     0.294917    0.036755     8.024 1.11e-15 ***
omega   0.011334    0.009021     1.256    0.209
alpha1  0.121625    0.024040     5.059 4.21e-07 ***
beta1   0.892568    0.018457    48.358 < 2e-16 ***
skew    0.861831    0.047754    18.047 < 2e-16 ***
shape   6.470892    1.532304     4.223 2.41e-05 ***
---
Standardised Residuals Tests:
                                     Statistic p-Value
 Ljung-Box Test       R      Q(10)   17.27285   0.0685404
 Ljung-Box Test       R      Q(20)   24.92514   0.2043099
 Ljung-Box Test       R^2    Q(10)    4.370475  0.9290906
 Ljung-Box Test       R^2    Q(20)   10.72815   0.95297
> plot(m5)
% Backtesting
> M3=arima(adjcp,order=c(3,0,0),include.mean=F)
> source("backtest.R")
> M3F=backtest(M3,adjcp,650,2,inc.mean=F)
[1] "RMSE of out-of-sample forecasts"
[1] 2.368055 2.557732
> source("backtestGarch.R")
> M4F=backtestGarch(adjcp,650,2,inc.mean=F,cdist="sstd")
[1] "RMSE of out-of-sample forecasts"
[1] 2.270438 2.435757
```

EXERCISES

1. Consider the weekly world crude oil prices from January 3, 1997 to September 24, 2010. The data are available from the US Energy Information Administration and in the file w-petroprice.txt. Focus on the change series of the price.

   - Build a pure ARMA model for the change series of the world oil prices.
   - Is there any seasonality in the price change series? If yes, perform simple seasonal adjustment by removing seasonal impact.
   - Build a pure ARMA model for the seasonally adjusted series of price change.

- Build an ARMA-GARCH model for the seasonally adjusted series of price change.
- Compare the models with and without GARCH components using backtesting on the last 56 data points.

2. Consider the daily log returns of the Abbott Laboratories stock from January 2, 2001 to December 31, 2010. The simple return of the stock is given in d-a2a-0110.txt. Employ a GARCH(1,1) model with Gaussian innovations to obtain the volatility term structure from $t = 2011$ to $t = 2515$ for $h = 1$, 5, 10, 20, 25, 30, 35, and 40, where h denotes the number of trading days.

3. Use the S&P 500 index to represent the US Market. Obtain the time-varying betas for the Abbott Laboratories stock. The sample period is from January 2, 2001 to December 31, 2010. The data are available from CRSP or Yahoo Finance, or the file d-abtsp-0110.txt.

4. Suppose that a portfolio consists of three US stocks, namely, Alcoa, American Express, and Abbott Laboratories. The daily simple returns are available from the file d-a2a-0110.txt. Use GARCH models to obtain the conditional covariance matrices of the stocks for the period from December 29, 2008 to December 31, 2010. Obtain the weights and the resulting volatility of the minimum variance portfolio for the period. Plot the weights and the volatilities similar to those in Figures 5.7 and 5.8.

5. Consider the daily returns of Apple stock from January 2, 2001 to December 31, 2010. The data are available from the file d-a2a-0110.txt. Build a Gaussian GARCH(1,1) model for the daily log returns. Assume that the risk free interest rate is 1% per annum and the current price of the stock is $350. Use simulation to compute the prices of a European call and an Asian call if the strike price is $355 and the time to expiration is 10 trading days.

REFERENCES

Amin K, Ng V. Option valuation with systematic stochastic volatility. J Finance 1993; 48: 881–910.

Black F. Capital market equilibrium with restricted borrowing. J Bus 1972; 45: 444–454.

Duan JC. The GARCH option pricing model. Math Finance 1995; 5: 13–32.

Engle RE, Rosenberg J. GARCH gamma. J Deriv 1995; 2: 47–59.

Hafner CM, Härdle W. Discrete time option pricing with flexible volatility estimation. Finance and Stochastics 2000; 4: 189–207.

Hull J. Options, Futures & Other Derivatives with Derivagem CD Value Package. 8th ed. Upper Saddle River (NJ): Prentice Hall; 2011.

Hull J, White A. The pricing of options on assets with stochastic volatilities. J Finance 1987; 42: 281–300.

Markovitz H. Portfolio Selection: Efficient Diversification of Investments. New York: John Wiley & Sons; 1959.

Tsay RS. Analysis of Financial Time Series. 3rd ed. Hoboken (NJ): John Wiley; 2010.

6

HIGH FREQUENCY FINANCIAL DATA

High frequency data in finance typically denote observations taken daily or at a finer time scale. These data become available primarily because of advances in information technology and the trend of moving toward electronic trading. They have attracted much attention in recent years because the data are important in empirical study of market microstructure and high frequency trading. Extreme events such as the Flash Crash of May 6, 2010 highlight the need for a deeper understanding of market operation in real time. As a matter of fact, financial markets have witnessed ever increasing interest in direct market access (DMA) in recent years.

The ultimate high frequency data in finance are the transaction-by-transaction or tick-by-tick data in security markets. Here, time is often measured in seconds or fractions of a second. The Trades and Quotes (TAQ) database of the New York Stock Exchange (NYSE) contains all quotes and transactions of equities reported on the *Consolidated Tape*, which includes transactions on the NYSE, AMEX, NASDAQ, and the regional exchanges. For options data, see the Web site of Chicago Board Options Exchange (CBOE). Transactions data for many other securities and markets, both domestic and foreign, are continuously collected and processed. Wood (2000) provides some historical perspective of high frequency financial study.

An Introduction to Analysis of Financial Data with R, First Edition. Ruey S. Tsay.
© 2013 John Wiley & Sons, Inc. Published 2013 by John Wiley & Sons, Inc.

High frequency financial data are important in studying a variety of issues related to the trading process and market microstructure. They can be used to compare the efficiency of different trading systems in price discovery (e.g., the open out-cry system of the NYSE and the computer trading system of NASDAQ). They can also be used to study the dynamics of bid and ask quotes of a particular stock (Hasbrouck, 1999; Zhang et al., 2008). In an order-driven stock market (e.g., the Taiwan Stock Exchange), high frequency data can be used to study the order dynamics and, more interesting, to investigate the question of "who provides the market liquidity." Cho et al. (2003) use intraday 5-min returns of more than 340 stocks traded on the Taiwan Stock Exchange to study the impact of daily stock price limits and find significant evidence of magnet effects toward the price ceiling. High frequency financial data also play an important role in algorithmic trading and DMA. They can be used to design trading strategies and monitor inventory and risk. See Johnson (2010) and Hasbrouck (2007) for further information.

However, high frequency data have some unique characteristics that do not appear in lower frequencies. Analysis of these data thus introduces new challenges to financial economists and statisticians. In this chapter, we study some of the special characteristics, consider methods for analyzing high frequency data, and discuss implications of the results obtained. In particular, we discuss nonsynchronous trading, bid–ask spread, diurnal pattern of high frequency data, movements of trading prices, and trading intensity. We also introduce some algorithmic trading.

6.1 NONSYNCHRONOUS TRADING

We begin with nonsynchronous trading. Stock trading such as those on the NYSE do not occur in a synchronous manner; different stocks have different trading frequencies, and for a single stock, the trading intensity varies from time to time. Yet, we often analyze a return series in a fixed time interval such as daily, weekly, or monthly. For daily series, price of a stock is its *closing* price, which is the last transaction price of the stock in a trading day. The actual time of the last transaction of the stock varies from day to day. As such, we incorrectly assume daily returns as an equally spaced time series with a 24-h interval. It turns out that such an assumption can lead to erroneous conclusions about the predictability of stock returns even if the true return series are serially independent.

For daily stock returns, nonsynchronous trading can introduce (i) lag-1 cross-correlation between stock returns, (ii) lag-1 serial correlation in a portfolio return, and (iii) negative serial correlations of the return series of a single stock in some situations. Consider stocks A and B. Assume that the two stocks are independent and stock A is traded more frequently than stock B. For special news affecting the market, which arrives near the closing hour on one day, stock A is more likely than B to show the effect of the news on the same day simply because A is traded more frequently. The effect of the news on B will eventually appear, but it may be delayed until the following trading day. If this situation indeed happens, return of stock A appears to lead that of stock B. Consequently, the return series may show a significant lag-1 cross-correlation from A to B even though the two stocks are independent. For a portfolio that holds stocks A and B, the prior cross-correlation would become a significant lag-1 serial correlation.

In a more complicated manner, nonsynchronous trading can also induce erroneous negative serial correlations for a single stock. There are several models available in the literature to study this phenomenon; see Campbell et al. (1997) and the references therein. Here, we adopt a simplified version of the model proposed in Lo and MacKinlay (1990). Let r_t be the continuously compounded return of a security at the time index t. For simplicity, assume that $\{r_t\}$ is a sequence of independent and identically distributed random variables with mean $E(r_t) = \mu$ and variance $\text{Var}(r_t) = \sigma^2$. For each time period, the probability that the security is not traded is π, which is time invariant and independent of r_t. Let r_t^o be the observed return. When there is no trade at time index t, we have $r_t^o = 0$ because there is no information available. Yet when there is a trade at time index t, we define r_t^o as the cumulative return from the previous trade (i.e., $r_t^o = r_t + r_{t-1} + \cdots + r_{t-k_t}$, where k_t is the largest nonnegative integer such that no trade occurred in the periods $t - k_t, t - k_t + 1, \ldots, t - 1$). Mathematically, the relationship between r_t and r_t^o is

$$
r_t^o = \begin{cases}
0 & \text{with probability } \pi \\
r_t & \text{with probability } (1 - \pi)^2 \\
r_t + r_{t-1} & \text{with probability } (1 - \pi)^2 \pi \\
r_t + r_{t-1} + r_{t-2} & \text{with probability } (1 - \pi)^2 \pi^2 \\
\vdots & \vdots \\
\sum_{i=0}^{k} r_{t-i} & \text{with probability } (1 - \pi)^2 \pi^k \\
\vdots & \vdots
\end{cases}
\tag{6.1}
$$

These probabilities are easy to understand. For example, $r_t^o = r_t$ if and only if there are trades at both t and $t - 1$; $r_t^o = r_t + r_{t-1}$ if and only if there are trades at t and $t - 2$, but no trade at $t - 1$; $r_t^o = r_t + r_{t-1} + r_{t-2}$ if and only if there are trades at t and $t - 3$, but no trades at $t - 1$ and $t - 2$; and so on. As expected, the total probability is 1 given by

$$
\pi + (1 - \pi)^2 [1 + \pi + \pi^2 + \cdots] = \pi + (1 - \pi)^2 \frac{1}{1 - \pi} = \pi + 1 - \pi = 1.
$$

We are ready to consider the moment equations of the observed return series $\{r_t^o\}$. First, the expectation of r_t^o is

$$
E(r_t^o) = (1 - \pi)^2 E(r_t) + (1 - \pi)^2 \pi E(r_t + r_{t-1}) + \cdots
$$
$$
= (1 - \pi)^2 \mu + (1 - \pi)^2 \pi 2\mu + (1 - \pi)^2 \pi^2 3\mu + \cdots
$$

$$= (1 - \pi)^2 \mu [1 + 2\pi + 3\pi^2 + 4\pi^3 + \cdots]$$

$$= (1 - \pi)^2 \mu \frac{1}{(1 - \pi)^2} = \mu. \tag{6.2}$$

In the prior derivation, we use the result $1 + 2\pi + 3\pi^2 + 4\pi^3 + \cdots = 1/(1 - \pi)^2$, which can be derived by taking the derivative of the geometric series $g(\pi) = 1 + \pi + \pi^2 + \cdots = 1/(1 - \pi)$. Next, for the variance of r_t^o, we use $\text{Var}(r_t^o) = E[(r_t^o)^2] - [E(r_t^o)]^2$ and

$$E(r_t^o)^2 = (1 - \pi)^2 E[(r_t)^2] + (1 - \pi)^2 \pi E[(r_t + r_{t-1})^2] + \cdots$$

$$= (1 - \pi)^2 [(\sigma^2 + \mu^2) + \pi(2\sigma^2 + 4\mu^2) + \pi^2(3\sigma^2 + 9\mu^2) + \cdots] \tag{6.3}$$

$$= (1 - \pi)^2 \{\sigma^2 [1 + 2\pi + 3\pi^2 + \cdots] + \mu^2 [1 + 4\pi + 9\pi^2 + \cdots]\} \tag{6.4}$$

$$= \sigma^2 + \mu^2 \left[\frac{2}{1 - \pi} - 1 \right]. \tag{6.5}$$

In Equation (6.3), we use

$$E \left(\sum_{i=0}^{k} r_{t-i} \right)^2 = \text{Var} \left(\sum_{i=0}^{k} r_{t-i} \right) + \left[E \left(\sum_{i=0}^{k} r_{t-i} \right) \right]^2$$

$$= (k + 1)\sigma^2 + [(k + 1)\mu]^2$$

under the serial independence assumption of r_t. Using techniques similar to that of Equation (6.2), we can show that the first term of Equation (6.4) reduces to σ^2. For the second term of Equation (6.4), we use the identity

$$1 + 4\pi + 9\pi^2 + 16\pi^3 + \cdots = \frac{2}{(1 - \pi)^3} - \frac{1}{(1 - \pi)^2},$$

which can be obtained as follows. Let

$$H = 1 + 4\pi + 9\pi^2 + 16\pi^3 + \cdots \quad \text{and} \quad G = 1 + 3\pi + 5\pi^2 + 7\pi^3 + \cdots.$$

Then $(1 - \pi)H = G$ and

$$(1 - \pi)G = 1 + 2\pi + 2\pi^2 + 2\pi^3 + \cdots$$

$$= 2(1 + \pi + \pi^2 + \cdots) - 1 = \frac{2}{(1 - \pi)} - 1.$$

Consequently, from Equations (6.2) and (6.5), we have

$$\text{Var}(r_t^o) = \sigma^2 + \mu^2 \left[\frac{2}{1 - \pi} - 1 \right] - \mu^2 = \sigma^2 + \frac{2\pi\mu^2}{1 - \pi}. \tag{6.6}$$

Consider next the lag-1 autocovariance of $\{r_t^o\}$. Here, we use $\text{Cov}(r_t^o, r_{t-1}^o) = E(r_t^o r_{t-1}^o) - E(r_t^o)E(r_{t-1}^o) = E(r_t^o r_{t-1}^o) - \mu^2$. The question then reduces to finding $E(r_t^o r_{t-1}^o)$. Notice that $r_t^o r_{t-1}^o$ is 0 if there is no trade at t, no trade at $t-1$, or no trade at both t and $t-1$. Therefore, we have

$$
r_t^o r_{t-1}^o = \begin{cases}
0 & \text{with probability } 2\pi - \pi^2 \\
r_t r_{t-1} & \text{with probability } (1-\pi)^3 \\
r_t(r_{t-1} + r_{t-2}) & \text{with probability } (1-\pi)^3 \pi \\
r_t(r_{t-1} + r_{t-2} + r_{t-3}) & \text{with probability } (1-\pi)^3 \pi^2 \\
\quad \vdots & \quad \vdots \\
r_t\left(\sum_{i=1}^{k} r_{t-i}\right) & \text{with probability } (1-\pi)^3 \pi^{k-1} \\
\quad \vdots & \quad \vdots
\end{cases}
\tag{6.7}
$$

Again the total probability is unity. To understand the prior result, notice that $r_t^o r_{t-1}^o = r_t r_{t-1}$ if and only if there are three consecutive trades at $t-2$, $t-1$, and t. Using Equation (6.7) and the fact that $E(r_t r_{t-j}) = E(r_t)E(r_{t-j}) = \mu^2$ for $j > 0$, we have

$$
E(r_t^o r_{t-1}^o) = (1-\pi)^3 \left\{ E(r_t r_{t-1}) \right.
$$

$$
\left. + \pi E[r_t(r_{t-1} + r_{t-2})] + \pi^2 E\left[r_t\left(\sum_{i=1}^{3} r_{t-i}\right)\right] + \cdots \right\}
$$

$$
= (1-\pi)^3 \mu^2 (1 + 2\pi + 3\pi^2 + \cdots) = (1-\pi)\mu^2.
$$

The lag-1 autocovariance of $\{r_t^o\}$ is then

$$
\text{Cov}(r_t^o, r_{t-1}^o) = -\pi\mu^2.
\tag{6.8}
$$

Provided that μ is not 0, the nonsynchronous trading induces a *negative* lag-1 autocorrelation in r_t^o given by

$$
\rho_1(r_t^o) = \frac{-(1-\pi)\pi\mu^2}{(1-\pi)\sigma^2 + 2\pi\mu^2}.
$$

In general, we can extend the prior result and show that

$$
\text{Cov}(r_t^o, r_{t-j}^o) = -\mu^2 \pi^j, \quad j \geq 1.
$$

The magnitude of the lag-1 ACF (autocorrelation function) depends on the choices of μ, π, and σ and can be substantial. Thus, when $\mu \neq 0$, the nonsynchronous trading induces negative autocorrelations in an observed security return series.

The previous discussion can be generalized to the return series of a portfolio that consists of N securities Campbell et al. (1997, Chapter 3). In the time series literature, effects of nonsynchronous trading on the return of a single security are equivalent to those of random temporal aggregation on a stationary time series, with the trading probability π governing the mechanism of aggregation (Tsay and Yeh, 2011).

6.2 BID–ASK SPREAD OF TRADING PRICES

In some stock exchanges (e.g., NYSE), market makers play an important role in facilitating trades. They provide market liquidity by standing ready to buy or sell whenever the public wishes to sell or buy. By market liquidity, we mean the ability to buy or sell significant quantities of a security quickly, anonymously, and with little price impact. In return for providing liquidity, market makers are granted monopoly rights by the exchange to post different prices for purchases and sales of a security. They buy at the *bid* price P_b and sell at a higher ask price P_a. (For the public, P_b is the sale price and P_a is the purchase price.) The difference $P_a - P_b$ is called the *bid–ask spread*, which is the primary source of compensation for market makers. Typically, the bid–ask spread is small—namely, 1 or 2 cents.

The existence of a bid–ask spread, although small in magnitude, has several important consequences in time series properties of asset returns. We briefly discuss the bid–ask bounce—namely, the bid–ask spread introduces *negative* lag-1 serial correlation in an asset return. Consider the simple model of Roll (1984). The observed market price P_t of an asset is assumed to satisfy

$$P_t = P_t^* + I_t \frac{S}{2}, \tag{6.9}$$

where $S = P_a - P_b$ is the bid–ask spread, P_t^* is the time-t fundamental value of the asset in a frictionless market, and $\{I_t\}$ is a sequence of independent binary random variables with equal probabilities (i.e., $I_t = 1$ with probability 0.5 and $I_t = -1$ with probability 0.5). The I_t can be interpreted as an order-type indicator, with 1 signifying buyer-initiated transaction and -1 seller-initiated transaction. Alternatively, the model can be written as

$$P_t = P_t^* + \begin{cases} +S/2 & \text{with probability } 0.5, \\ -S/2 & \text{with probability } 0.5. \end{cases}$$

If there is no change in P_t^*, then the observed process of price changes is

$$\Delta P_t = (I_t - I_{t-1}) \frac{S}{2}. \tag{6.10}$$

Under the assumption of I_t in Equation (6.9), $E(I_t) = 0$ and $\text{Var}(I_t) = 1$, and we have $E(\Delta P_t) = 0$ and

$$\text{Var}(\Delta P_t) = S^2/2, \tag{6.11}$$

$$\text{Cov}(\Delta P_t, \Delta P_{t-1}) = -S^2/4, \tag{6.12}$$

$$\text{Cov}(\Delta P_t, \Delta P_{t-j}) = 0, \quad j > 1. \tag{6.13}$$

Therefore, the autocorrelation function of ΔP_t is

$$\rho_j(\Delta P_t) = \begin{cases} -0.5 & \text{if } j = 1, \\ 0 & \text{if } j > 1. \end{cases} \tag{6.14}$$

The bid–ask spread thus introduces a negative lag-1 serial correlation in the series of observed price changes. This is referred to as the *bid–ask bounce* in the finance literature. Intuitively, the bounce can be seen as follows. Assume that the fundamental price P_t^* is equal to $(P_a + P_b)/2$. Then P_t assumes the value P_a or P_b. If the previously observed price is P_a (the higher value), then the current observed price is either unchanged or lower at P_b. Thus, ΔP_t is either 0 or $-S$. However, if the previous observed price is P_b (the lower value), then ΔP_t is either 0 or S. The negative lag-1 correlation in ΔP_t becomes apparent. However, the bid–ask spread does not introduce any serial correlation beyond lag 1.

A more realistic formulation is to assume that P_t^* follows a random walk so that $\Delta P_t^* = P_t^* - P_{t-1}^* = \epsilon_t$, which forms a sequence of independent and identically distributed random variables with mean 0 and variance σ^2. In addition, $\{\epsilon_t\}$ is independent of $\{I_t\}$. In this case, $\text{Var}(\Delta P_t) = \sigma^2 + S^2/2$, but $\text{Cov}(\Delta P_t, \Delta P_{t-j})$ remains unchanged. Therefore,

$$\rho_1(\Delta P_t) = \frac{-S^2/4}{S^2/2 + \sigma^2}, \quad \text{satisfying} \quad -0.5 < \rho_1(\Delta P_t) \le 0.$$

The magnitude of the lag-1 autocorrelation of ΔP_t is reduced but remains negative when $S = P_a - P_b > 0$. In finance, it might be of interest to study the components of the bid–ask spread. Interested readers are referred to Campbell et al. (1997) and the references therein.

To illustrate, consider the tick-by-tick trading data of the Caterpillar stock from January 4 to January 8, 2010. The data are obtained from the NYSE TAQ database with 155,267 transactions. We focus on 30-s intraday log returns within the normal trading hours with effective sample size of 3895. Figure 6.1a shows the trading prices with 3900 observations, whereas Figure 6.1b shows the corresponding log returns with 3895 observations. Here, the price is defined as the last transaction price within the 30-s interval if trading occurred and the previous price if there was no trade in the interval. Figure 6.2 shows the sample ACF of the 30-s intraday log returns. The two horizontal lines denote the asymptotic 95% interval for the ACF. Clearly, lag-1 ACF

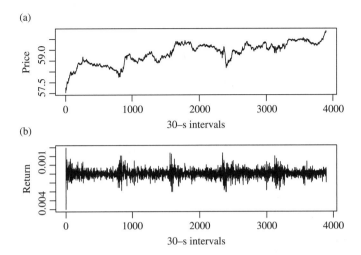

Figure 6.1. Intraday trading prices and log returns of Caterpillar stock from January 4 to January 8, 2010: (a) prices and (b) log returns both in 30-s interval.

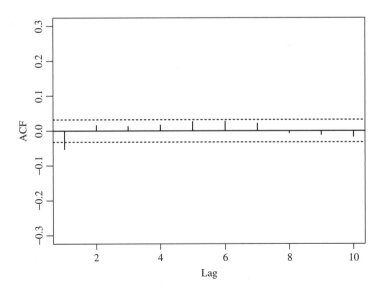

Figure 6.2. Sample ACF of the intraday 30-s long returns for Caterpillar stock from January 4 to January 8, 2010. There are 3895 returns.

is negative and statistically significant at the 5% level. In fact, we have $\hat{\rho}_1 = -0.052$, $\hat{\rho}_2 = 0.016$, and $\hat{\rho}_3 = 0.013$. The effect of bid–ask spread is evident.

The effect of bid–ask spread continues to exist in portfolio returns and in multivariate financial time series. Consider the bivariate case. Denote the bivariate order-type indicator by $I_t = (I_{1t}, I_{2t})'$, where I_{1t} is for the first security and I_{2t} for the second security. If I_{1t} and I_{2t} are contemporaneously positively correlated, then the bid–ask spreads can introduce negative lag-1 cross-correlations.

6.3   EMPIRICAL CHARACTERISTICS OF TRADING DATA

Let t_i be the calendar time, measured in seconds from midnight, at which the ith trade of an asset takes place. Associated with the transaction are several variables such as the trading price, the transaction volume, the prevailing bid and ask quotes, and so on. The collection of t_i and the associated measurements are referred to as the *transactions data*. These data have several important characteristics that do not exist when the observations are aggregated over time. Some of the characteristics are given as follows:

1. *Unequally Spaced Time Intervals.* Transactions such as stock trading on an exchange do not occur at equally spaced time intervals. As such, the actual trading prices of an asset do not form an equally spaced time series. The time duration between trades becomes important and might contain useful information about market microstructure (e.g., trading intensity).

2. *Discrete-Valued Prices.* In United States, the price change of an asset from one transaction to the next only occurred in multiples of tick size before January 29, 2001. On the NYSE, the tick size was 1/8th of a dollar before June 24, 1997 and was 1/16th of a dollar before January 29, 2001. Therefore, the price was a discrete-valued variable in transactions data. Although all equity markets in United States now use the decimal system, the price change in consecutive trades tends to occur in multiples of 1 cent and can be treated approximately as a discrete-valued variable. In some markets, price change may also be subject to limit constraints set by regulators.

3. *Existence of a Daily Periodic or Diurnal Pattern.* Under the normal trading conditions, transaction activity can exhibit a periodic pattern. For instance, on the NYSE, transactions are "heavier" at the beginning and closing of the trading hours and "thinner" during lunch hour, resulting in a U-shaped transaction intensity. Consequently, time durations between transactions also exhibit a daily cyclical pattern. A similar pattern is also observed in stock volatility.

4. *Multiple Transactions within a Single Second.* It is possible that multiple transactions, even with different prices, occur at the same time. This is partly due to the fact that time is measured in seconds, which may be too long a time scale in computer trading. The TAQ database now provides transactions data in milliseconds.

To demonstrate these characteristics, we consider the transactions data of Johnson and Johnson (JNJ) stock from October 4 to October 15, 2010. There are 10 trading days and the total number of transactions is 419,565. Our analysis focuses on transactions occurred during the normal trading hours from 9:30 AM to 16:00 PM Eastern time. It is well known that overnight stock returns differ substantially from intraday returns; see Stoll and Whaley (1990) and the references therein. However, for simplicity, we do not consider overnight returns in the demonstration. Table 6.1 gives a snapshot of the transactions data. From the table, it is clear that (i) multiple transactions may occur in the same second, (ii) price change in consecutive trades tends to be small,

TABLE 6.1. A Snapshot of the Transactions Data for Johnson and Johnson Stock on October 5, 2011

Date	Hour	Minute	Second	Price	Volume
20101005	9	29	30	62.1000	100
20101005	9	29	45	62.1000	100
20101005	9	30	00	62.0800	185
20101005	9	30	00	62.0800	185
20101005	9	30	00	62.1000	100
20101005	9	30	00	62.1000	100
20101005	9	30	00	62.1000	100
20101005	9	30	00	62.1000	100
20101005	9	30	00	62.1000	100
20101005	9	30	00	62.1000	100
20101005	9	30	00	62.1000	100
20101005	9	30	00	62.1000	100
20101005	9	30	00	62.1000	100
20101005	9	30	00	62.0800	550
20101005	9	30	00	62.1000	100
20101005	9	30	00	62.1000	100
⋮	⋮	⋮	⋮	⋮	⋮
20101005	9	30	00	62.0800	100
20101005	9	30	03	62.0900	100
20101005	9	30	07	62.0700	100
20101005	9	30	07	62.0800	100
20101005	9	30	08	62.1000	100
20101005	9	30	08	62.1000	100
20101005	9	30	08	62.1000	100
20101005	9	30	08	62.1000	100
20101005	9	30	08	62.1000	100
20101005	9	30	08	62.1000	100
20101005	9	30	09	62.1000	100
20101005	9	30	09	62.1000	100
20101005	9	30	10	62.0900	126753
20101005	9	30	11	62.0900	120

and (iii) most transactions have small volume, but there exist occasional transactions with large volume.

Within the normal trading hours from October 4 to October 15, 2010, there are 418,855 intraday price changes for the JNJ stock. Figure 6.3a shows the time plot of the intraday price changes, whereas Figure 6.3b gives the associated histogram. Since large price changes are rare, the histogram is shown for price changes between −4 cents to 4 cents only. The time plot and histogram show a symmetric pattern between increases and decreases in trading prices of the JNJ stock. In fact, large price changes seem to occur in pairs consisting of two large movements in the opposite directions. From the histogram, most of consecutive trades are without any price changes.

Table 6.2 gives the counts and frequencies of price change for the JNJ stock. The changes are classified into seven categories. From the table, we make the following observations:

1. About 73% of JNJ intraday transactions were without price change.
2. Approximately 26% of the intraday transactions result in a price change that is less than or equal to 1 cent.

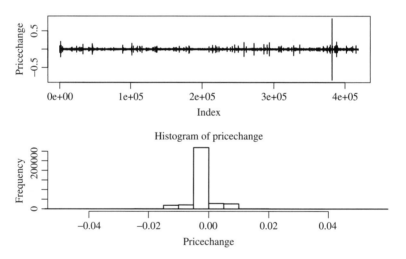

Figure 6.3. Time plot and histogram of intraday price changes in consecutive trades for Johnson and Johnson stock from October 4 to October 15, 2010. Only transactions occurred in the normal trading hours are used. There are 418,855 price changes in 10 trading days.

TABLE 6.2. Frequencies of Price Change in Consecutive Trades for Johnson and Johnson Stock From October 4 to October 15, 2010

Cents	< − 2	[−2, −1)	[−1, 0)	0	(0, 1]	(1, 2]	> 2
Counts	540	1,794	55,325	304,067	54,860	1,711	558
Percentage	0.128	0.428	13.209	72.595	13.098	0.408	0.132

[a] Only Transactions occurred in the normal trading hours are used. Total number of price changes is 418,855.

3. Only 0.83% of the transactions were associated with a price change between $(1, 2]$ cents.

4. Only about 0.26% of the transactions resulted in price changes of 2 cents or more.

5. The empirical distribution of price changes is approximately symmetric with respect to 0.

Consider next the time duration, measured in seconds, between two consecutive trades of the JNJ stock. Figure 6.4 shows the time plot of the first 3000 durations from October 4 to October 15, 2010. The plot confirms that trading did not occur at equally spaced time intervals and there exist zero durations, that is, multiple trades in a second.

For diurnal pattern, we consider the number of transactions in a 5-min time interval for the JNJ stock. Denote the series by x_t. That is, x_1 is the number of JNJ transactions from 09:30 AM to 09:35 AM on October 4, 2010 Eastern time, x_2 is the number of transactions from 09:35 AM to 09:40 AM, and so on. The time gaps between trading days are ignored. Figure 6.5a shows the time plot of x_t, and Figure 6.5b shows the sample ACF of x_t for lags 1–234. The time plot exhibits roughly a cyclical U-shaped pattern with 10 cycles, and the ACF shows a seasonal pattern with periodicity 78, which is the number of 5-min intervals in a trading day. The number of transactions thus exhibits a daily pattern.

Finally, we consider the transactions data of Caterpillar stock on January 4, 2010. There are 37,716 transactions in the regular trading hours. Figure 6.6a shows the transaction prices versus the calendar time measured in seconds from the midnight, and Figure 6.6b shows the time plot of price changes. In this particular instance, the price increased gradually until approximately 4250 s from midnight then it decreased slowly to about $58.5 per share when the market is closed. The price changes continue to exhibit patterns similar to those of the JNJ stock. Figure 6.7 shows the histogram of the price changes for the Caterpillar stock. The histogram shows some distinct characteristics. First, similar to JNJ stock, the price changes appear to be symmetric with respective to 0. Second, the price changes concentrate on multiples of 1 cent. Out of the 37,715 transactions, 64.98% has no price change; see the big spike of the histogram. Details of the summary of price changes for the Caterpillar stock are given in Table 6.3.

6.4 MODELS FOR PRICE CHANGES

Consider the trading of an asset. Let t_i be the calendar time, measured in seconds from midnight, when the ith transaction took place. Let P_{t_i} be the transaction price. The price change from the $(i - 1)$th to the ith trade is $y_i \equiv \Delta P_{t_i} = P_{t_i} - P_{t_{i-1}}$ and the time duration is $\Delta t_i = t_i - t_{i-1}$. Here, it is understood that the subscript i in Δt_i and y_i denotes the time sequence of transactions, not the calendar time. In the following text, we consider models for y_i and Δt_i both individually and together.

The discreteness and concentration on "no change" make it difficult to model the intraday price changes. Campbell et al. (1997) discuss several econometric models

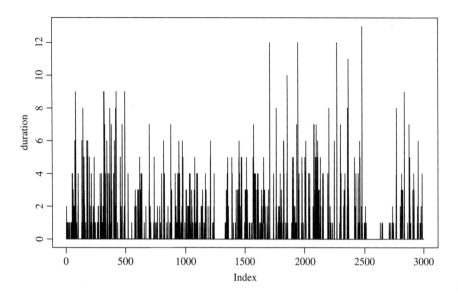

Figure 6.4. Time plot of durations, measured in seconds, between consecutive trades for Johnson and Johnson stock from October 4 to October 15, 2010. Only the first 3000 durations in the normal trading hours are shown.

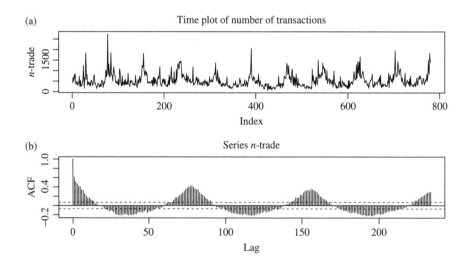

Figure 6.5. Johnson and Johnson intraday transactions data from October 4 to October 15, 2010: (a) the number of transactions in 5-min time intervals and (b) sample ACF of the series in part (a).

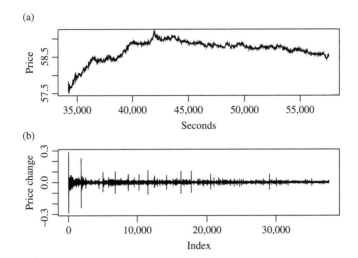

Figure 6.6. Transactions data of Caterpillar stock on January 4, 2010. (a) Price series over calendar time measured in seconds from midnight and (b) time plot of price changes in consecutive trades measured in cents. Only data during the normal trading hours are included.

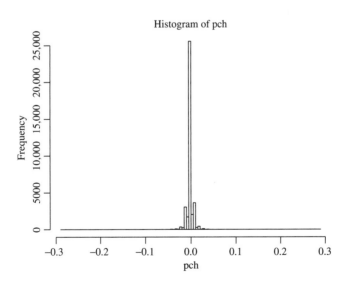

Figure 6.7. The histogram of price changes for Caterpillar stock on January 4, 2010.

that have been proposed in the literature. Here, we mention two models that have the advantage of employing explanatory variables to study the intraday price movements. The first model is the ordered probit model used by Hauseman et al. (1992) to study the price movements in transactions data. The second model has been considered recently by McCulloch and Tsay (2000) and is a simplified version of the model proposed by Rydberg and Shephard (2003); see also Ghysels (2000).

6.4.1 Ordered Probit Model

Let y_i^* be the unobservable price change of the asset under study (i.e., $y_i^* = P_{t_i}^* - P_{t_{i-1}}^*$), where P_t^* is the *virtual* price of the asset at time t. The ordered probit model assumes that y_i^* is a continuous random variable and follows the model

$$y_i^* = x_i \beta + \epsilon_i, \qquad (6.15)$$

where x_i is a p-dimensional row vector of explanatory variables available at time t_{i-1}, β is a $p \times 1$ parameter vector, $E(\epsilon_i | x_i) = 0$, $\text{Var}(\epsilon_i | x_i) = \sigma_i^2$, and $\text{Cov}(\epsilon_i, \epsilon_j) = 0$ for $i \neq j$. The conditional variance σ_i^2 is assumed to be a positive function of the explanatory variable w_i—that is,

$$\sigma_i^2 = g(w_i), \qquad (6.16)$$

where $g(.)$ is a positive function. For financial transactions data, w_i may contain the time interval $t_i - t_{i-1}$ and some conditional heteroscedastic variables. Typically, one also assumes that the conditional distribution of ϵ_i given x_i and w_i is Gaussian.

Suppose that the observed price change y_i may assume k possible values. In theory, k can be infinity, but countable. In practice, k is finite and may involve combining several categories into a single value. For example, we have $k = 7$ in Table 6.3, where the first value "< -2 cents" means that the price drops more than 2 cents. We denote the k possible values as $\{s_1, \ldots, s_k\}$. The ordered probit model postulates the relationship between y_i and y_i^* as

$$y_i = s_j \quad \text{if} \quad \alpha_{j-1} < y_i^* \leq \alpha_j, \quad j = 1, \ldots, k, \qquad (6.17)$$

TABLE 6.3. Frequencies of Price Change for Caterpillar Stock on January 4, 2010

Category	1	2	3	4	5	6	7
Cents	< -2	$[-2,-1)$	$[-1,0)$	0	$(0,1]$	$(1,2]$	> 2
Percentage	0.605	1.692	15.20	64.98	15.04	1.832	0.655

where α_j are real numbers satisfying $-\infty = \alpha_0 < \alpha_1 < \cdots < \alpha_{k-1} < \alpha_k = \infty$. Under the assumption of conditional Gaussian distribution, we have

$$P(y_i = s_j | x_i, w_i) = P(\alpha_{j-1} < x_i\beta + \epsilon_i \le \alpha_j | x_i, w_i)$$

$$= \begin{cases} P(x_i\beta + \epsilon_i \le \alpha_1 | x_i, w_i) & \text{if } j = 1, \\ P(\alpha_{j-1} < x_i\beta + \epsilon_i \le \alpha_j | x_i, w_i) & \text{if } j = 2, \ldots, k-1, \\ P(\alpha_{k-1} < x_i\beta + \epsilon_i | x_i, w_i) & \text{if } j = k, \end{cases}$$

$$= \begin{cases} \Phi\left[\dfrac{\alpha_1 - x_i\beta}{\sigma_i(w_i)}\right] & \text{if } j = 1, \\ \Phi\left[\dfrac{\alpha_j - x_i\beta}{\sigma_i(w_i)}\right] - \Phi\left[\dfrac{\alpha_{j-1} - x_i\beta}{\sigma_i(w_i)}\right] & \text{if } j = 2, \ldots, k-1, \\ 1 - \Phi\left[\dfrac{\alpha_{k-1} - x_i\beta}{\sigma_i(w_i)}\right] & \text{if } j = k, \end{cases} \tag{6.18}$$

where $\Phi(x)$ is the cumulative distribution function of the standard normal random variable evaluated at x, and we write $\sigma_i(w_i)$ to denote that σ_i^2 is a positive function of w_i. From the definition, an ordered probit model is driven by an unobservable continuous random variable. The observed values, which have a natural ordering, can be regarded as categories representing the underlying process.

The ordered probit model contains parameters β, α_i ($i = 1, \ldots, k-1$), and those in the conditional variance function $\sigma_i(w_i)$ in Equation (6.16). These parameters can be estimated by the maximum likelihood or Markov chain Monte Carlo methods. In this chapter, we use the command `polr` of the R package MASS to estimate ordered probit models.

Example 6.1. To illustrate, we consider again the intraday price changes of Caterpillar stock on January 4, 2010. There are 37,716 transactions during the normal trading hours so that we have 37,715 price changes. For simplicity, we classify the price change into seven categories as shown in Table 6.3. Our analysis focuses on the dynamic dependence of intraday price changes. As such, we define indicator (or dummy) variables for lagged price changes:

$$y_{\ell,j} = \begin{cases} 1 & \text{if } y_{i-\ell} = s_j \\ 0 & \text{otherwise,} \end{cases}$$

where s_j denotes the jth category of price change and $y_{i-\ell}$ is the $(i - \ell)$th price change at time $t_{i-\ell}$, where $j = 2, \ldots, 7$ and $\ell = 1$ and 2. In other words, we employ the classifications of price changes for the previous two consecutive trades. As usual, with seven categories, only six indicator variables are needed.

We also employ the observed price changes $y_{i-\ell}$ for $\ell = 1, 2, 3$ and the lag-2 transaction volume defined as $v_{i-2} = V_{i-2}/100$, where V_{i-2} is the actual volume. We do not use price volume because price is relatively stable in a trading day. Consequently, the model entertained is

$$x_i \beta = \beta_1 v_{i-2} + \sum_{\ell=1}^{3} \beta_{1+\ell} y_{i-\ell} + \sum_{j=2}^{7} \gamma_{1,j} y_{1,j} + \sum_{j=2}^{7} \gamma_{2,j} y_{2,j}. \qquad (6.19)$$

For simplicity, we start with $\sigma_i^2(w_i) = \sigma^2$, a constant. Parameter estimates of the model are given in Table 6.4, where all estimates of Equation (6.19) are negative. Clearly, all estimates but one are statistical significant at the usual 5% level. As a matter of fact, the model shown is a simplified one after removing some explanatory variables that were not statistically significant. For instance, we also included the time duration $\Delta t_i = t_i - t_{i-1}$ in the preliminary analysis and decided to drop the variable because its estimate is not statistically significant at the 5% level. The significance of the indicator variables shows that there exists dynamic dependence in intraday price change. The fitted model thus can be used to provide probability forecasts for the next transaction price change. Indeed, the model provides probability for each category of price change at each transaction. Figure 6.8 shows the time plots of fitted probabilities for categories 1, 4, and 7. As expected, the fitted probability for category 4, which denotes no price change, is high, but there are cases in which the probabilities of category 1 or 7 are high.

Finally, it is interesting to study the fitted boundary partitions of the ordered probit model in Table 6.4. First, because the explanatory variables may have nonzero means, the estimates of boundary parameters α_i are not symmetric with respect to 0. Second, $\hat{\alpha}_2 - \hat{\alpha}_1 = 0.577$ and $\hat{\alpha}_6 - \hat{\alpha}_5 = 0.601$. The two intervals roughly have the same length. Similarly, $\hat{\alpha}_3 - \hat{\alpha}_2 = 1.157$, which is close to $\hat{\alpha}_5 - \hat{\alpha}_4 = 1.140$. These results are consistent with the empirical observation that price changes appear to be roughly symmetric with respect to 0 as shown in Table 6.3.   □

TABLE 6.4. Estimation Results of an Ordered Probit Model for the Intraday Price Changes of Caterpillar Stock on January 4, 2010 with 37,716 Transactions[a]

(a)	Boundary Partitions of the Probit Model					
Parameter	α_1	α_2	α_3	α_4	α_5	α_6
Estimate	−4.594	−4.017	−2.860	−0.853	0.287	0.888
t	−31.48	−27.80	−19.89	−5.944	2.000	6.188

(b)	Equation Prameters of Probit Model (Estimates are Negative)						
Parameter	β_1	β_2	β_3	β_4	$\gamma_{1,2}$	$\gamma_{1,3}$	$\gamma_{1,4}$	$\gamma_{1,5}$
Estimate	0.004	7.837	10.86	12.28	0.274	0.743	1.331	1.858
t	3.983	5.363	7.098	15.93	2.971	8.173	13.81	17.83
Parameter	$\gamma_{1,6}$	$\gamma_{1,7}$	$\gamma_{2,2}$	$\gamma_{2,3}$	$\gamma_{2,4}$	$\gamma_{2,5}$	$\gamma_{2,6}$	$\gamma_{2,7}$
Estimate	2.262	2.493	0.099	0.307	0.531	0.745	0.933	0.859
t	18.57	15.95	1.053	3.324	5.419	7.009	7.528	5.381

[a]The model is in Equation (6.19) and t denotes t-ratio.

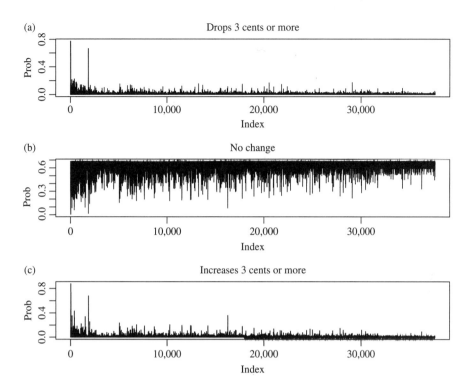

Figure 6.8. Time plots of fitted probabilities of price change for Caterpillar stock on January 4, 2010: the three plots are for (a) decrease 2 cents or more, (b) no change, and (c) increase 2 cents or more.

Discussion. The command `polr` allows for predetermined weights to handle heteroscedasticity, but it cannot perform simultaneous estimation of the volatility and probit equations. See Hauseman et al. (1992) and Tsay (2010) for some examples with time-varying $\sigma_i^2(w_i)$ function. Finally, as usual, only six indicator variables are needed for each lagged value of y_i.

R Demonstrations for Ordered Probit Models. Output edited.

```
> da=read.table("taq-cat-t-jan042010.txt",header=T)
> head(da)
      date hour minute second price size
1 20100104    9     30      0 57.65 3910
 . . . . .
6 20100104    9     30      1 57.65  462
> vol=da$size/100
> da1=read.table("taq-cat-cpch-jan042010.txt")
> cpch=da1[,1]   % category of price change
> pch=da1[,2]   % price change
> cf=as.factor(cpch)   % create categories in R
```

```
> length(cf)
[1] 37715

> y=cf[4:37715]
> y1=cf[3:37714]   % create indicator variables for lag-1 cpch
> y2=cf[2:37713]   % create indicator variables for lag-2 cpch

> vol=vol[2:37716]
> v2=vol[2:37713] % create lag-2 volume

> cp1=pch[3:37714] % select lagged price changes
> cp2=pch[2:37713]; cp3=pch[1:37712]

> library(MASS) % load package
> m1=polr(y~v2+cp1+cp2+cp3+y1+y2,method="probit")
> summary(m1)
Call:
polr(formula = y ~ v2 + cp1 + cp2 + cp3 + y1 + y2, method = "probit")
Coefficients:
        Value      Std. Error    t value
v2    -0.003765    0.0009453    -3.983
cp1   -7.836883    1.4613047    -5.363
cp2  -10.864394    1.5306456    -7.098
cp3  -12.283682    0.7710955   -15.930
y12   -0.274407    0.0923566    -2.971
y13   -0.742792    0.0908854    -8.173
y14   -1.330665    0.0963540   -13.810
y15   -1.858199    0.1042257   -17.829
y16   -2.261587    0.1218013   -18.568
y17   -2.493321    0.1563177   -15.950
y22   -0.098542    0.0935908    -1.053
y23   -0.307034    0.0923725    -3.324
y24   -0.531115    0.0980150    -5.419
y25   -0.744706    0.1062435    -7.009
y26   -0.932655    0.1238918    -7.528
y27   -0.858858    0.1596219    -5.381

Intercepts:
       Value    Std.Error    t value
1|2   -4.5941    0.1459     -31.4803
2|3   -4.0170    0.1445     -27.7989
3|4   -2.8599    0.1438     -19.8926
4|5   -0.8528    0.1435      -5.9437
5|6    0.2868    0.1434       1.9996
6|7    0.8882    0.1435       6.1883

Residual Deviance: 74802.56
AIC: 74846.56
> names(m1)
 [1] "coefficients"   "zeta"         "deviance"       "fitted.values"
 [5] "lev"            "terms"        "df.residual"    "edf"
 [9] "n"              "nobs"         "call"           "method"
[13] "convergence"    "niter"        "lp"             "model"
[17] "contrasts"      "xlevels"
```

```
> yhat=m1$fitted.values
> print(yhat[1:5,],digits=3)
        1        2       3     4      5      6        7
1 1.11e-03 0.005420 0.08605 0.660 0.2134 0.0266 0.007696
2 1.55e-02 0.041461 0.27883 0.608 0.0535 0.0028 0.000444
3 8.99e-06 0.000094 0.00522 0.287 0.4311 0.1605 0.116298
4 1.87e-04 0.001251 0.03267 0.539 0.3343 0.0658 0.027144
5 6.41e-04 0.003470 0.06457 0.630 0.2527 0.0365 0.011836
```

6.4.2   A Decomposition Model

An alternative approach to modeling price change is to decompose it into three components and use conditional specifications for the components (Rydberg and Shephard, 2003). The three components are indicators for price change, the direction of price movement if there is a change, and the size of price change if a change occurs. Specifically, the price change at the ith transaction can be written as

$$y_i \equiv P_{t_i} - P_{t_{i-1}} = A_i D_i S_i, \tag{6.20}$$

where A_i is a binary variable defined as

$$A_i = \begin{cases} 1 & \text{if there is a price change at the } i \text{ th trade,} \\ 0 & \text{if price remains the same at the} i \text{ th trade,} \end{cases} \tag{6.21}$$

D_i is also a discrete variable signifying the *direction* of the price change if a change occurs — that is,

$$D_i | (A_i = 1) = \begin{cases} 1 & \text{if price increases at the } i\text{th trade,} \\ -1 & \text{if price drops at the } i\text{th trade,} \end{cases} \tag{6.22}$$

where $D_i | (A_i = 1)$ means that D_i is defined under the condition of $A_i = 1$, and S_i is the size of the price change. We set $S_i = 0$ if there is no price change at the ith trade. When there is a price change, S_i is positive. If price change is measured in tick size, S_i becomes a positive integer-valued random variable when there is a price change at the ith trade.

Note that D_i is not needed when $A_i = 0$, and there is a natural ordering in the decomposition. D_i is well defined only when $A_i = 1$ and S_i is meaningful when $A_i = 1$ and D_i is given. Model specification under the decomposition makes use of the ordering.

Let F_i be the information set available at the ith transaction. Examples of elements in F_i are Δt_{i-j}, A_{i-j}, D_{i-j}, and S_{i-j} for $j \geq 0$. The evolution of price change under model (Eq. 6.20) can then be partitioned as

$$P(y_i | F_{i-1}) = P(A_i D_i S_i | F_{i-1}) = P(S_i | D_i, A_i, F_{i-1}) P(D_i | A_i, F_{i-1}) P(A_i | F_{i-1}). \tag{6.23}$$

Since A_i is a binary variable, it suffices to consider the evolution of the probability $p_i = P(A_i = 1)$ over time. We assume that

$$\ln\left(\frac{p_i}{1 - p_i}\right) = x_i\beta \quad \text{or} \quad p_i = \frac{e^{x_i\beta}}{1 + e^{x_i\beta}}, \tag{6.24}$$

where x_i is a finite-dimensional vector consisting of elements of F_{i-1} and β is a parameter vector. Conditioned on $A_i = 1$, D_i is also a binary variable, and we use the following model for $\delta_i = P(D_i = 1|A_i = 1)$:

$$\ln\left(\frac{\delta_i}{1 - \delta_i}\right) = z_i\gamma \quad \text{or} \quad \delta_i = \frac{e^{z_i\gamma}}{1 + e^{z_i\gamma}}, \tag{6.25}$$

where z_i is a finite-dimensional vector consisting of elements of F_{i-1} and γ is a parameter vector. To allow for asymmetry between positive and negative price changes, we assume that

$$S_i|(D_i, A_i = 1) \sim 1 + \begin{cases} g(\lambda_{u,i}) & \text{if } D_i = 1, A_i = 1, \\ g(\lambda_{d,i}) & \text{if } D_i = -1, A_i = 1, \end{cases} \tag{6.26}$$

where $g(\lambda)$ is a geometric distribution with parameter λ and the parameters $\lambda_{j,i}$ evolve over time as

$$\ln\left(\frac{\lambda_{j,i}}{1 - \lambda_{j,i}}\right) = w_i\theta_j \quad \text{or} \quad \lambda_{j,i} = \frac{e^{w_i\theta_j}}{1 + e^{w_i\theta_j}}, \quad j = u, d, \tag{6.27}$$

where w_i is again a finite-dimensional explanatory variable in F_{i-1} and θ_j is a parameter vector.

In Equation (6.26), the probability mass function of a random variable x, which follows the geometric distribution $g(\lambda)$, is

$$p(x = m) = \lambda(1 - \lambda)^m, \quad m = 0, 1, 2, \ldots$$

We added 1 to the geometric distribution so that the price change, if it occurs, is at least 1 tick. In Equation (6.27), we take the logistic transformation to ensure that $\lambda_{j,i} \in [0, 1]$.

The previous specification classifies the ith trade, or transaction, into one of three categories:

1. No price change: $A_i = 0$ and the associated probability is $(1 - p_i)$.
2. A price increase: $A_i = 1$, $D_i = 1$, and the associated probability is $p_i\delta_i$. The size of the price increase is governed by $1 + g(\lambda_{u,i})$.
3. A price drop: $A_i = 1$, $D_i = -1$, and the associated probability is $p_i(1 - \delta_i)$. The size of the price drop is governed by $1 + g(\lambda_{d,i})$.

Let $I_i(j)$ for $j = 1, 2, 3$ be the indicator variables of the prior three categories. That is, $I_i(j) = 1$ if the jth category occurs and $I_i(j) = 0$ otherwise. The log likelihood function of Equation (6.23) becomes

$$\ln[P(y_i|F_{i-1})]$$
$$= I_i(1)\ln[(1-p_i)] + I_i(2)[\ln(p_i) + \ln(\delta_i) + \ln(\lambda_{u,i}) + (S_i - 1)\ln(1 - \lambda_{u,i})]$$
$$+ I_i(3)[\ln(p_i) + \ln(1-\delta_i) + \ln(\lambda_{d,i}) + (S_i - 1)\ln(1 - \lambda_{d,i})],$$

and the overall log likelihood function is

$$\ln[P(y_1,\ldots,y_n|F_0)] = \sum_{i=1}^{n} \ln[P(y_i|F_{i-1})], \tag{6.28}$$

which is a function of parameters β, γ, θ_u, and θ_d. For simplicity, one can estimate parameters of the three categories separately with properly identified observations.

Example 6.2. We illustrate the decomposition model by analyzing, again, the intraday transactions of Caterpillar stock on January 4, 2010. As mentioned in Example 6.1, there are 37,715 price changes during the normal trading hours, and we classify the price changes into seven categories. To keep the model simple, we employ the following explanatory variables:

1. A_{i-1}: the action indicator of the previous trade (i.e., the $(i-1)$th trade).
2. D_{i-1}: the direction indicator of the previous trade.
3. S_{i-1}: the size of the previous trade.

Other variables available following the $(i-1)$ trade can also be used. Because we use lag-1 explanatory variables, the actual sample size is 37,714, and the model employed is

$$\ln\left(\frac{p_i}{1-p_i}\right) = \beta_0 + \beta_1 A_{i-1},$$

$$\ln\left(\frac{\delta_i}{1-\delta_i}\right) = \gamma_0 + \gamma_1 D_{i-1}, \tag{6.29}$$

$$\ln\left(\frac{\lambda_{u,i}}{1-\lambda_{u,i}}\right) = \theta_{u,0} + \theta_{u,1} S_{i-1},$$

$$\ln\left(\frac{\lambda_{d,i}}{1-\lambda_{d,i}}\right) = \theta_{d,0} + \theta_{d,1} S_{i-1}.$$

The parameter estimates of the model are given in Table 6.5. The estimated simple model shows some dynamic dependence in the price change. In particular, the trade-by-trade price changes of CAT stock exhibit some of the following appealing features:

TABLE 6.5. Parameter Estimates of the ADS Model in Equation (6.29) for Caterpillar Stock Traded on January 4, 2010

Parameter	β_0	β_1	γ_0	γ_1
Estimate	−1.073	1.183	−0.010	−1.241
standard error	0.015	0.023	0.019	0.029
Parameter	$\theta_{u,0}$	$\theta_{u,1}$	$\theta_{d,0}$	$\theta_{d,1}$
Estimate	1.649	−0.297	1.534	−0.162
standard error	0.041	0.035	0.039	0.037

1. The probability of a price change depends on the previous price change. Specifically, we have

$$P(A_i = 1|A_{i-1} = 0) = \frac{\exp(-1.073)}{1 + \exp(-1.073)} = 0.255$$

$$P(A_i = 1|A_{i-1} = 1) = \frac{\exp(-1.073 + 1.183)}{1 + \exp(-1.073 + 1.183)} = 0.527.$$

The result indicates that (non) price changes may occur in clusters and, as expected, most transactions are without price change. When no price change occurred at the $(i - 1)$th trade, then only about one out of four trades in the subsequent transaction has a price change. When there is a price change at the $(i - 1)$th transaction, the probability of a price change in the ith trade increases to about 0.5.

2. The direction of price change is governed by

$$P(D_i = 1|F_{i-1}, A_i) = \begin{cases} 0.500 & \text{if } D_{i-1} = 0 \text{ (i.e., } A_{i-1} = 0), \\ 0.223 & \text{if } D_{i-1} = 1, A_i = 1, \\ 0.774 & \text{if } D_{i-1} = -1, A_i = 1. \end{cases}$$

This result says that (i) if no price change occurred at the $(i - 1)$th trade, then the chances for a price increase or decrease at the ith trade are about even; and (ii) the probabilities of consecutive price increases or decreases are relatively small. The probability of a price increase at the ith trade given that a price change occurs at the ith trade, and there was a price increase at the $(i - 1)$th trade is only 22.3%. However, the probability of a price increase is about 77.4% given that a price change occurs at the ith trade, and there was a price decrease at the $(i - 1)$th trade. Consequently, this result shows the effect of bid–ask bounce and supports price reversals in high frequency trading.

3. There is weak evidence suggesting that big price changes have a higher probability to be followed by another big price change. Consider the size of a price increase. We have

$$S_i|(D_i = 1) \sim 1 + g(\lambda_{u,i}), \quad \lambda_{u,i} = 1.649 - 0.297S_{i-1}.$$

Using the probability mass function of a geometric distribution, we obtain that the probability of a price increase by 1 cent is 0.794 at the ith trade if the transaction results in a price increase and $S_{i-1} = 1$. The probability reduces to 0.742 if $S_{i-1} = 2$ and to 0.681 if $S_{i-1} = 3$. Similarly, the probability of a price increase by 2 cents is 0.163 at the ith trade if the transaction results in a price increase and $S_{i-1} = 1$. The probability increases to 0.192 if $S_{i-1} = 2$ and to 0.217 if $S_{i-1} = 3$. Consequently, the probability of a large S_i is proportional to S_{i-1} given that there is a price increase at the ith trade.

□

R Demonstration for ADS Models with Geometric Distribution. Output edited.

```
> da=read.table("taq-cat-cpch-jan042010.txt")
> dim(da)
[1] 37715      2
> pch=da[,2]   % create Ai, Di, and Si and their lagged variables
> idx=c(1:37715)[pch > 0]
> jdx=c(1:37715)[pch < 0]
> A=rep(0,37715); A[idx]=1; A[jdx]=1
> D=rep(0,37715); D[idx]=1; D[jdx]=-1
> S=abs(da[,1]-4)
> Ai=A[2:37715]; Aim1=A[1:37714]
> Di=D[2:37715]; Dim1=D[1:37714]
> Si=S[2:37715]; Sim1=S[1:37714]
> m1=glm(Ai~Aim1,family="binomial")
> summary(m1)
Call: glm(formula = Ai ~ Aim1, family = "binomial")

Coefficients:
              Estimate Std. Error z value Pr(>|z|)
(Intercept) -1.07342     0.01466  -73.22   <2e-16 ***
Aim1         1.18316     0.02277   51.95   <2e-16 ***
---
Residual deviance: 46085  on 37712  degrees of freedom
AIC: 46089

> di=Di[Ai==1]
> dim1=Dim1[Ai==1]
> di=(di+abs(di))/2 % transform di to binary
> m2=glm(di~dim1,family="binomial")
> summary(m2)
Call: glm(formula = di ~ dim1, family = "binomial")

Coefficients:
              Estimate Std. Error z value Pr(>|z|)
(Intercept) -0.009755   0.018994  -0.514    0.608
dim1        -1.241364   0.028731 -43.207   <2e-16 ***
```

```
---
Residual deviance: 16069   on 13207   degrees of freedom
AIC: 16073

> si=Si[Di==1]
> sim1=Sim1[Di==1]
> source("GeoSize.R") % R script to fit Geometric dist.
> m3=GeoSize(si,sim1)
Coefficient(s):
          Estimate   Std. Error   t value    Pr(>|t|)
omega1    1.6489885   0.0406517   40.5639  < 2.22e-16 ***
omega2   -0.2966793   0.0354231   -8.3753  < 2.22e-16 ***
---
> nsi=Si[Di==-1]
> nsim1=Sim1[Di==-1]
> m4=GeoSize(nsi,nsim1)
Coefficient(s):
          Estimate   Std. Error   t value    Pr(>|t|)
omega1    1.5339270   0.0392826   39.04847  < 2.22e-16 ***
omega2   -0.1617696   0.0367264   -4.40472  1.0592e-05 ***
```

6.5 DURATION MODELS

Duration models are concerned with time intervals between trades. Longer durations indicate lack of trading activities, which in turn signify a period of no new information. The dynamic behavior of durations thus contains useful information about intraday market activities. Using concepts similar to the ARCH models for volatility, Engle and Russell (1998) propose an autoregressive conditional duration (ACD) model to describe the evolution of time durations for (heavily traded) stocks. Zhang et al. (2005) extend the ACD model to account for nonlinearity and structural breaks in the data. In this section, we introduce some simple duration models.

A special feature of duration data is that all observations are positive. This feature also occurs in data of other applications. For instance, the daily price range of a stock is typically positive. Thus, applications of the duration models discussed are not limited to transactions data. They can be used in many other situations.

To illustrate, we consider the intraday transactions of the Caterpillar stock from January 4 to January 8, 2010. Focusing on the normal trading hours, we have 155,077 transactions during the sample period. Thus, there are 155,076 time durations between trades. As many trades can occur in the same second, we ignore the zero durations. Consequently, we have 37,674 nonzero intraday durations. Figure 6.9 shows the time plot of those nonzero durations. From the plot, we see that there exists certain diurnal pattern in the durations.

Because intraday transactions exhibit some diurnal pattern, we focus on the adjusted time duration

$$x_i = \Delta t_i / f(t_i), \tag{6.30}$$

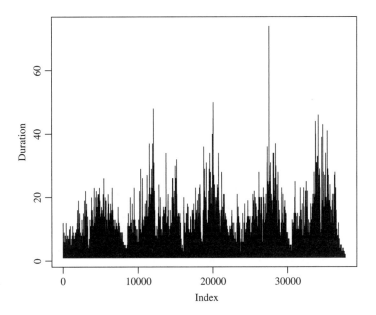

Figure 6.9. Time plot of nonzero durations between trades for Caterpillar stock from January 4 to January 8, 2010.

where $f(t_i)$ is a deterministic function consisting of the cyclical component of Δt_i. Obviously, $f(t_i)$ depends on the underlying asset and the systematic behavior of the market. In practice, there are many ways to estimate $f(t_i)$, but no single method dominates the others in terms of statistical properties. A common approach is to use smoothing spline. Here, we use simple quadratic functions to take care of the deterministic component of daily trading activities.

6.5.1 Diurnal Component

For the CAT transactions data, we assume

$$f(t_i) = \exp[d(t_i)], \quad d(t_i) = \beta_0 + \beta_1 f_1(t_i) + \beta_2 f_2(t_i), \tag{6.31}$$

where

$$f_1(t_i) = \frac{t_i - 43,200}{23,400} \quad \text{and} \quad f_2(t_i) = f_1^2(t_i),$$

where 43,200 denotes the 12:00 noon and 23,400 is number of trading hours measured in seconds. We use these two functions based on the daily pattern shown in Figure 6.9.

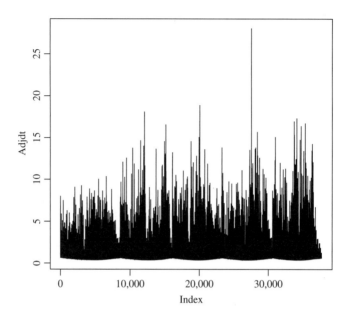

Figure 6.10. Time plot of adjusted nonzero durations between trades for Caterpillar stock from January 4 to January 8, 2010.

The coefficients β_j of Equation (6.31) are obtained by the least squares method of the linear regression

$$\ln(\Delta t_i) = \beta_0 + \sum_{j=1}^{2} \beta_j f_j(t_i) + \epsilon_i.$$

The fitted model is

$$\ln(\widehat{\Delta t_i}) = 0.9396 + 0.5702 f_1(t_i) - 2.1655 f_2(t_i),$$

where standard errors of the coefficient estimates are 0.0061, 0.0180, and 0.0511, respectively. These estimates are highly significant. Figure 6.10 shows the time plot of adjusted durations based on the model in Equation (6.31). From the plot, the diurnal pattern is largely removed.

R Demonstrations for Diurnal Component. Output edited.

```
> da=read.table("taq-cat-t-jan04t082010.txt",header=T)
> head(da)
       DATE hour minute second PRICE SIZE
1 20100104    5     34     26 57.56  200
. . . . . . .
> sec=3600*dahour+60*daminute+da$second % time in seconds
```

```
> ist=3600*9+30*60;  end=3600*16
> lunch=3600*12
> length(sec)
[1] 155267
> idx=c(1:155267)[sec < ist]   % before market opens
> jdx=c(1:155267)[sec > end]   % after market closes
> sec=sec[-c(idx,jdx)]   % normal trading hours only.
> length(sec)
[1] 155077
> dt=diff(sec)
> kdx=c(1:length(dt))[dt > 0] % Positive durations only
> length(kdx)
[1] 37674
> ti=sec[2:155077]
> dt=dt[kdx]
> ti=ti[kdx]
> plot(dt,type='l',xlab='index',ylab='duration')
> st=3600*6.5
> f1=(ti-lunch)/st
> ft=cbind(f1,f1^2)
> m2=lm(log(dt)~ft)   % Linear model for log(durations)
> summary(m2)
Call:  lm(formula = log(dt) ~ ft)

Coefficients:
            Estimate Std. Error t value Pr(>|t|)
(Intercept)  0.939622   0.006126  153.39  <2e-16 ***
ftf1         0.570215   0.017971   31.73  <2e-16 ***
ft          -2.165498   0.051115  -42.37  <2e-16 ***
---
Residual standard error: 0.7874 on 37671 degrees of freedom
Multiple R-squared: 0.04566,    Adjusted R-squared: 0.04561

> names(m2)
 [1] "coefficients"  "residuals"    "effects"       "rank"
 [5] "fitted.values" "assign"       "qr"            "df.residual"
 [9] "xlevels"       "call"         "terms"         "model"
> fit=m2$fitted.values
> adjdt=dt/exp(fit)
```

6.5.2 The ACD Model

The ACD model uses the idea of GARCH models to study the dynamic structure of the adjusted duration x_i of Equation (6.30).

Let $\psi_i = E(x_i|F_{i-1})$ be the conditional expectation of the adjusted duration between the $(i-1)$th and ith trades, where F_{i-1} is the information set available at the $(i-1)$th trade. In other words, ψ_i is the expected adjusted duration given F_{i-1}. The basic ACD model is defined as

$$x_i = \psi_i \epsilon_i, \tag{6.32}$$

where $\{\epsilon_i\}$ is a sequence of independent and identically distributed nonnegative random variables such that $E(\epsilon_i) = 1$. In Engle and Russell (1998), ϵ_i follows a standard exponential or a standardized Weibull distribution, and ψ_i assumes the form

$$\psi_i = \omega + \sum_{j=1}^{r} \gamma_j x_{i-j} + \sum_{j=1}^{s} \omega_j \psi_{i-j}. \tag{6.33}$$

Such a model is referred to as an ACD(r,s) model. When the distribution of ϵ_i is exponential, the resulting model is called an EACD(r,s) model. Similarly, if ϵ_i follows a Weibull distribution, the model is a WACD(r,s) model. If necessary, readers are referred to Appendix A for a quick review of exponential and Weibull distributions.

Similar to GARCH models, the process $\eta_i = x_i - \psi_i$ is a martingale difference sequence (i.e., $E(\eta_i | F_{i-1}) = 0$), and the ACD(r,s) model can be written as

$$x_i = \omega + \sum_{j=1}^{\max(r,s)} (\gamma_j + \omega_j) x_{i-j} - \sum_{j=1}^{s} \omega_j \eta_{i-j} + \eta_j, \tag{6.34}$$

which is in the form of an ARMA process with non-Gaussian innovations. It is understood here that $\gamma_j = 0$ for $j > r$ and $\omega_j = 0$ for $j > s$. Such a representation can be used to obtain the basic conditions for weak stationarity of the ACD model. For instance, taking expectation on both sides of Equation (6.34) and assuming weak stationarity, we have

$$E(x_i) = \frac{\omega}{1 - \sum_{j=1}^{\max(r,s)} (\gamma_j + \omega_j)}.$$

Therefore, we assume that $\omega > 0$ and $1 > \sum_j (\gamma_j + \omega_j)$ because the expected duration is positive. As another application of Equation (6.34), we study properties of the EACD(1,1) model.

EACD(1,1) Model. An EACD(1,1) model can be written as

$$x_i = \psi_i \epsilon_i, \quad \psi_i = \omega + \gamma_1 x_{i-1} + \omega_1 \psi_{i-1}, \tag{6.35}$$

where ϵ_i follows the standard exponential distribution. Using the moments of a standard exponential distribution in Appendix A, we have $E(\epsilon_i) = 1$, $\text{Var}(\epsilon_i) = 1$, and $E(\epsilon_i^2) = \text{Var}(x_i) + [E(x_i)]^2 = 2$. Assuming that x_i is weakly stationary (i.e., the first two moments of x_i are time invariant), we derive the variance of x_i. First, taking the expectation of Equation (6.35), we have

$$E(x_i) = E[E(\psi_i \epsilon_i | F_{i-1})] = E(\psi_i), \quad E(\psi_i) = \omega + \gamma_1 E(x_{i-1}) + \omega_1 E(\psi_{i-1}). \tag{6.36}$$

Under weak stationarity, $E(\psi_i) = E(\psi_{i-1})$ so that Equation (6.36) gives

$$\mu_x \equiv E(x_i) = E(\psi_i) = \frac{\omega}{1 - \gamma_1 - \omega_1}. \tag{6.37}$$

Next, because $E(\epsilon_i^2) = 2$, we have $E(x_i^2) = E[E(\psi_i^2 \epsilon_i^2 | F_{i-1})] = 2E(\psi_i^2)$.
Taking the square of ψ_i in Equation (6.35) and the expectation and using weak stationarity of ψ_i and x_i, we have, after some algebra, that

$$E(\psi_i^2) = \mu_x^2 \times \frac{1 - (\gamma_1 + \omega_1)^2}{1 - 2\gamma_1^2 - \omega_1^2 - 2\gamma_1 \omega_1}. \tag{6.38}$$

Finally, using $\text{Var}(x_i) = E(x_i^2) - [E(x_i)]^2$ and $E(x_i^2) = 2E(\psi_i^2)$, we have

$$\text{Var}(x_i) = 2E(\psi_i^2) - \mu_x^2 = \mu_x^2 \times \frac{1 - \omega_1^2 - 2\gamma_1 \omega_1}{1 - \omega_1^2 - 2\gamma_1 \omega_1 - 2\gamma_1^2},$$

where μ_x is defined in Equation (6.37). This result shows that, to have time-invariant unconditional variance, the EACD(1,1) model in Equation (6.35) must satisfy $1 > 2\gamma_1^2 + \omega_1^2 + 2\gamma_1 \omega_1$. The variance of a WACD(1,1) model can be obtained using the same techniques and the first two moments of a standardized Weibull distribution.

ACD Models with a Generalized Gamma Distribution. In the statistical literature, intensity function is often expressed in terms of hazard function. As shown in Appendix B, the hazard function of an EACD model is constant over time and that of a WACD model is a monotonous function. These hazard functions are rather restrictive in application, as the intensity function of stock transactions might not be constant or monotone over time. To increase the flexibility of the associated hazard function, Zhang et al. (2005) employ a (standardized) generalized gamma distribution for ϵ_i. See Appendix A for some basic properties of a generalized gamma distribution. The resulting hazard function may assume various patterns, including U shape or inverted U shape. We refer to an ACD model with innovations that follow a generalized gamma distribution as a GACD(r, s) model.

6.5.3 Estimation

For an ACD(r, s) model, let $i_o = \max(r, s)$ and $x_t = (x_1, \ldots, x_t)'$. The likelihood function of the durations $x_1, \ldots, x_T$ is

$$f(x_T | \theta) = \left[\prod_{i=i_o+1}^{T} f(x_i | F_{i-1}, \theta) \right] \times f(x_{i_o} | \theta),$$

where θ denotes the vector of model parameters and T is the sample size. The marginal probability density function (pdf) $f(x_{i_o} | \theta)$ of the previous equation is rather complicated for a general ACD model. Because its impact on the likelihood function is diminishing as the sample size T increases, this marginal density is often ignored, resulting in use of the conditional likelihood method. For a WACD model, we use the pdf of Equation (6.48) and obtain the conditional log likelihood function

$$\ell(x|\theta, x_{i_o}) = \sum_{i=i_0+1}^{T} \alpha \ln\left[\Gamma\left(1 + \frac{1}{\alpha}\right)\right] + \ln\left(\frac{\alpha}{x_i}\right) + \alpha \ln\left(\frac{x_i}{\psi_i}\right) - \left(\frac{\Gamma\left(1 + \frac{1}{\alpha}\right)x_i}{\psi_i}\right)^{\alpha},$$

(6.39)

where $\psi_i = \omega + \sum_{j=1}^{r} \gamma_j x_{i-j} + \sum_{j=1}^{s} \omega_j \psi_{i-j}$, $\theta = (\omega, \gamma_1, \ldots, \gamma_r, \omega_1, \ldots, \omega_s, \alpha)'$ and $x = (x_{i_o+1}, \ldots, x_T)'$. When $\alpha = 1$, the (conditional) log likelihood function reduces to that of an EACD(r, s) model.

For a GACD(r, s) model, the conditional log likelihood function is

$$\ell(x|\theta, x_{i_o}) = \sum_{i=i_0+1}^{T} \ln\left(\frac{\alpha}{\Gamma(\kappa)}\right) + (\kappa\alpha - 1)\ln(x_i) - \kappa\alpha \ln(\lambda\psi_i) - \left(\frac{x_i}{\lambda\psi_i}\right)^{\alpha}, \quad (6.40)$$

where $\lambda = \Gamma(\kappa)/\Gamma(\kappa + 1/\alpha)$ and the parameter vector θ now also includes κ. As expected, when $\kappa = 1$, $\lambda = 1/\Gamma(1 + 1/\alpha)$, and the log likelihood function in Equation (6.40) reduces to that of a WACD(r, s) model in Equation (6.39). This log likelihood function can be rewritten in many ways to simplify the estimation.

Under some regularity conditions, the conditional maximum likelihood estimates are asymptotically normal; see Engle and Russell (1998) and the references therein. In practice, simulation can be used to obtain finite-sample reference distributions for the problem of interest once a duration model is specified.

Example 6.3. As an illustration of duration models, we consider the intraday transaction durations of CAT stock from January 4 to January 8, 2010. As stated in the previous section, there are 37,674 positive durations during the 5 trading days, and we use the simple quadratic function of Equation (6.31) to remove the diurnal pattern of the data. Our analysis, thus, employs the adjusted durations of Figure 6.10.

Let x_i be the adjusted duration series. Figure 6.11a shows the sample ACF of x_i. Clearly, there are strong and persistent serial correlations in the data. If an EACD(1,1) model is entertained, we obtain the model

$$x_i = \psi_i \epsilon_i, \quad \psi_i = 0.0124 + 0.0411 x_{i-1} + 0.9503 \psi_{i-1},$$

where all estimates are highly significant with a minimum t-ratio of 6.59. The ACF of the residual series $\hat{\epsilon}_i = x_i/\hat{\psi}_i$ shows some small, but significant, serial correlations at lower-order lags. The model can be improved.

We entertain a WACD(1,2) model for the data and obtain the result

$$x_i = \psi_i \epsilon_i, \quad \psi_i = 0.0128 + 0.0573 x_{i-1} + 0.541 \psi_{i-1} + 0.393 \psi_{i-2}, \quad (6.41)$$

where $\{\epsilon_i\}$ is a sequence of independent and identically distributed random variates that follow the standardized Weibull distribution with parameter $\hat{\alpha} = 1.234(0.004)$,

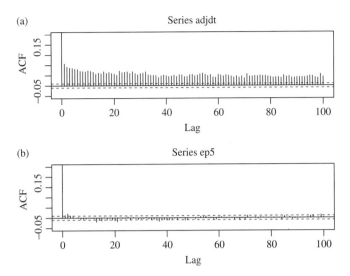

Figure 6.11. Sample ACF of adjusted trading durations for Caterpillar stock from January 4 to January 8, 2010: (a) the adjusted durations and (b) innovations of a WACD(1,2) model.

where 0.005 is the estimated standard error. Standard errors of the estimates in Equation (6.41) are 0.002, 0.004, 0.069, and 0.067, respectively. All t-ratios of the estimates are greater than 5.9, indicating that the estimates are highly significant. Figure 6.12 shows the time plot of $\hat{\epsilon}_i = x_i/\hat{\psi}_i$ for the WACD(1,2) model in Equation (6.41), and Figure 6.11b provides the sample ACF of $\hat{\epsilon}_i$. The time plot shows a few large outlying innovations and the ACF indicates that some minor serial correlations remain in the data. The sample mean and standard deviation of $\hat{\epsilon}_i$ are 0.99 and 0.95, respectively. The mean is close to 1, but the standard deviation is larger than that of a Weibull distribution with $\alpha = 1.234$. This is consistent with the outlying innovations observed from the time plot. Overall, the fitted model describes the data reasonably well, but it can be further improved.

In model (6.41), the estimated coefficients show $\hat{\gamma}_1 + \hat{\omega}_1 + \hat{\omega}_2 \approx 0.992$, indicating certain persistence in the adjusted durations. The expected adjusted duration is $0.0128/(1 - 0.992) = 1.60$ s, which is close to the sample mean 1.42 of the adjusted durations. The estimated α of the standardized Weibull distribution is 1.234, which is greater than 1. Thus, the conditional hazard function is monotonously increasing.

Finally, it is common in statistical modeling to see that more sophisticated models are needed to adequately describe the data when the sample size is large. The analysis can be simplified when a subsample is used. For the adjusted durations of the CAT stock, we reanalyze the data using the first 1200 observations. In this subsample, a WACD(1,1) model appears to be adequate. The fitted model is

$$x_i = \psi_i \epsilon_i, \quad \psi_i = 0.162 + 0.068x_{i-1} + 0.788\psi_{i-1},$$

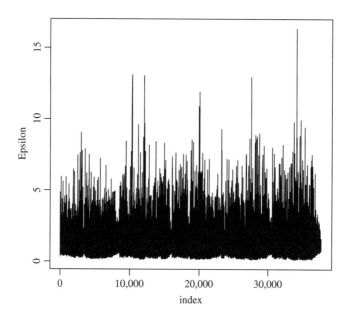

Figure 6.12. Time plot of innovations of a WACD(1,2) model fitted to adjusted trading durations of Caterpillar stock from January 4 to January 8, 2010.

where ϵ_i follows a standardized Weibull distribution with parameter $\alpha = 1.478$ (0.029), where the number in parentheses is the standard error. The standard errors of the coefficient estimates are 0.067, 0.019, and 0.071. Figure 6.13a shows the time plot of the innovations $\hat{\epsilon}_i$, whereas Figure 6.13b gives the ACF of $\hat{\epsilon}_i$. From the plots, the model seems to be adequate. As a matter of fact, the Ljung–Box statistics of $\hat{\epsilon}_i$ gives $Q(10) = 9.60(0.47)$ and $Q(20) = 14.95(0.78)$, where the number in parentheses denotes the p-value. For the ϵ_i^2 series, we obtain $Q(10) = 5.69(0.84)$ and $Q(20) = 10.38(0.96)$. These statistics confirm that the innovations of the WACD(1,1) model have no serial correlations or conditional heteroscedasticity.   □

Remark. Estimation of EACD models can be carried out using programs for ARCH models with some minor modification (Engle and Russell, 1998). In this book, we use a simple R script.   □

R Demonstrations of Duration Models. Output edited.

```
> source("acd.R")
> m2=acd(adjdt,order=c(1,1),cond.dist="exp")
Coefficient(s):
          Estimate   Std. Error    t value    Pr(>|t|)
omega 0.01247473   0.00189210    6.59305  4.3087e-11  ***
alpha 0.04106574   0.00273273   15.02735  < 2.22e-16  ***
```

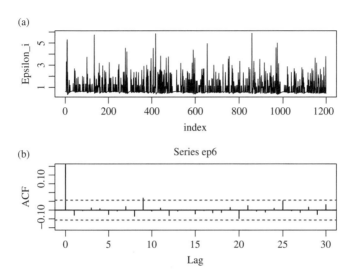

Figure 6.13. Model checking for a WACD(1,1) model fitted to the first 1200 observations of adjusted trading durations of Caterpillar stock from January 4 to January 8, 2010: (a) innovation series and (b) ACF of the innovations.

```
beta  0.95029295  0.00364684 260.57992 < 2.22e-16 ***
---
> names(m2)
[1] "estimates" "Hessian"    "epsilon"
> m3=acd(adjdt,order=c(1,1),cond.dist="weibull")
> m5=acd(adjdt,order=c(1,2),cond.dist="weibull")
Coefficient(s):
        Estimate  Std. Error   t value    Pr(>|t|)
omega 0.01275520  0.00183458   6.95264 3.5849e-12 ***
alpha 0.05729856  0.00359740  15.92776 < 2.22e-16 ***
beta1 0.54141462  0.06912212   7.83273 4.8850e-15 ***
beta2 0.39333491  0.06667232   5.89952 3.6455e-09 ***
shape 1.23368806  0.00450529 273.83124 < 2.22e-16 ***
---
> ep5=m5$epsilon
> acf(ep5,ylim=c(-0.05,0.25))

> adt1=adjdt[1:1200]   % Subsample
> plot(adt1,type='l')
> m6=acd(adt1,order=c(1,1),cond.dist="weibull")
Coefficient(s):
        Estimate  Std. Error  t value   Pr(>|t|)
omega 0.1616787   0.0670298   2.41204  0.0158635 *
alpha 0.0677561   0.0194223   3.48857  0.0004856 ***
beta  0.7881004   0.0710153  11.09761 < 2.22e-16 ***
```

```
shape 1.4783354    0.0292454 50.54928 < 2.22e-16 ***
---
> ep6=m6$epsilon
> Box.test(ep6,lag=10,type='Ljung')
          Box-Ljung test
data:   ep6
X-squared = 9.6035, df = 10, p-value = 0.4759
> Box.test(ep6,lag=20,type='Ljung')
> Box.test(ep6^2,lag=10,type='Ljung')
> Box.test(ep6^2,lag=20,type='Ljung')
> par(mfcol=c(2,1))
> plot(ep6,type='l',xlab='index',ylab='epsilon_t')
> acf(ep6,ylim=c(-0.1,.25))
```

One can apply the ACD model to study asset volatility (Chou, 2005). Here volatility is measured by the daily price range (or range of log prices). Similar to duration, daily price range is nonnegative and can be used as an alternative approach to quantify price variability. Tsay (2009) and Tsay (2010, Chapter 6) apply ACD models with intervention analysis to investigate the impact on stock volatility of switching from tick size to the decimal system on January 29, 2001.

6.6 REALIZED VOLATILITY

In this section, we return to the topic of using high frequency data to compute volatility. In Equation (4.45), if we further assume that the sample mean $\bar{r}_t$ is 0, then we have $\hat{\sigma}_m^2 \approx \sum_{i=1}^n r_{t,i}^2$. In this case, the cumulative sum of squares of daily log returns in a month is used as an estimate of monthly volatility. This concept has been generalized to estimate daily volatility of an asset using intraday log returns. Let r_t be the daily log return of an asset. Suppose that there are n equally spaced intraday log returns available such that $r_t = \sum_{i=1}^n r_{t,i}$. The quantity

$$\text{RV}_t = \sum_{i=1}^n r_{t,i}^2$$

is called the *realized* volatility of r_t (Andersen et al., 2001a,b).

Mathematically, realized volatility assumes that the log price x_t of a security follows the model

$$dx_t = \mu_t dt + \sigma_t dw_t, \qquad (6.42)$$

where w_t is a standard Brownian motion and μ_t and σ_t are the drift and diffusion of x_t, respectively. This stochastic diffusion equation is known as an *Ito process*. See Tsay (2010, Chapter 6) for more information. The quantity of interest is the integrated variance $\int_0^T \sigma_t^2 dt$ for some time interval $[0, T]$. For instance, in studying realized volatility, T is often chosen to be 1 day. For most securities, the drift μ_t is close to 0 when the time interval is small so that Equation (6.42) reduces to $dx_t = \sigma_t dw_t$. In discrete-time, the model says that the log return of the security within a small time interval Δ can be approximated by $\sigma_t \epsilon_t$, where ϵ_t is a normal random variable with mean 0 and variance Δ. Now, divide the interval $[0, T]$ into $n = T/\Delta$ subintervals and compute the log return $r_t = x_{t\Delta} - x_{(t-1)\Delta}$ for $t = 1, \ldots, n$. Thus, r_t is the intraday log return for time tth interval of length Δ. A natural way to estimate the integrated variance is based on the following statistical property:

$$\lim_{n \to \infty} \sum_{t=1}^{n} r_t^2 \to_p \int_0^T \sigma_t^2 dt, \tag{6.43}$$

where $\to_p$ denotes convergence in probability. Since $n \to \infty$ is equivalent to $\Delta \to 0$, the prior equation states that one can use the quadratic variation to estimate the integrated variance, and the estimate should be more accurate if the time interval Δ is small. Thus, realized volatility is an integrated volatility. Note that it is not necessary to have equally spaced intraday log returns for Equation (6.43) to hold. As long as the length of the largest interval approaches 0, the result holds.

Equation (6.43) implies that one would use as many intraday returns as possible to compute the daily realized volatility. In practice, this means using the tick-by-tick log returns, because they are the intraday log returns at the finest possible intervals. However, the observed returns are contaminated by market microstructure noises, say $r_t^o = r_t + e_t$. Two well-known examples of the microstructure noises are the bid–ask bounce and nonsynchronous trading (Chapter 6). Such noises introduce bias when one uses the empirical quadratic variation to estimate the integrated variance of the underlying asset price. As a matter of fact, the bias becomes more serious as the time interval Δ becomes finer. One can even show that, under certain assumptions, $\sum_{t=1}^{n} (r_t^o)^2/(2n)$ converges to the variance of the noise e_t instead of the integrated variance of x_t (Zhang et al., 2001; Bandi and Russell, 2008). Consequently, methods must be sought to obtain bias-corrected estimate of the realized volatility.

To demonstrate, we consider the tick-by-tick data for the Caterpillar stock from January 4 to May 28, 2010. The data are from the Trade-and-Quote (TAQ) database of the New York Stock Exchange. The numbers of trading days for the 5 months are 19, 19, 23, 21, and 20, respectively. Thus, we analyze tick-by-tick data for 102 trading days. For simplicity, we only employ transactions in the normal trading hours from 9:30 AM to 4:00 PM Eastern time. Figure 6.14 shows the time plot of daily numbers of transactions in the sample. It is clear that the trading intensity varies markedly from one day to another. Figure 6.15 shows the intraday 5-min log returns of the stock for May 2010, whereas Figure 6.16 gives the histogram of the returns. To compute the log returns, we follow the convention using the last transaction price in the interval as the stock price for that interval. As expected, the returns vary around 0, show volatility clustering, and have heavy tails. As will be seen later, the extreme returns have a substantial impact on realized volatility.

Turn to realized volatility. We compute the realized volatility using different time intervals. In particular, time interval of length "0" denotes the case in which we simply

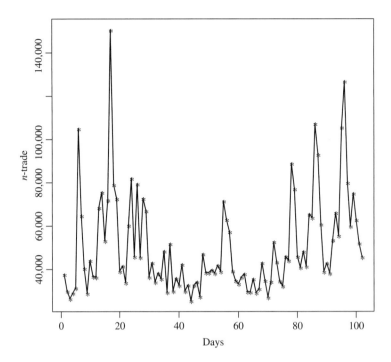

Figure 6.14. Time plot of the daily number of transactions in the normal trading hours of Caterpillar stock from January 04, 2010 to May 28, 2010.

employ the sum of squared log returns of consecutive trades in each trading day. This means that we treat the observed sequence as the true sequence of transactions when there are multiple transactions within a second. We employ this extreme case to highlight the impact of microstructure noises on realized volatility. Figure 6.17 shows the time plot of realized volatility of Caterpillar stock with different time intervals. The volatility is annualized and the time intervals used are 0, 1, 2, 3, 4, 5, 10, 15, 20 and 30 min. Therefore, we consider 10 estimates for the realized volatility. The solid line represents the realized volatility with time interval 0. From the plot, we make the following observations. First, realized volatility for the case of zero time interval differs substantially from those of other time intervals. In fact, the solid line is truncated. This is in agreement with the theory mentioned earlier as the realized volatility in this particular case is estimating $2n \times \text{Var}(e_t)$, which is unbounded as n increases. Second, the realized volatility varies substantially from one time interval to another. On the other hand, they do show similar characteristics. Thus, the plot raises the important issue of how to obtain a reasonable realized volatility.

To gain further insight, we obtain the boxplot for the 10 realized volatility series (Fig. 6.18). The plot ignores the serial dependence of the volatility, but it provides a summary for the size of the realized volatility. Figure 6.19 shows the mean and

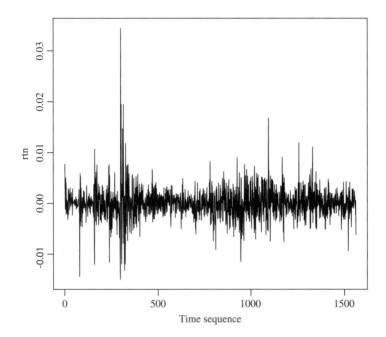

Figure 6.15. Time plot of intraday 5-min log returns for Caterpillar stock from May 03 to May 28, 2010. The last transaction price of each interval is used.

median of the realized volatility for each time interval. These plots indicate that an interval of 3–5 min might work well for the Caterpillar stock considered.

Remark. We use R scripts to obtain various realized volatility series. These scripts are available on the book web page. □

R Demonstration

```
> source("hfanal.R")
% Process January data
> da=read.table("taq-cat-jan2010.txt",header=T)
> m1=hfanal(da,1)
> names(m1)
[1] "returns"  "Ytot"       "realized" "ntrad"
> Ytot=m1$Ytot
> Ntrad=m1$ntrad
> Rv=cbind(Ytot,m1$realized)
> m2=hfanal(da,2)
> Rv=cbind(Rv,m2$realized)
> m3=hfanal(da,3)
> Rv=cbind(Rv,m3$realized)
```

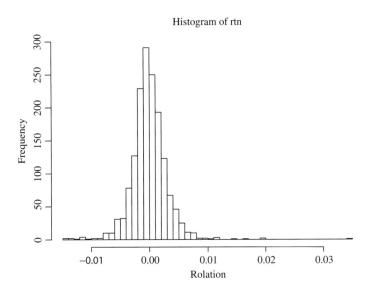

Figure 6.16. Histogram of intraday 5-min log returns for Caterpillar stock from May 03 to May 28, 2010. The last transaction price of each interval is used.

```
> m4=hfanal(da,4)
> Rv=cbind(Rv,m4$realized)
> m5=hfanal(da,5)
> Rv=cbind(Rv,m5$realized)
> m6=hfanal(da,10)
> Rv=cbind(Rv,m6$realized)
> m7=hfanal(da,15)
> Rv=cbind(Rv,m7$realized)
> m8=hfanal(da,20)
> Rv=cbind(Rv,m8$realized)
> m9=hfanal(da,30)
> Rv=cbind(Rv,m9$realized)
%% Process February data
> da=read.table("taq-cat-feb2010.txt",header=T)
> m1=hfanal(da,1)
> Rv2=cbind(m1$Ytot,m1$realized)
> Ytot=c(Ytot,m1$Ytot)
> Ntrad=c(Ntrad,m1$ntrad)
> m2=hfanal(da,2)
> Rv2=cbind(Rv2,m2$realized)
> m3=hfanal(da,3)
> Rv2=cbind(Rv2,m3$realized)
> m4=hfanal(da,4)
> Rv2=cbind(Rv2,m4$realized)
> m5=hfanal(da,5)
> Rv2=cbind(Rv2,m5$realized)
```

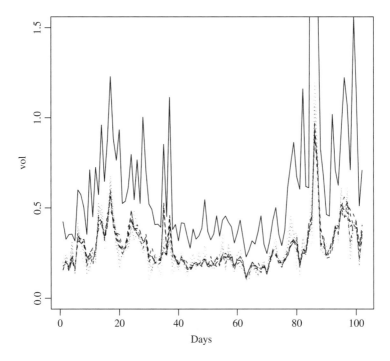

Figure 6.17. Realized volatilities for Caterpillar stock from January 04, 2010 to May 28, 2010. The solid line is based on tick-by-tick log returns, whereas the dashed lines are based on log returns for time intervals of 1, 2, 3, 4, 5, 10, 15, 20, and 30 min.

```
> m6=hfanal(da,10)
> Rv2=cbind(Rv2,m6$realized)
> m7=hfanal(da,15)
> Rv2=cbind(Rv2,m7$realized)
> m8=hfanal(da,20)
> Rv2=cbind(Rv2,m8$realized)
> m9=hfanal(da,30)
> Rv2=cbind(Rv2,m9$realized)

> RV=rbind(Rv,Rv2)   % Combine Jan and Feb results
% repeat the same process for March, April and May data.
```

6.6.1 Handling Microstructure Noises

There are two methods commonly used in the literature to handle the microstructure noises in calculating realized volatility. The first method is used to obtain an optimal sampling interval (Zhang et al., 2001; Bandi and Russell, 2008). The idea here is to

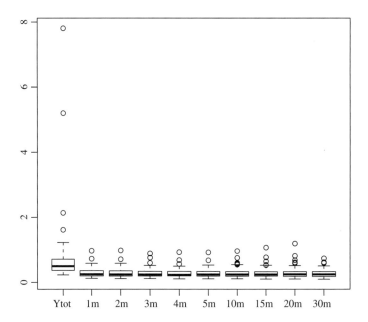

Figure 6.18. Boxplot of realized volatilities for Caterpillar stock from January 04, 2010 to May 28, 2010. The realized volatilities are obtained by intraday log returns of intervals with length 0, 1, 2, 3, 4, 5, 10, 15, 20, and 30 min, where 0 denotes the result obtained using tick-by-tick log returns.

minimize the mean square of errors in estimating the integrated volatility. It represents a compromise between accuracy in approximating the true realized volatility and reducing the bias caused by microstructure noises. Roughly speaking, one uses the tick-by-tick log returns to estimate the variance of the noise, then obtains an optimal sampling interval to minimize the mean square of errors in estimation. The resulting optimal interval varies from stock to stock and period to period. For heavily traded stocks, the optimal interval is often between 1 to 5 min.

The second method to calculating realized volatility is to use subsampling and to correct the bias. This results in a two-scale procedure (Zhang et al., 2001). We use a simple example to introduce the idea of subsampling. Suppose that we are interested in 5-min intraday log returns of an asset. Typically, we use the time intervals (9:30, 9:35], (9.35, 9:40], ... to construct a series of 5-min log returns. In this way, we drop many transaction prices of the asset because only the last transaction price of each interval is used. One can also use the intervals (9:31, 9:36], (9.36, 9:41], ... to compute another series of 5-min log returns. This process continues until we use the intervals (9:34,9:39], (9.39,9:44], ... to obtain yet another series of 5-min log returns. In this manner, we have five time series of 5-min intraday log returns, each of them can be used to construct a realized volatility for the asset. A simple average of these five estimates should provide a more accurate realized volatility. As a matter of fact, we can use other increment, instead of 1 min, to construct more sequences of 5-min

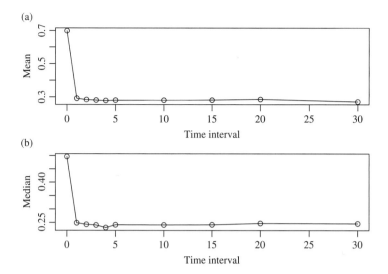

Figure 6.19. Mean and median of realized volatilities for Caterpillar stock from January 04, 2010 to May 28, 2010. The realized volatilities ate obtained by intraday log returns of intervals with length 0, 1, 2, 3, 4, 5, 10, 15, 20, and 30 min, where 0 denotes the result obtained using tick-by-tick log returns.

time intervals, for example, (9:30:30, 9:35:30], (9:35:30, 9:40:30], ... In practice, a choice must be made to keep the procedure simple. Here we use 1 min as the time increment to construct sequences of time intervals.

Mathematically, the two-scale estimator of realized volatility for the tth trading day of a security can be written as

$$RV_t = \left(1 - \frac{\bar{n}}{n}\right)^{-1}\left(RV_t^{\text{ave}} - \frac{\bar{n}}{n}RV_{0,t}\right),$$

where

$$RV_t^{\text{ave}} = \frac{1}{m}\sum_{i=1}^{m}\sum_{j=1}^{n_i}r_{i,j}^2, \quad RV_{0,t} = \sum_{j=1}^{n}r_j^2,$$

where m is the number of subsamplings, $r_{i,j}$ is the jth log returns of the ith subsampling, n_i is the number of log returns for the ith subsampling, $\bar{n}$ is the average of $\{n_i\}$, r_j is the log return of the jth transaction (from the $(j-1)$th transaction) of the security, and n is the total number of log returns in day t. The quantity $\bar{n}RV_{0,t}/n$ is the bias correction to remove the impact of microstructure noises and the average is used to improve the estimation because it uses more transaction prices.

Figure 6.20 shows the two-scale estimator of realized volatility for the Caterpillar stock from January 04 to May 28, 2010. Here we use 5-min log returns and the

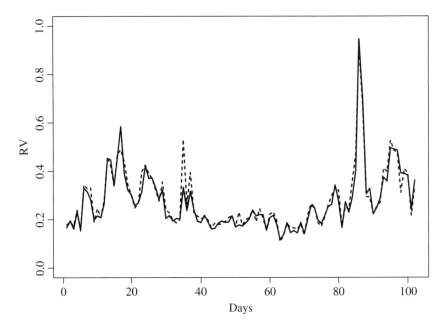

Figure 6.20. Realized volatilities for the Caterpillar stock from January 04 to May 28, 2010. The solid line denotes the two-scale estimator using 5-min log returns and 1-min increment in subsampling. The dashed line is the result using the usual 5-min log return series.

increment in subsampling is 1 min. Thus, RV_t^{ave} is based on five realized volatilities. All volatilities are annualized. Also shown in the plot in dashed line is the realized volatility based on the usual 5-min log returns. There are some differences between the two estimates. Figure 6.21 is the scatter plot between the two-scale estimator and the average estimator, that is, RV_t^{ave}. The two estimates are close. Thus, in this particular instance, the bias correction only has a small effect in estimating the realized volatility.

R Demonstration

```
> source("hf2ts.R")
> da=read.table("taq-cat-may2010.txt",header=T)
> m5=hf2ts(da,int=5)
> names(m5)
[1] "Ytot"     "realized" "ave.RV"   "ntrad"
> da=read.table("taq-cat-apr2010.txt",header=T)
> m4=hf2ts(da,int=5)
> da=read.table("taq-cat-mar2010.txt",header=T)
> m3=hf2ts(da,int=5)
> da=read.table("taq-cat-feb2010.txt",header=T)
> m2=hf2ts(da,int=5)
> da=read.table("taq-cat-jan2010.txt",header=T)
> m1=hf2ts(da,int=5)
```

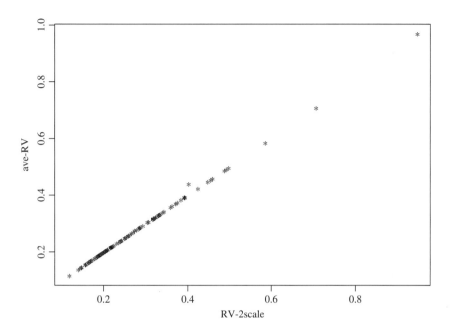

Figure 6.21. Scatter plot of realized volatilities for the Caterpillar stock from January 04 to May 28, 2010. The horizontal axis denotes the two-scale estimator and the *y*-axis denotes the average estimator of five 5-min log return series.

```
% Combine the results
> Ytot=c(m1$Ytot,m2$Ytot,m3$Ytot,m4$Ytot,m5$Ytot)   % Consecutive trades
% 2-scale method (Zhang et al. method)
> RV=c(m1$realized,m2$realized,m3$realized,m4$realized,m5$realized)
% average of 5-m RV.
> mRV=c(m1$ave.RV,m2$ave.RV,m3$ave.RV,m4$ave.RV,m5$ave.RV)
```

6.6.2 Discussion

Once a realized volatility series is obtained, it can be analyzed to produce volatility forecasts. We use the realized volatility series constructed by 5-min intraday log returns of the Caterpillar stock from January 04 to May 28, 2010 to demonstrate the analysis. The realized volatility and its log series are shown in Figure 6.22, with the log series in the panel (b). Let V_t be the realized volatility. Applying the modeling process of Chapter 2, we obtain the AR(1) model

$$(1 - 0.6527B)(V_t - 0.2794) = a_t, \quad \sigma_a^2 = 0.008742, \quad (6.44)$$

where the standard errors of the estimates are 0.0741 and 0.0262, respectively. AIC of the model is -187.42. Model checking shows that this simple AR(1) model is adequate

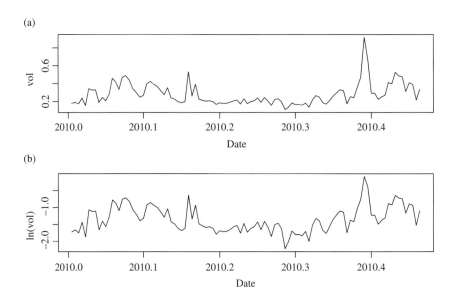

Figure 6.22. Time plots of realized volatility and its log series for Caterpillar stock from January 04 to May 28, 2010. The realized volatility is based on 5-min intraday log returns and shown in panel (a).

TABLE 6.6. Forecasts For Realized Volatility Of Caterpillar Stock. The Sampling Period is From January 04 to May 28, 2010, and The Realized Volatility Is Constructed Using 5-min Intraday Log Returns. The Forecast Origin Is May 28, 2010

Model	1-step	2-step	3-step	4-step	5-step
AR(1)	0.3154	0.3029	0.2947	0.2894	0.2860
AR(6)	0.3569	0.3381	0.3443	0.3447	0.3058
ARIMA(0,1,3)	0.3163	0.3753	0.3623	0.3639	0.3654

(Fig. 6.23). We then use the model to produce 1-step to 5-step ahead forecasts. The results are given in Table 6.6.

Alternatively, we also consider the log series v_t of the realized volatility. Two models are found to fit the data well. The first one is an AR(6) model:

$$(1 - 0.6296B - 0.2032B^6)(v_t + 1.3507) = a_t, \quad \sigma_a^2 = 0.07161, \qquad (6.45)$$

where the standard errors of the estimates are 0.0705, 0.0726, and 0.1437, respectively. The AIC of the model is 29.62. Model checking, not shown, also indicates that the model is adequate. The second model for the log volatility series is

$$(1 - B)v_t = (1 - 0.4287B - 0.2426B^3)a_t, \quad \sigma_a^2 = 0.07662, \qquad (6.46)$$

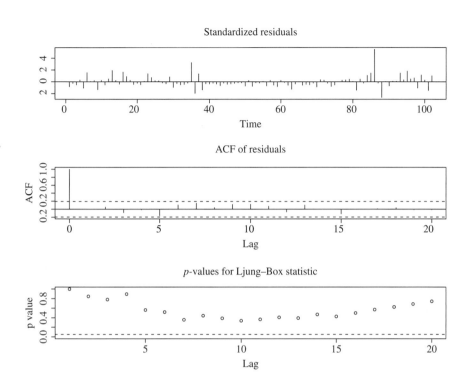

Figure 6.23. Model checking plots for an AR(1) model fitted to the realized volatility of Caterpillar stock from January 04 to May 28, 2010.

where the standard errors of the MA coefficients are 0.0906 and 0.0729, respectively. The AIC of the model is 33.67. Again, model checking, not shown, fails to indicate any model inadequacy. If AIC criterion is used to compare the models, then one would select the AR(6) model in Equation (6.45) for the log volatility series. We include the ARIMA(0,1,3) model here because limited experience indicates that $\ln(RV_t)$ often follows approximately a Gaussian ARIMA(0,1,q) model. Note that to obtain volatility forecasts for the models in Equations (6.45) and (6.46), we make use of the relationship between log-normal and normal distributions. In other words, we use $E(V_t) = \exp[E(v_t) + 0.5\text{Var}(v_t)]$. The resulting volatility forecasts are also given in Table 6.6. Finally, one cannot compare the AIC or other information criterion between models for V_t and models for v_t. On the other hand, as realized volatility can be directly calculated from the data, one can use backtesting to compare the models. We do not perform backtesting here because there are only 102 data points in the series.

Advantages of realized volatility include simplicity and making use of intraday returns. On the other hand, realized volatility for stock returns does not include the overnight volatility of the asset. By overnight volatility, we mean the volatility of the asset when the market is close, for example, from the close of day $t - 1$ to the open of day t. In some instances, this overnight volatility could be substantial. For instance, most quarterly earnings are announced after the market close, and earning surprises tend to have marked impact on stock volatility. Consequently, realized volatility is likely to underestimate the true volatility for stock returns. A possible solution to this problem is to use a weighted average between overnight variance and realized volatility (Hansen and Lunde, 2005). However, the square of overnight return is not an accurate estimate of overnight volatility. Our limited experience shows that overnight returns appear to be small for index returns or foreign exchange returns. Finally, in a series of recent articles, Barndorff-Nielsen and Shephard (2004) have used high frequency returns to study bipower variations of an asset return and developed some methods to detect jumps in volatility.

APPENDIX A: SOME PROBABILITY DISTRIBUTIONS

Exponential Distribution. A random variable X has an exponential distribution with parameter $\beta > 0$ if its probability density function (pdf) is given by

$$f(x|\beta) = \begin{cases} \dfrac{1}{\beta}e^{-x/\beta} & \text{if } x \geq 0, \\ 0 & \text{otherwise.} \end{cases}$$

Denoting such a distribution by $X \sim \exp(\beta)$, we have $E(X) = \beta$ and $\text{Var}(X) = \beta^2$. The cumulative distribution function (CDF) of X is

$$F(x|\beta) = \begin{cases} 0 & \text{if } x < 0, \\ 1 - e^{-x/\beta} & \text{if } x \geq 0. \end{cases}$$

When $\beta = 1$, X is said to have a standard exponential distribution.

Gamma Function. For $\kappa > 0$, the gamma function $\Gamma(\kappa)$ is defined by

$$\Gamma(\kappa) = \int_0^\infty x^{\kappa-1}e^{-x}dx.$$

The most important properties of the gamma function are:

1. For any $\kappa > 1$, $\Gamma(\kappa) = (\kappa - 1)\Gamma(\kappa - 1)$.
2. For any positive integer m, $\Gamma(m) = (m - 1)!$.
3. $\Gamma(\frac{1}{2}) = \sqrt{\pi}$.

The integration

$$\Gamma(y|\kappa) = \int_0^y x^{\kappa-1}e^{-x}dx, \quad y > 0$$

is an *incomplete* gamma function. Its values have been tabulated in the literature. Computer programs are now available to evaluate the incomplete gamma function.

Gamma Distribution. A random variable X has a gamma distribution with parameter κ and β $(\kappa > 0, \beta > 0)$ if its pdf is given by

$$f(x|\kappa, \beta) = \begin{cases} \dfrac{1}{\beta^\kappa \Gamma(\kappa)} x^{\kappa-1} e^{-x/\beta} & \text{if } x \geq 0, \\ 0 & \text{otherwise.} \end{cases}$$

By changing variable $y = x/\beta$, one can easily obtain the moments of X:

$$E(X^m) = \int_0^\infty x^m f(x|\kappa, \beta) dx = \frac{1}{\beta^\kappa \Gamma(\kappa)} \int_0^\infty x^{\kappa+m-1} e^{-x/\beta} dx$$

$$= \frac{\beta^m}{\Gamma(\kappa)} \int_0^\infty y^{\kappa+m-1} e^{-y} dy = \frac{\beta^m \Gamma(\kappa+m)}{\Gamma(\kappa)}.$$

In particular, the mean and variance of X are $E(X) = \kappa\beta$ and $\text{Var}(X) = \kappa\beta^2$. When $\beta = 1$, the distribution is called a *standard gamma distribution* with parameter κ. We use the notation $G \sim \text{gamma}(\kappa)$ to denote that G follows a standard gamma distribution with parameter κ. The moments of G are

$$E(G^m) = \frac{\Gamma(\kappa+m)}{\Gamma(\kappa)}, \quad m > 0. \tag{6.47}$$

Weibull Distribution. A random variable X has a Weibull distribution with parameters α and β $(\alpha > 0, \beta > 0)$ if its pdf is given by

$$f(x|\alpha, \beta) = \begin{cases} \dfrac{\alpha}{\beta^\alpha} x^{\alpha-1} e^{-(x/\beta)^\alpha} & \text{if } x \geq 0, \\ 0 & \text{if } x < 0, \end{cases}$$

where β and α are the scale and shape parameters, respectively, of the distribution. The mean and variance of X are

$$E(X) = \beta\Gamma\left(1 + \frac{1}{\alpha}\right), \quad \text{Var}(X) = \beta^2 \left\{ \Gamma\left(1 + \frac{2}{\alpha}\right) - \left[\Gamma\left(1 + \frac{1}{\alpha}\right)\right]^2 \right\},$$

respectively, and the CDF of X is

$$F(x|\alpha, \beta) = \begin{cases} 0 & \text{if } x < 0, \\ 1 - e^{-(x/\beta)^\alpha} & \text{if } x \geq 0. \end{cases}$$

When $\alpha = 1$, the Weibull distribution reduces to an exponential distribution.

Define $Y = X/[\beta\Gamma(1 + 1/\alpha)]$. We have $E(Y) = 1$ and the pdf of Y is

$$
f(y|\alpha) = \begin{cases} \alpha \left[\Gamma\left(1 + \frac{1}{\alpha}\right)\right]^{\alpha} y^{\alpha-1} \exp\left\{-\left[\Gamma\left(1 + \frac{1}{\alpha}\right)y\right]^{\alpha}\right\} & \text{if } y \geq 0, \\ 0 & \text{otherwise}, \end{cases}
\tag{6.48}
$$

where the scale parameter β disappears because of standardization. The CDF of the standardized Weibull distribution is

$$
F(y|\alpha) = \begin{cases} 0 & \text{if } y < 0, \\ 1 - \exp\left\{-\left[\Gamma\left(1 + \frac{1}{\alpha}\right)y\right]^{\alpha}\right\} & \text{if } y > 0, \end{cases}
$$

and we have $E(Y) = 1$ and $\mathrm{Var}(Y) = \Gamma(1 + 2/\alpha)/[\Gamma(1 + 1/\alpha)]^2 - 1$. For a duration model with Weibull innovations, the pdf in Equation (6.48) is used in the maximum likelihood estimation.

Generalized Gamma Distribution. A random variable X has a generalized gamma distribution with parameter α, β, κ ($\alpha > 0$, $\beta > 0$, and $\kappa > 0$) if its pdf is given by

$$
f(x|\alpha, \beta, \kappa) = \begin{cases} \dfrac{\alpha x^{\kappa\alpha-1}}{\beta^{\kappa\alpha}\Gamma(\kappa)} \exp\left[-\left(\dfrac{x}{\beta}\right)^{\alpha}\right] & \text{if } x \geq 0, \\ 0 & \text{otherwise}, \end{cases}
$$

where β is a scale parameter, and α and κ are shape parameters. This distribution can be written as

$$
G = \left(\frac{X}{\beta}\right)^{\alpha},
$$

where G is a standard Gamma random variable with parameter κ. The pdf of X can be obtained from that of G by the technique of changing variables. Similarly, the moments of X can be obtained from that of G in Equation (6.47) by

$$
E(X^m) = E[(\beta G^{1/\alpha})^m] = \beta^m E(G^{m/\alpha}) = \beta^m \frac{\Gamma(\kappa + m/\alpha)}{\Gamma(\kappa)} = \frac{\beta^m \Gamma(\kappa + m/\alpha)}{\Gamma(\kappa)}.
$$

When $\kappa = 1$, the generalized gamma distribution reduces to that of a Weibull distribution. Thus, the exponential and Weibull distributions are special cases of the generalized gamma distribution.

The expectation of a generalized gamma distribution is $E(X) = \beta\Gamma(\kappa + 1/\alpha)/\Gamma(\kappa)$. In duration models, we need a distribution with unit expectation.

Therefore, defining a random variable $Y = \lambda X/\beta$, where $\lambda = \Gamma(\kappa)/\Gamma(\kappa + 1/\alpha)$, we have $E(Y) = 1$ and the pdf of Y is

$$f(y|\alpha, \kappa) = \begin{cases} \dfrac{\alpha y^{\kappa\alpha-1}}{\lambda^{\kappa\alpha}\Gamma(\kappa)} \exp\left[-\left(\dfrac{y}{\lambda}\right)^{\alpha}\right] & \text{if } y > 0, \\ 0 & \text{otherwise,} \end{cases} \tag{6.49}$$

where again the scale parameter β disappears and $\lambda = \Gamma(\kappa)/\Gamma(\kappa + \frac{1}{\alpha})$.

APPENDIX B: HAZARD FUNCTION

A useful concept in modeling duration is the *hazard function* implied by a distribution function. For a random variable X, the *survival function* is defined as

$$S(x) \equiv P(X > x) = 1 - P(X \le x) = 1 - \text{CDF}(x), \quad x > 0,$$

which gives the probability that a subject, which follows the distribution of X, survives at the time x. The hazard function (or intensity function) of X is then defined by

$$h(x) = \frac{f(x)}{S(x)}, \tag{6.50}$$

where $f(.)$ and $S(.)$ are the pdf and survival function of X, respectively.

Example. For the Weibull distribution with parameters α and β, the survival function and hazard function are

$$S(x|\alpha, \beta) = \exp\left[-\left(\frac{x}{\beta}\right)^{\alpha}\right], \quad h(x|\alpha, \beta) = \frac{\alpha}{\beta^{\alpha}}x^{\alpha-1}, \quad x > 0.$$

□

    In particular, when $\alpha = 1$, we have $h(x|\beta) = 1/\beta$. Therefore, for an exponential distribution, the hazard function is constant. For a Weibull distribution, the hazard is a monotone function. If $\alpha > 1$, then the hazard function is monotonously increasing. If $\alpha < 1$, the hazard function is monotonously decreasing. For the generalized gamma distribution, the survival function and, hence, the hazard function involve the incomplete gamma function. Yet the hazard function may exhibit various patterns, including U shape or inverted U shape. Thus, the generalized gamma distribution provides a flexible approach to modeling the duration of stock transactions.

    For the standardized Weibull distribution, the survival and hazard functions are

$$S(y|\alpha) = \exp\left\{-\left[\Gamma\left(1 + \frac{1}{\alpha}\right)y\right]^{\alpha}\right\}, \quad h(y|\alpha) = \alpha\left[\Gamma\left(1 + \frac{1}{\alpha}\right)\right]^{\alpha}y^{\alpha-1}, \quad y > 0.$$

EXERCISES

1. Let r_t be the log return of an asset at time t. Assume that $\{r_t\}$ is a Gaussian white noise series with mean 0.02 and variance 0.04. Suppose that the probability of a trade at each time point is 50% and is independent of r_t. Denote the observed return by r_t^o. Is r_t^o serially correlated? If yes, calculate the first three lags of autocorrelations of r_t^o.

2. Let P_t be the observed market price of an asset, which is related to the fundamental value of the asset P_t^* via Equation (6.9). Assume that $\Delta P_t^* = P_t^* - P_{t-1}^*$ forms a Gaussian white noise series with mean 0 and variance 1.0. Suppose that the bid–ask spread is 2 cents. What is the lag-1 autocorrelation of the price change series $\Delta P_t = P_t - P_{t-1}$ when the tick size is 1 cent.

3. The file `taq-aa-t-june72010.txt` contains the tick-by-tick trading data of Alcoa stock on June 7, 2010. It has seven columns, namely, date, hour, minute, second, price, and volume. Focus on the transactions that occurred during the normal hours.
   (a) Obtain the nonzero time durations between trades. What are sample mean and variance of the series.
   (b) Remove the diurnal pattern using

   $$f(t_i) = \exp[d(t_i)], \quad d(t_i) = \beta_0 + \beta_1 f_1(t_i) + \beta_2 f_1^2(t_i) + \beta_3 f_3(t_i),$$

   where $f_1(t_i)$ is defined as in Equation (6.31) and $f_3(t_i) = \ln(t_i)$. Write down the fitted model for diurnal pattern.
   (c) Build an EACD model for the adjusted duration and check the fitted model.
   (d) Build a WACD model for the adjusted duration and check the fitted model.
   (e) Compare the prior two duration models.

4. Consider, again, the transactions data of Alcoa stock on June 7, 2010. Focus on the transactions occurred during the normal trading hours.
   (a) Obtain the price change series and its histogram.
   (b) Divide the price changes into seven categories as those in Table 6.3. What is the percentage of transactions with no price change?
   (c) Fit an ordered probit model similar to that of Example 6.1 to the categorical price changes. Write down the fitted model.
   (d) Fit an ADS model to the categorical price changes similar to Example 6.2. Write down the fitted model.

5. Consider the questions of the prior problem, but using the transactions data of Alcoa stock on June 8, 2010. The data are in the file `taq-aa-t-june82010.txt`.

6. The file `taq-aa-t-june7t112010.txt` contains the transactions data of Alcoa stock from June 7 to June 11, 2010. Focus on the transactions during the normal trading hours. Consider the number of transactions within a 5-min

time interval. Denote the resulting series by x_t. Is there any diurnal pattern in x_t? Why?

7. Consider the tick-by-tick transactions data of the stock of Starbucks from July 25 to July 29, 2011. The data are in `taq-sbux-jul2011.txt`. Focus on trades occurred in the normal trading hours.

   (a) Obtain the intraday 5-min log returns of the stock. Plot the histogram of the returns.

   (b) Obtain daily realized volatility of the stock using 5-min intraday log returns.

   (c) Obtain daily realized volatility of the stock using 1-min intraday log returns using subsampling method with the average estimator.

   (d) Obtain daily realized volatility of the stock using 1-min intraday log returns using subsampling method with the two-scale estimator.

REFERENCES

Andersen TG, Bollerslev T, Diebold FX, Labys P. The distribution of realized exchange rate volatility. J Am Stat Assoc 2001a;96:42–55.

Andersen TG, Bollerslev T, Diebold FX, Labys P. The distribution of realized stock return volatility. J Financ Econ 2001b;61:43–76.

Bandi F, Russell JR. Microstructure noise, realized volatility, and optimal sampling. Rev Econ Studies 2008;2:339–369.

Barndorff-Nielsen OE, Shephard N. Power and bi-power variations with stochastic volatility and jumps (with discussion). J Financ Econometrics 2004;2:1–48.

Campbell JY, Lo AW, MacKinlay AC. The Econometrics of Financial Markets. Princeton (NJ): Princeton University Press; 1997.

Cho D, Russell JR, Tiao GC, Tsay RS. The magnet effect of price limits: evidence from high frequency data on Taiwan stock exchange. J Empir Finance 2003;10:133–168.

Chou RY. Forecasting financial volatilities with extreme values: the conditional autoregressive range (CARR) model. J Money Credit Bank 2005;37:561–582.

Engle RF, Russell JR. Autoregressive conditional duration: a new model for irregularly spaced transaction data. Econometrica 1998;66:1127–1162.

Ghysels E. Some econometric recipes for high-frequency data cooking. J Bus Econ Stat 2000;18:154–163.

Hansen PR, Lunde A. A realized variance for the whole day based on intermittent high-frequency data. J Financ Econometrics 2005;3:525–554.

Hasbrouck J. The dynamics of discrete bid and ask quotes. J Finance 1999;54:2109–2142.

Hasbrouck J. Empirical Market Microstructure: The Institutions, Economics, and Econometrics of Securities Trading. New York: Oxford University Press; 2007.

Hauseman J, Lo A, MacKinlay C. An ordered probit analysis of transaction stock prices. J Financ Econ 1992;31:319–379.

Johnson B. Algorithmic Trading & DMA: An Introduction to Direct Access Trading Strategies. London: 4Myeloma Press; 2010.

Lo A, MacKinlay AC. An econometric analysis of nonsynchronous trading. J Econometrics 1990;45:181–212.

McCulloch RE, Tsay RS. Nonlinearity in high frequency data and hierarchical models. Stud Nonlinear Dynam Econometrics 2000;5:1–17.

Roll R. A simple implicit measure of the effective bid-ask spread in an efficient market. J Finance 1984;39:1127–1140.

Rydberg TH, Shephard N. Dynamics of trade-by-trade price movements: decomposition and models. J Financ Econometrics 2003;1:2–25.

Stoll H, Whaley R. Stock market structure and volatility. Rev Financ Stud 1990;3:37–71.

Tsay RS. Autoregressive conditional duration models. In: Mills TC, Patterson K, editors. Palgrave Handbook in Econometrics, Volume 2: Applied Econometrics; Macmillan: Hampshire, UK; 2009.

Tsay RS. Analysis of Financial Time Series. 3rd ed. Hoboken (NJ): John Wiley & Sons; 2010.

Tsay RS, Yeh JH. Random aggregation with applications in high-frequency finance. J Forecast 2011;30:72–103.

Wood RA. Market microstructure research databases: history and projections. J Bus Econ Stat 2000;18:140–145.

Zhang L, Mykland PA, Aït- Sahalia Y. A tale of two time scales: determining integrated volatility with noisy high-frequency data. J Am Stat Assoc 2005;100:1394–1411.

Zhang MY, Russell JR, Tsay RS. A nonlinear autoregressive conditional duration model with applications to financial transaction data. J Econometrics 2001;104:179–207.

Zhang MY, Russell JR, Tsay RS. Determinants of bid and ask quotes and implications for the cost of trading. J Empir Finance 2008;15:656–678.

7

VALUE AT RISK

One of the lessons we learned from repeated financial crises is that *risk* is real and should be an integral part of any financial decision. Assessing financial risk thus becomes important in asset pricing and allocation. In this chapter, we introduce some risk measures for quantifying financial risk, discuss statistical methods for calculating financial risk and the theory behind them, and demonstrate risk assessment via real examples. Similar to other topics discussed in the previous chapters, our goal here is to provide readers with basic knowledge about financial risk and risk management. We use real examples in the demonstration, and all computations are carried out step-by-step with R.

Following the framework of Basel Accords, financial risk can be classified into three categories. They are market risk, credit risk, and operational risk. If necessary, one can treat liquidity (or refinancing) risk as an additional category. Market risk is concerned with loss arising from changes in stock prices, interest rates, foreign exchange rates, and commodity prices. It includes equity risk, interest rate risk, currency risk, commodity risk, and volatility risk. Because equity prices and interest rates are widely available and of high quality, market risk is the most well-studied and understood financial risk. It is also the main focus of this chapter.

An Introduction to Analysis of Financial Data with R, First Edition. Ruey S. Tsay.
© 2013 John Wiley & Sons, Inc. Published 2013 by John Wiley & Sons, Inc.

Credit risk is also known as *default risk* or *counterparty risk*. It occurs when a borrower fails to make a payment as promised. It covers consumer credit risk, concentration risk, securitization, and credit derivatives. Because credit data are harder to obtain and are available mainly to large rating agencies, credit risk has been less investigated. However, it has attracted much attention in recent years because of the failure of several large financial institutions in the United States, for example, Lehman Brothers, Merry Lynch, Wachovia, and Washington Mutual, and the increases in use of structured financial products such as credit default swaps (CDS) and collateralized debt obligations (CDO). According to Basel-II, operational risk is concerned with risk of loss resulting from inadequate or failed internal processes, people and systems, or external events. Legal and political risks are examples of operational risk. Because it covers a wide range of risks and often involves people and processes, operational risk is the least studied and understood financial risk. However, it has started to gain attention in recent years as more data have been collected and become available.

As financial products and trading become more and more sophisticated and the financial markets around the world become more integrated, understanding financial risk becomes more important. Much research is needed to provide insight into financial risk in today's global economy. Nevertheless, there exist some fundamental properties about financial risk and some general concepts of risk management. The goal of this chapter is therefore to provide readers with the basic concept to quantify and assess risk of a financial position. In particular, we provide a comprehensive treatment of market risk.

Even though we focus on market risk, some of the concepts and methods discussed apply equally well to credit and operational risks. For instance, RiskMetrics of J.P. Morgan for assessing value at risk (VaR) has been generalized to CreditMetrics for evaluating credit risk.

7.1   RISK MEASURE AND COHERENCE

Loss of a given financial position for a specified holding period can be represented by a random variable, say X. For example, suppose that one bought 100 shares of Stock A at \$50 per share today. Then, the potential loss of the position for tomorrow is $X = \$100(Y - 50)$, where Y is the tomorrow's share price of the stock. Here, the holding period is a trading day and X is a random variable because Y is unknown today. We shall use random variables to describe financial losses. All inferences concerning the loss of a financial position are based on the distribution of the associated loss random variable.

As the distributions of losses are unknown and it is hard to adequately estimate them based on the available data, we often employ some summary statistics to quantify the loss distributions in real applications. A risk measure is simply one of these summary statistics. In short, a risk measure is a mapping from the loss random variable X into the real line. It provides an estimate of the potential risk. Keep in mind, however, a risk measure usually does not provide a complete description of the possible losses, because it is just a summary statistic. The goal then is to select a risk measure that is widely applicable and can best describe the losses we often encounter in finance.

To begin with, a sensible risk measure in finance must be consistent with the basic theory in finance. Let η be a risk measure. We say that η is *coherent* if it satisfies the following four conditions for any two loss random variables X and Y:

1. *Subadditivity*: $\eta(X + Y) \leq \eta(X) + \eta(Y)$.
2. *Monotonicity*: If $X \leq Y$ for all possible outcomes, then $\eta(X) \leq \eta(Y)$.
3. *Positive homogeneity*: For any positive constant c, $\eta(cX) = c\eta(X)$.
4. *Translation invariance*: For any positive constant c, $\eta(X + c) = \eta(X) + c$.

See Artzner et al. (1997) for more details.

The subadditivity states that the risk measure for a combined position should not be greater than risks of the two positions treated separately. In finance, this is related to diversification. The risk of a diversified portfolio should not be greater than risks of the individual components. Thus, subadditivity simply expresses the fact that there should be some diversification benefit from combining risks. Without the subadditivity, companies would find it to be advantageous to divide into smaller companies.

The monotonicity is easily understandable. It simply states that if one financial position always has greater losses than another position under all circumstances, then its risk measure should always be greater. The positive homogeneity also has important implications. First, it states that doubling a financial position should also double its risk. Second, it implies that the risk does not depend on the currency in which the risk is measured. The translation invariance means that there is no additional risk if there is no additional uncertainty, because in statistics adding a constant to a random variable does not affect its variability.

There are many coherent risk measures available. In what follows, we shall introduce some of the commonly used risk measures.

7.1.1 Value at Risk (VaR)

VaR is perhaps the most well-known risk measure. It is a single estimate of the amount by which an institution's position in a risk category could decline because of general market movements during a given holding period; see Duffie and Pan (1997) and Jorion (2006) for a general exposition of VaR. The measure can be used by financial institutions to assess their risks or by a regulatory committee to set margin requirements. In either case, VaR is used to ensure that the financial institutions can still be in business after a catastrophic event.

In what follows, we define VaR using the loss random variable of a financial position for a given holding period. Suppose that at the time index t we are interested in the risk of a financial position for the next ℓ periods. Let $L_t(\ell)$ be the loss random variable of the position. The loss is typically measured in dollars. Let V_t be the value of the position at time t. Then, $L_t(\ell)$ is either a positive or negative function of $V_{t+\ell} - V_t$, depending on the financial position being long or short. Denote the cumulative distribution function (CDF) of $L_t(\ell)$ by $F_\ell(x)$. Here, we drop the subscript t from $F_\ell(x)$. However, it is understood that $F_\ell(x)$ depends on the time index t.

As big loss occurs less frequently, we assess loss using a small probability, for example, 5% or 1% or 0.1%. Denote the probability by p. We define the VaR of the financial position over the time horizon ℓ with given probability p as

$$\text{VaR}_{1-p} = \inf\{x | F_\ell(x) \geq 1 - p\} \qquad (7.1)$$

where inf denotes the smallest real number x satisfying the condition. From the definition, $F_\ell(\text{VaR}_{1-p}) \geq 1 - p$, which says

$$\Pr[L_t(\ell) \leq \text{VaR}_{1-p}] \geq 1 - p.$$

Thus, with probability $(1 - p)$, the potential loss encountered by the holder of the financial position from time t to time $t + \ell$ is less than or equal to VaR_{1-p}.

Using the property $\Pr[L_t(\ell) \leq x] = 1 - \Pr[L_t(\ell) > x]$, we have

$$\Pr[L_t(\ell) > \text{VaR}_{1-p}] \leq p.$$

Therefore, the probability that the position holder would encounter a loss greater than VaR_{1-p} over the period from t to $t + \ell$ is at most p.

The previous definition shows that VaR is concerned with the upper tail probability of the loss CDF $F_\ell(x)$. For a univariate CDF $F_\ell(x)$ and a given probability q satisfying $0 < q < 1$, the quantity

$$x_q = \inf\{x | F_\ell(x) \geq q\}$$

is called the qth *quantile* of $F_\ell(x)$. If the random variable $L_t(\ell)$ of $F_\ell(x)$ is continuous, then $q = \Pr[L_t(\ell) \leq x_q]$. Thus, letting $q = 1 - p$, we see that VaR is simply the $(1 - p)$th quantile of the loss distribution, where p is a small tail probability. For this reason, we used the subscript $1 - p$ in Equation (7.1) to define VaR. In the statistical literature, the $(1 - p)$th quantile of a distribution is also referred to as the $100(1 - p)$th percentile of the distribution.

Figure 7.1 shows the VaR of a continuous loss random variable based on its CDF. The upper horizontal line denotes the probability $1 - p$ so that the upper tail probability is p. The VaR is then the X-coordinate of the vertical line, confirming that VaR is just the $(1 - p)$th quantile of the loss distribution. Figure 7.2 shows the VaR based on the probability density function (pdf) of a continuous loss random variable. The area of the upper tail under the density is p. In statistics, this means

$$\int_{\text{VaR}}^{\infty} f(x)dx = p \quad \text{or equivalently} \quad \int_{-\infty}^{\text{VaR}} f(x)dx = 1 - p,$$

where $f(x)$ denotes the pdf of X and, for simplicity, we drop the subscript $1 - p$ from VaR. Again, the prior equation says that VaR is the $100(1 - p)$th percentile of the loss variable X.

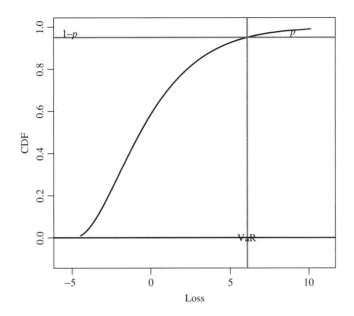

Figure 7.1. Definition of value at risk (VaR) for a continuous loss random variable based on the cumulative distribution function.

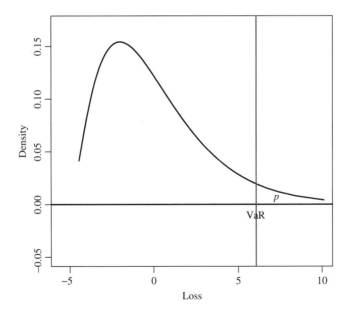

Figure 7.2. Definition of value at risk (VaR) for a continuous loss random variable based on the probability density function.

Discussion. Arguably, VaR has become the standard measure to quantify the risk exposure of a financial institution. Its popularity arises because it has some good properties. First, it is simple and easy to calculate for some well-known distributions.

Case 1: (Normal distribution). If the loss random variable X is normally distributed, say $X \sim N(\mu_t, \sigma_t^2)$, then

$$\text{VaR}_{1-p} = \mu_t + z_{1-p}\sigma_t, \tag{7.2}$$

where z_{1-p} denotes the $(1-p)$th quantile of the standard normal distribution and the subscript t is used to signify that VaR is time varying. The quantile can be obtained from the normal probability table or from any statistical software package. For instance, $z_{0.95} \approx 1.645$ and $z_{0.99} \approx 2.326$. In R, one can use qnorm(0.95) and qnorm(0.99) to obtain these two quantiles.

Case 2: (Student-t distribution). If the loss random variable X satisfies that $Y = (X - \mu_t)/\sigma_t$ is a Student-t distribution with v degrees of freedom, then

$$\text{VaR}_{1-p} = \mu_t + t_{1-p,v}\sigma_t, \tag{7.3}$$

where $t_{1-p,v}$ is the $(1-p)$th quantile of a Student-t distribution with v degrees of freedom. For instance, for $v = 5$, we have $t_{0.95,5} \approx 2.015$ and $t_{0.99,5} \approx 3.365$. In R, one can use the command qt(0.95,5) and qt(0.99,5) to obtain these two quantiles.

Case 3: (Standardized Student-t distribution). The variance of the Student-t random variable Y of Case 2 is $v/(v-2)$ provided that $v > 2$. In volatility modeling, we may employ the standardized version of Y as innovations; see Chapter 4. This is achieved by

$$Y^* = \frac{Y}{\sqrt{v/(v-2)}} = \frac{X - \mu_t}{\sigma_t \sqrt{v/(v-2)}}.$$

Consequently, we have

$$\text{VaR}_{1-p} = \mu_t + t_{1-p,v}^*\sigma_t\sqrt{v/(v-2)}, \tag{7.4}$$

where $t_{1-p,v}^*$ is the $(1-p)$th quantile of a standardized Student-t distribution with v degrees of freedom. For instance, we have $t_{0.95,5}^* \approx 1.561$ and $t_{0.99,5}^* \approx 2.606$. In R with package fGarch, one can use qstd(0.95,nu = 5) and qstd(0.99,nu = 5) to obtain these two quantiles. It is easy to verify that $t_{1-p,v}^*\sqrt{v/(v-2)} = t_{1-p,v}$.

Second, VaR is a coherent risk measure if the loss random variable is normally distributed (or more generally, it follows a spherical distribution). To see this, we assume, without of loss of generality, that the means of the two loss random variables X and Y are zero so that $X \sim N(0, \sigma_x^2)$ and $Y \sim N(0, \sigma_y^2)$ and the $(1-p)$th VaR are $z_{1-p}\sigma_x$ and $z_{1-p}\sigma_y$, respectively. In this case, it is easy to see, from properties of the normal distribution, that monotonicity, positive homogeneity, and translation invariance hold. It remains to check the subadditivity of VaR. This can be seen as follows:

$$\mathrm{Var}(X+Y) = \mathrm{Var}(X) + \mathrm{Var}(Y) + 2\mathrm{Cov}(X,Y)$$
$$= \sigma_x^2 + \sigma_y^2 + 2\rho\sigma_x\sigma_y$$
$$\leq \sigma_x^2 + \sigma_y^2 + 2\sigma_x\sigma_y$$
$$= (\sigma_x + \sigma_y)^2,$$

where ρ denotes the correlation between X and Y and we have $\rho \leq 1$. Therefore, $\sigma_{x+y} \leq \sigma_x + \sigma_y$. This implies that $z_{1-p}\sigma_{x+y} \leq z_{1-p}\sigma_x + z_{1-p}\sigma_y$, or equivalently, VaR of $X+Y$ is less than or equal to the sum of VaR of X and VaR of Y. Consequently, the subadditivity of VaR holds under the normality assumption.

VaR, however, is not a coherent risk measure in general. We give a simple counterexample, which is similar to Example 3.13 of Klugman et al. (2008).

Example 7.1. Suppose the CDF $F_\ell(x)$ of a continuous loss random variable X satisfies the following probabilities:

$$F_\ell(80) = 0.9215$$
$$F_\ell(90) = 0.95$$
$$F_\ell(100) = 0.97.$$

For $p = 0.05$, the VaR of X is 90 because 90 is the 0.95th quantile of X. We denote this by $\mathrm{VaR}_{0.95}^x = 90$. Now, define two loss random variables X_1 and X_2 by

$$X_1 = \begin{cases} X, & \text{if } X \leq 100 \\ 0, & \text{if } X > 100 \end{cases}$$

and

$$X_2 = \begin{cases} 0, & \text{if } X \leq 100 \\ X, & \text{if } X > 100. \end{cases}$$

These two loss variables are simply truncated versions of X and we have $X = X_1 + X_2$. As the total probability must be 1, the CDF $F_\ell^1(X)$ of X_1 satisfies

$$F_\ell^1(80) = 0.9215/0.97 = 0.95$$
$$F_\ell^1(90) = 0.95/0.97 = 0.9794$$
$$F_\ell^1(100) = 0.97/0.97 = 1.$$

The 0.95th quantile of X_1 is 80. Therefore, $\text{VaR}^1_{0.95} = 80$, where the superscript 1 is used to denote X_1. On the other hand, $Pr(X_2 \leq 0) = P(X \leq 100) = 0.97$. Therefore, the 0.95th quantile of X_2 is less than or equal to 0. We denote this by $\text{VaR}^2_{0.95} \leq 0$. Taking the sum, we have $\text{VaR}^1_{0.95} + \text{VaR}^2_{0.95} \leq 80$.

In this particular instance, $X = X_1 + X_2$, yet $\text{VaR}^x_{0.95} = 90 > \text{VaR}^1_{0.95} + \text{VaR}^2_{0.95}$. Therefore, the subadditivity of VaR fails. ☐

Finally, VaR is simply a quantile with the upper tail probability p. It does not describe the actual tail behavior of the loss random variable. One can easily construct two loss random variables that share the same VaR for a given probability p but have very different tail behavior. In this case, the actual risks of the two loss variables are different even though they have the same VaR. To demonstrate, consider Figure 7.3 that shows the density functions of two loss random variables. The two density functions are identical for $x \leq \text{VaR}$ but different for $x > \text{VaR}$ with the dash line indicating a heavier upper tail. These two loss variables have the same VaR, but the one associated with dash line has higher probabilities of bigger losses. This demonstration shows that VaR is not a perfect risk measure.

7.1.2 Expected Shortfall

To overcome the drawbacks of VaR, a new risk measure, called *expected shortfall* (ES), is introduced in financial econometrics. The measure, however, is not new in

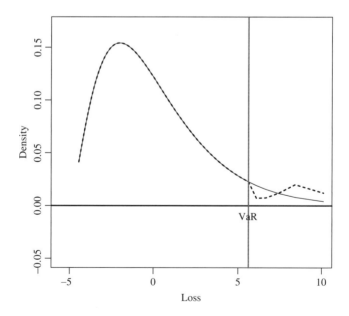

Figure 7.3. Density functions of two loss random variables that have the same VaR but different loss implications.

the literature. It is called the *tail value at risk* (TVaR) in the actuarial science. Simply put, ES is the expected loss of a financial position after a catastrophic event. Consider the loss random variable X of a financial position with holding period ℓ. Denote the pdf and CDF of X by $f(x)$ and $F(x)$, respectively. For a given tail probability p, let VaR be the value at risk of X. Here, for simplicity, we drop the subscript $1 - p$ from VaR. Then, ES of X is defined as

$$ES_{1-p} = E(X|X > VaR) = \frac{\int_{VaR}^{\infty} xf(x)dx}{Pr(X > VaR)}. \tag{7.5}$$

From the definition, ES is the expected loss of X given that X exceeds its VaR. For this reason, ES is also called the conditional value at risk (CVaR) in the literature. It is called TVaR because it focuses on the upper tail behavior of the loss distribution.

For simplicity, assume that X is continuous. We shall use change of variable in integration to rewrite Equation (7.5). Let $u = F(x)$ for VaR $\leq x \leq \infty$. Then, we have $du = f(x)dx$, $F(VaR) = 1 - p$, $F(\infty) = 1$, and $x = F^{-1}(u) = VaR_u$. Equation (7.5) can then be rewritten as

$$ES_{1-p} = \frac{\int_{1-p}^{1} VaR_u \, du}{p}.$$

Thus, ES can be seen to average all VaR$_u$ for $1 - p \leq u \leq 1$. This averaging feature enables ES to better reflect the tail behavior of the loss random variable X than VaR. As a matter of fact, it can be shown that ES is a coherent risk measure. For the two loss densities in Figure 7.3, their ESs are different with the dashed line corresponding to a higher value.

Closed-form solutions for ES are also available for some loss distributions.

Case 1: (Normal distribution). Suppose that the loss random variable X is distributed as $N(\mu_t, \sigma_t^2)$. In this case, ES is simply the expectation of a left-truncated normal random variable for which a closed-form solution is available:

$$ES_{1-p} = \mu_t + \frac{f(z_{1-p})}{p}\sigma_t, \tag{7.6}$$

where $f(z)$ is the pdf of a standard normal random variable, z_{1-p} is the $(1 - p)$th quantile of $f(z)$, and p is a small tail probability. This result can also be obtained by integration by parts. For $p = 0.05$, we have $z_{1-p} \approx 1.645$ and $f(1.645) \approx 0.103$. In R, these quantities are qnorm(0.95) and dnorm(qnorm(0.95)), respectively.

Case 2: (Student-t distribution with v degrees of freedom). If the loss random variable X satisfies that $Y = (X - \mu_t)/\sigma_t$ is a Student-t distribution with v

degrees of freedom, then we have

$$ES_{1-p} = \mu_t + \sigma_t \frac{f_v(x_{1-p})}{p} \left(\frac{v + x_{1-p}^2}{v - 1} \right), \tag{7.7}$$

where $f_v(x)$ denotes the pdf of a Student-t distribution with v degrees of freedom and x_{1-p} is the $(1-p)$th quantile of $f_v(x)$. This result can be obtained by direct integration. For $p = 0.01$ and $v = 5$, we have $x_{1-p} \approx 3.365$ and $f_5(x_{1-p}) \approx 0.0109$. In R, these two quantities are `qt(0.99,5)` and `dt(qt(0.99,5),5)`, respectively.

Case 3: (Standardized Student-t distribution with v degrees of freedom). Suppose the loss random variable X satisfies that

$$Y^* = \frac{X - \mu_t}{\sigma_t \sqrt{v/(v-2)}},$$

follows a standardized Student-t distribution with v degrees of freedom, where $v > 2$. In this case, we have

$$ES_{1-p} = \mu_t + \sigma_t \sqrt{v/(v-2)} \frac{f_v^*(x_{1-p}^*)}{p} \left(\frac{(v-2) + [x_{1-p}^*]^2}{v - 1} \right), \tag{7.8}$$

where $f_v^*(x)$ is the pdf of a Standardized Student-t distribution with v degrees of freedom and x_{1-p}^* is the $(1-p)$th quantile of $f_v^*(x)$. For $p = 0.01$ and $v = 5$, we have $x_{1-p}^* \approx 2.606$ and $f_v^*(x_{1-p}^*) \approx 0.014$. In R with the package fGarch, we can obtain these quantities using `qstd(0.99,nu = 5)` and `dstd(qstd(0.99,nu = 5),nu = 5)`, respectively.

7.2   REMARKS ON CALCULATING RISK MEASURES

Before introducing methods for calculating VaR and ES, we discuss some practical issues in assessing financial risk.

Calculation of VaR involves several factors:

1. The probability of interest p, such as $p = 0.01$ for risk management and $p = 0.001$ in stress testing. The choice of p is somewhat arbitrary.
2. The time horizon ℓ. It might be set by a regulatory committee, such as 1 day or 10 days for market risk and 1 year or 5 years for credit risk.
3. The frequency of the data, which might not be the same as the time horizon ℓ. Daily observations are often used in market risk analysis. Monthly or quarterly data are common for credit risk modeling.
4. The CDF $F_\ell(x)$ or its quantiles of the loss random variable.
5. The amount of the financial position or the mark-to-market value of the portfolio.

Among these factors, the CDF $F_\ell(x)$ is the focus of econometric modeling. Different methods for estimating the CDF give rise to different approaches to VaR and ES calculation.

The definition of VaR in Equation (7.1) is in dollar amount. As log returns correspond approximately to percentage changes in value of a financial asset, we use log returns r_t in data analysis. For a long position, loss occurs when r_t is negative. On the contrary, for a short position, loss occurs when r_t is positive. Therefore, in this chapter, we define the loss random variable as

$$x_t = \begin{cases} r_t, & \text{if the position is short,} \\ -r_t, & \text{if the position is long.} \end{cases} \qquad (7.9)$$

The VaR calculated from the upper quantile of the loss distribution given information available at time t is therefore in percentage. The dollar amount of VaR is then the cash value of the financial position times the VaR of the loss variable. That is, VaR = Value $\times$ VaR(of x_t). If necessary, one can also use the approximation VaR = Value $\times$ [exp(VaR of x_t) $-$ 1].

VaR is a prediction concerning possible loss of a portfolio over a holding period. It should be computed using the *predictive distribution* of the loss variable over the specified holding period. For example, the VaR for a 1-day horizon of a portfolio using daily returns r_t should be calculated using the predictive distribution of r_{t+1} given information available at time t. From a statistical viewpoint, predictive distribution takes into account the parameter uncertainty of the entertained econometric model. However, predictive distribution is hard to obtain in general, and most of the available methods for VaR calculation ignore the effects of parameter uncertainty. Furthermore, when the entertained model is uncertain, the VaR should also take into consideration the model uncertainty. In short, because of some simplicity considerations, the commonly used methods for calculating VaR and ES overlook both the parameter and model uncertainties.

7.3 RISKMETRICS

J. P. Morgan developed the RiskMetrics™ methodology to VaR calculation; see Longerstaey and More (1995). In its simple form, RiskMetrics assumes that the continuously compounded daily return r_t of a portfolio follows a conditional normal distribution. Let x_t denote the daily loss random variable defined in Equation (7.9) and denote the information set available at time $t - 1$ by F_{t-1}. RiskMetrics assumes that $x_t | F_{t-1} \sim N(0, \sigma_t^2)$, where σ_t^2 is the conditional variance of x_t and it evolves over time according to the simple model:

$$\sigma_t^2 = \alpha \sigma_{t-1}^2 + (1 - \alpha)x_{t-1}^2, \quad 1 > \alpha > 0. \qquad (7.10)$$

Therefore, the method assumes that the logarithm of the daily price, $p_t = \ln(P_t)$, of the portfolio satisfies the difference equation $p_t - p_{t-1} = a_t$, where $a_t = \sigma_t \epsilon_t$ is an

IGARCH(1,1) process without drift. The value of α is often in the interval (0.9, 1) with a typical value of 0.94.

A good property of such a special random-walk IGARCH model is that the conditional distribution of a multiperiod loss is easily available. Specifically, for a k-period horizon, the loss variable from time $t + 1$ to time $t + k$ (inclusive) is $x_t[k]$ $= x_{t+1} + \cdots + x_{t+k-1} + x_{t+k}$. We use the square bracket $[k]$ to denote the k holding periods. Under the special IGARCH(1,1) model in Equation (7.10), the conditional distribution $x_t[k]|F_t$ is normal with mean zero and variance $\sigma_t^2[k]$, where $\sigma_t^2[k]$ can be computed using the forecasting method discussed in Chapter 4. Using the independence assumption of ϵ_t and model (Eq. 7.10), we have

$$\sigma_t^2[k] = \text{Var}(x_t[k]|F_t) = \sum_{i=1}^{k} \text{Var}(a_{t+i}|F_t),$$

where $\text{Var}(a_{t+i}|F_t) = E(\sigma_{t+i}^2|F_t)$ can be obtained recursively. Using $x_{t-1} = a_{t-1} = \sigma_{t-1}\epsilon_{t-1}$, we can rewrite the volatility equation of the IGARCH(1,1) model in Equation (7.10) as

$$\sigma_t^2 = \sigma_{t-1}^2 + (1 - \alpha)\sigma_{t-1}^2(\epsilon_{t-1}^2 - 1) \quad \text{for all } t.$$

In particular, we have

$$\sigma_{t+i}^2 = \sigma_{t+i-1}^2 + (1 - \alpha)\sigma_{t+i-1}^2(\epsilon_{t+i-1}^2 - 1) \quad \text{for} \quad i = 2, \cdots, k.$$

As $E(\epsilon_{t+i-1}^2 - 1|F_t) = 0$ for $i \geq 2$, the prior equation shows that

$$E(\sigma_{t+i}^2|F_t) = E(\sigma_{t+i-1}^2|F_t) \quad \text{for} \quad i = 2, \cdots, k. \tag{7.11}$$

For the 1-step ahead volatility forecast, Equation (7.10) shows that $\sigma_{t+1}^2 = \alpha\sigma_t^2 + (1 - \alpha)x_t^2$. Therefore, Equation (7.11) shows that $\text{Var}(x_{t+i}|F_t) = \sigma_{t+1}^2$ for $i \geq 1$ and, hence, $\sigma_t^2[k] = k\sigma_{t+1}^2$. The results show that $x_t[k]|F_t \sim N(0, k\sigma_{t+1}^2)$. Consequently, under the special IGARCH(1,1) model in Equation (7.10), the conditional variance of $x_t[k]$ is proportional to the time horizon k. The conditional standard deviation of a k-period horizon loss variable is then $\sqrt{k}\sigma_{t+1}$, which is $\sqrt{k}$ times σ_{t+1}.

Given a tail probability, RiskMetrics uses the result $x_t[k]|F_t \sim N(0, k\sigma_{t+1}^2)$ to calculate VaR of the loss random variable. If the tail probability is $p = 0.05$, then VaR $= 1.65\sigma_{t+1}$ for the next trading day (Section 7.1.1). For the next k trading days, VaR$[k] = 1.65\sqrt{k}\sigma_{t+1}$, which is the 95th percentile of $N(0, k\sigma_{t+1}^2)$. Similarly, if the tail probability is $p = 0.01$, then VaR $= 2.326\sigma_{t+1}$ for the next trading day and VaR$[k]$ $= 2.326\sqrt{k}\sigma_{t+1}$ for the next k trading days.

Consider the case of $p = 0.01$. The VaR for the portfolio under RiskMetrics is then

$$\text{VaR} = \text{Amount of position} \times 2.326\sigma_{t+1},$$

for the next trading day and that of a k-day horizon is

$$\text{VaR}(k) = \text{Amount of position} \times 2.326\sqrt{k}\sigma_{t+1},$$

where the argument (k) of VaR is used to denote the time horizon and the portfolio value is measured in dollars. Consequently, under RiskMetrics, we have

$$\text{VaR}(k) = \sqrt{k} \times \text{VaR}.$$

This is referred to as the *square root of time rule* in VaR calculation under RiskMetrics.

If the loss variables are in percentages, then the 1% VaR for the next trading day is VaR = Amount of position $\times 2.326\sigma_{t+1}/100$, where σ_{t+1} is the volatility of the loss variable.

Note that because RiskMetrics assumes log returns are normally distributed with mean zero, the loss function is symmetric and VaR are the same for long and short financial positions.

We can also use the results of Section 7.1.2 to compute ES under RiskMetrics. In particular, we have

$$\text{ES}_{0.95} = \frac{0.103}{0.05}\sigma_{t+1} = 2.063\sigma_{t+1}, \quad \text{and} \quad \text{ES}_{0.99} = \frac{0.0267}{0.01}\sigma_{t+1} = 2.67\sigma_{t+1},$$

for the next trading day. For the next k-trading days, we can apply the square root of time rule.

Example 7.2. Figure 7.4 shows the time plot of daily log returns of IBM stock from January 2, 2001 to December 31, 2010 for 2515 observations. We apply the RiskMetrics approach to calculate VaR and ES for a long position of 1 million on the stock. In this case, the loss variable is the negative daily log return. To begin with, we estimate the special IGARCH(1,1) model of Equation (7.10) to obtain an estimate of the parameter α. For IBM daily returns in percentages, we have $\hat{\alpha} = 0.943(0.007)$, where the number in parentheses is the estimated standard error. In addition, we have $x_{2515} = -0.061$ and $\hat{\sigma}_{2515} = 0.734$ from the data and the fitted model. The 1-step ahead prediction of volatility is then $\hat{\sigma}_{2516} = 0.7133$ obtained from $\hat{\sigma}_{2516}^2 = 0.943(0.734)^2 + (1 - 0.943)(-0.061)^2$. Consequently, using RiskMetrics, we have

$$\text{VaR}_{0.95} = 1.173, \quad \text{VaR}_{0.99} = 1.659, \quad \text{ES}_{0.95} = 1.471, \text{and} \quad \text{ES}_{0.99} = 1.901.$$

Using these results, we can obtain the VaR and ES for the long position. For instance,

$$\text{VaR}_{0.95} = \$1,000,000\frac{1.173}{100} = \$11,730 \quad \text{and} \quad \text{VaR}_{0.99} = \$16,590.$$

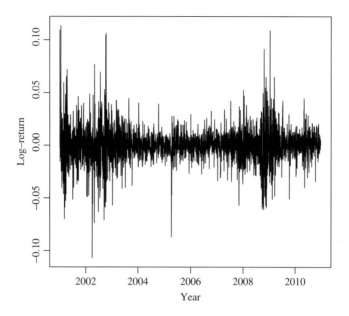

Figure 7.4. Daily log returns of IBM stock from January 2, 2001 to December 31, 2010.

Finally, suppose that we are interested in the risk for holding the position for the next 15 trading days. Then, by the square root of time rule, we have

$$\text{VaR}_{0.95}(15) = \sqrt{15} \times \$11730 = \$45,430 \quad \text{ES}_{0.95}(15) = \sqrt{15} \times \$14710 = \$56,972.$$

In this example, we use an R script RMfit to estimate the special IGARCH(1,1) model in Equation (7.10) and to calculate risk measures. The R script is available on the book web. □

R Demonstration

```
> da=read.table("d-ibm-0110.txt",header=T)
> head(da)
      date      return
1 20010102  -0.002206
  .....
> ibm=log(da[,2]+1)*100
> source("RMfit.R")
> mm=RMfit(ibm)
Coefficient(s):
        Estimate  Std. Error  t value   Pr(>|t|)
alpha   0.942857    0.007172  131.464 < 2.22e-16 ***
---
 Volatility prediction:
```

```
     Orig    Vpred
[1,] 2515 0.713303

 Risk measure based on RiskMetrics:
       prob       VaR          ES
[1,]  0.950 1.173279 1.471339
[2,]  0.990 1.659391 1.901105
[3,]  0.999 2.204272 2.401756
```

Example 7.3. Consider the daily log returns of dollar–euro exchange rate from January 5, 2001 to October 28, 2011 for 2721 observations. The exchange rate and its log return are shown in Figure 7.5. The exchange rates are downloaded from the Federal Reserve Bank at St. Louis, and we calculate log returns by taking the difference of the log exchange rate. For simplicity, assume that we hold 1 Euro and are interested in the associated risks for the next 1 period and 10 periods.

Similar to Example 7.2, we start with fitting the special IGARCH(1,1) model in Equation (7.10). The result is

$$\sigma_t^2 = 0.9698\sigma_{t-1}^2 + (1 - 0.9698)r_t^2.$$

The standard error of the coefficient is 0.004. Using the fitted model, the VaR and ES for the next trading day after October 28, 2011 are

$$\text{VaR}_{0.95} = 0.014, \quad \text{VaR}_{0.99} = 0.0199, \quad \text{ES}_{0.95} = 0.0177, \quad \text{and} \quad \text{ES}_{0.99} = 0.0229.$$

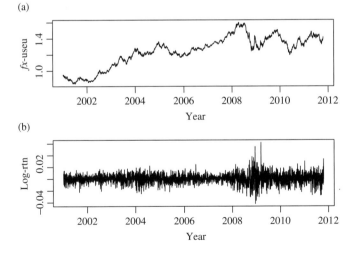

Figure 7.5. Daily log returns of dollar–euro exchange rate from January 4, 2001 to October 28, 2011.

Applying the square root of time rule, we have

$$\text{VaR}_{0.95}(10) = 0.0446 \quad \text{and} \quad \text{ES}_{0.95}(10) = 0.0559.$$

□

R Demonstration

```
> da1=read.table("d-useu0111.txt",header=T)
> head(da1)
   year mm dd    rate
1  2001  1   4 0.9448
......
> par(mfcol=c(2,1))
> rate=da1$rate; plot(rate,type='l')
> rt=diff(log(rate)); plot(rt,type='l')
> m2=RMfit(rt)
Coefficient(s):
         Estimate  Std. Error  t value   Pr(>|t|)
alpha  0.96975   0.003813    254.35 < 2.22e-16 ***
---

 Volatility prediction:
       Orig      Vpred
[1,] 2721 0.008574549


 Risk measure based on RiskMetrics:
        prob         VaR            ES
[1,] 0.950 0.01410388 0.01768683
[2,] 0.990 0.01994738 0.02285301
[3,] 0.999 0.02649735 0.02887128
```

7.3.1 Discussion

An advantage of RiskMetrics is simplicity. It is easy to understand and apply. Another advantage is that it makes risk more transparent in the financial markets. However, as security returns tend to have heavy tails (or fat tails), the normality assumption used often results in underestimation of VaR. Other approaches to VaR calculation avoid making such an assumption.

The square root of time rule is a consequence of the special model used by RiskMetrics. If either the zero mean assumption or the special IGARCH(1,1) model assumption of the log returns fails, then the rule is invalid. Consider the simple model

$$r_t = \mu + a_t, \quad a_t = \sigma_t \epsilon_t, \quad \mu \neq 0,$$
$$\sigma_t^2 = \alpha \sigma_{t-1}^2 + (1 - \alpha)a_{t-1}^2,$$

where $\{\epsilon_t\}$ is a standard Gaussian white noise series. The assumption that $\mu \neq 0$ holds for returns of many heavily traded stocks on the NYSE; see Chapter 1. For this simple model, the distribution of r_{t+1} given F_t is $N(\mu, \sigma_{t+1}^2)$. The 95% quantile used to calculate the 1-period horizon VaR becomes $\mu + 1.65\sigma_{t+1}$. For a k-period horizon, the distribution of $r_t[k]$ given F_t is $N(k\mu, k\sigma_{t+1}^2)$, where as before $r_t[k] = r_{t+1} + \cdots + r_{t+k}$. The 95% quantile used in the k-period horizon VaR calculation is $k\mu + 1.65\sqrt{k}\sigma_{t+1} = \sqrt{k}(\sqrt{k}\mu + 1.65\sigma_{t+1})$. Consequently, $\text{VaR}(k) \neq \sqrt{k} \times \text{VaR}$ when the mean return is not zero. It is also easy to show that the rule fails when the volatility model of the return is not an IGARCH(1,1) model without drift.

7.3.2 Multiple Positions

In some applications, an investor may hold multiple positions and needs to compute the overall VaR of the positions. RiskMetrics adopts a simple approach for doing such a calculation under the assumption that daily log returns of each position follow a random-walk IGARCH(1,1) model. The additional quantities needed are the cross-correlation coefficients between the returns. Consider the case of two positions. Let VaR_1 and VaR_2 be the VaR for the two positions and ρ_{12} be the correlation coefficient between the two returns — that is, $\rho_{12} = \text{Cov}(r_{1t}, r_{2t})/[\text{Var}(r_{1t})\text{Var}(r_{2t})]^{0.5}$. Then, the overall VaR of the investor is

$$\text{VaR} = \sqrt{\text{VaR}_1^2 + \text{VaR}_2^2 + 2\rho_{12}\text{VaR}_1\text{VaR}_2}.$$

The generalization of VaR to a position consisting of m instruments is straightforward as

$$\text{VaR} = \sqrt{\sum_{i=1}^{m} \text{VaR}_i^2 + 2\sum_{i<j}^{m} \rho_{ij}\text{VaR}_i\text{VaR}_j},$$

where ρ_{ij} is the cross-correlation coefficient between returns of the ith and jth instruments and VaR_i is the VaR of the ith instrument.

The prior formula is obtained using the assumption that the joint distribution of the log returns of assets involved in the portfolio is multivariate normal with mean zero and covariance matrix Σ_{t+1}. Under such an assumption, the log return of the portfolio is normal with mean zero and finite variance; see Appendix B of Chapter 8 of Tsay (2010) for properties of multivariate normal variables.

Example 7.4. Consider a simple portfolio consisting of 40% in AAA bonds and 60% on IBM stock. The market value of the portfolio is U.S. $ 1 million. To measure the bond returns, we employ the daily log return of the Bank of America Merrill Lynch U.S. Corp AAA total return index from January 2, 2001 to December 31, 2010. The data of bond index are obtained from the Federal Reserve Bank at St. Louis. The original bond index has 2612 observations. On the other hand, as given in Example 7.2, the daily stock returns only have 2515 observations within the same period. For simplicity, we remove those bond indices that correspond to no stock returns. The resulting bond index is then used to calculate the daily bond returns.

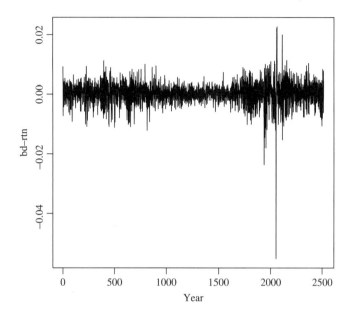

Figure 7.6. Daily log returns of bond index from January 2, 2001 to December 31, 2010. The bond index is the Bank of America Merrill Lynch U.S. Corp AAA total return index.

Figure 7.6 shows the log returns of the bond index. Similar to stock returns, bond returns also exhibit the pattern of volatility clustering and weak stationarity. Applying the RiskMetrics method, we obtain the following IGARCH(1,1) model for the log returns of bond index:

$$\sigma_t^2 = 0.9577\sigma_{t-1}^2 + (1 - 0.9577)r_{t-1}^2,$$

where r_t denotes the log bond return at time t and the standard error of the coefficient is 0.006. On the basis of this model, we have

$$\text{VaR}_{0.95} = 0.00705 \quad \text{and} \quad \text{VaR}_{0.99} = 0.00997.$$

Recall, from Example 7.2, that for the daily log returns of IBM stock, we have

$$\text{VaR}_{0.95} = 0.01173 \quad \text{and} \quad \text{VaR}_{0.99} = 0.01659.$$

Next, consider the VaR of the portfolio. We use $p = 0.05$ in the demonstration. The sample correlation coefficient of the log returns between IBM stock and AAA bond index is -0.2215. Consequently, for the portfolio, we have

$$\text{VaR}_{0.95}^e = 0.01173 \times 0.6 = 0.00704 \quad \text{and} \quad \text{VaR}_{0.95}^b = 0.00705 \times 0.4 = 0.00282,$$

where the superscripts e and b denote equity and bond returns, respectively. The $VaR_{0.95}$ for the portfolio is then

$$VaR_{0.95} = \sqrt{(VaR_{0.95}^e)^2 + (VaR_{0.95}^b)^2 + 2(-0.2215)VaR_{0.95}^e VaR_{0.95}^b}$$

$$= 0.006978.$$

For this particular instance, we see that with tail probability $p = 0.05$, the VaR of the portfolio is less than the VaR of each component. More specifically, with \$1 million investment, we have

1. Equity market only: $VaR_{0.95} = \$11{,}730$.
2. Bond market only: $VaR_{0.95} = \$7050$.
3. Portfolio (60-40): $VaR_{0.95} = \$6978$.

This result is expected because VaR is a coherent risk measure under the normality assumption. The example, thus, demonstrates the value of diversification.   □

7.4 AN ECONOMETRIC APPROACH

A general approach to VaR and ES calculation is to use the time series econometric models of Chapters 2 and 4. For a log return series, the time series models of Chapter 2 can be used for the mean equation, and the conditional heteroscedastic models of Chapter 4 are used to handle the volatility. For simplicity, we use GARCH models in our discussion and refer to the approach as an *econometric approach* to VaR and ES calculation. Other volatility models can also be used.

Consider the loss variable x_t of a financial position. A general time series model for x_t can be written as

$$x_t = \phi_0 + \sum_{i=1}^{p} \phi_i x_{t-i} + a_t - \sum_{j=1}^{q} \theta_j a_{t-j}, \tag{7.12}$$

$$a_t = \sigma_t \epsilon_t,$$

$$\sigma_t^2 = \alpha_0 + \sum_{i=1}^{u} \alpha_i a_{t-i}^2 + \sum_{j=1}^{v} \beta_j \sigma_{t-j}^2. \tag{7.13}$$

Equations (7.12) and (7.13) are the mean and volatility equations, respectively, for x_t. These two equations can be used to obtain 1-step ahead forecasts of the conditional mean and conditional variance of x_t assuming that the parameters are known. Specifically, we have

$$\hat{x}_t(1) = \phi_0 + \sum_{i=1}^{p} \phi_i x_{t+1-i} - \sum_{j=1}^{q} \theta_j a_{t+1-j},$$

$$\hat{\sigma}_t^2(1) = \alpha_0 + \sum_{i=1}^{u} \alpha_i a_{t+1-i}^2 + \sum_{j=1}^{v} \beta_j \sigma_{t+1-j}^2.$$

If one further assumes that ϵ_t follows a Gaussian or a standardized Student-t distribution, then the results of Sections 7.1.1 and 7.1.2 can be used to calculate VaR and ES for a given tail probability.

Example 7.5. Consider again the daily log returns of IBM stock employed in Example 7.2. We use two volatility models to calculate VaR and ES for a long position of $1 million dollars. The two econometric models are reasonable based on the modeling techniques of Chapters 2 and 4.

Because the position is long, the loss variable is $x_t = -r_t$, where r_t is the daily log return. In this example, all calculations use the last data point $T = 2515$ as the date of making inference.

Model 1: Here, we employ a GARCH(1,1) model with Gaussian innovations. The fitted model is

$$x_t = -6.01 \times 10^{-4} + a_t, \quad a_t = \sigma_t \epsilon_t, \quad \epsilon_t \sim N(0,1)$$
$$\sigma_t^2 = 4.378 \times 10^{-6} + 0.101 a_{t-1}^2 + 0.884 \sigma_{t-1}^2.$$

Except for the normality assumption, model checking statistics discussed in Chapter 4 show that this model is adequate in describing the mean and volatility of the loss variable. All coefficient estimates are statistically significant at the usual 5% level; see the attached R output. The 1-step ahead predictions of the mean and volatility of the loss variable x_t at $T = 2515$ are -6.01×10^{-4} and 7.82×10^{-3}, respectively. Consequently, we have

$$\text{VaR}_{0.95} = 0.01227, \quad \text{ES}_{0.95} = 0.01554,$$
$$\text{VaR}_{0.99} = 0.01760, \quad \text{ES}_{0.99} = 0.02025.$$

These results can then be used to calculate the risk measure of the financial position. For instance, $\text{VaR}_{0.95} = \$12,270$ and $\text{ES}_{095} = \$15,540$ for the next trading day.

Model 2: To handle the heavy tails of x_t, we employ a GARCH(1,1) model with standardized Student-t innovations. The fitted model is

$$x_t = -4.113 \times 10^{-4} + a_t, \quad a_t = \sigma_t \epsilon_t, \quad \epsilon_t \sim t_{5.751}^*$$
$$\sigma_t^2 = 1.922 \times 10^{-6} + 0.0645 a_{t-1}^2 + 0.9286 \sigma_{t-1}^2.$$

Except for the constant of the mean equation, all coefficient estimates are significant at the usual 5% level. The fitted degrees of freedom are 5.751 confirming the heavy tails of the loss variable x_t. Again, model checking statistics confirm the adequacy of this model. The 1-step ahead predictions of the model for the mean and volatility at

$T = 2515$ are -4.113×10^{-4} and 0.00801, respectively. Applying the results of risk measures discussed before, we obtain that, for x_t,

$$\text{VaR}_{0.95} = 0.01545, \quad \text{ES}_{0.95} = 0.02185, \quad \text{VaR}_{0.99} = 0.02542, \quad \text{ES}_{0.99} = 0.03295.$$

Consequently, using the Student-t innovations, the risk measures for the financial position are $\text{VaR}_{0.95} = \$15,450$ and $\text{ES}_{0.95} = \$21,850$.

From the results of Models 1 and 2, we see that the heavy-tailed innovations give rise to higher risk measures. This is understandable because Model 2 provides a better description of the tail behavior of x_t. As the normality assumption for x_t is rejected by the data, this simple example shows that the VaR under normality is likely to underestimate the true risk.   □

R Demonstration. Output edited.

```
> da=read.table("d-ibm-0110.txt",header=T)
> xt=-log(da$return+1) % calculate negative log returns.
> library(fGarch)
> m1=garchFit(~garch(1,1),data=xt,trace=F)
> m1
Title:  GARCH Modelling
Call: garch Fit(formula = ~garch(1, 1), data = xt, trace = F)

Mean and Variance Equation:
 data ~ garch(1, 1)   [data = xt]

Conditional Distribution: norm

Error Analysis:
         Estimate    Std. Error    t value    Pr(>|t|)
mu       -6.010e-04  2.393e-04     -2.511     0.012044   *
omega    4.378e-06   1.160e-06     3.774      0.000161   ***
alpha1   1.011e-01   1.851e-02     5.463      4.67e-08   ***
beta1    8.841e-01   1.991e-02     44.413     < 2e-16    ***
---
> predict(m1,3)
   meanForecast      meanError           standardDeviation
1 -0.0006009667    0.007824302          0.007824302
2 -0.0006009667    0.008043298          0.008043298
3 -0.0006009667    0.008253382          0.008253382
> source("RMeasure.R")
> m11=RMeasure(-.000601,.0078243)

 Risk Measures for selected probabilities:
      prob          VaR              ES
[1,]  0.950 0.01226883 0.01553828
[2,]  0.990 0.01760104 0.02025244
```

```
[3,] 0.999 0.02357790 0.02574412
>
> m2=garchFit(~garch(1,1),data=xt,trace=F,cond.dist="std")
> m2
Title: GARCH Modelling
Call:  garch-
Fit(formula =~garch(1,1), data=xt,cond.dist="std", trace=F)

Mean and Variance Equation:
 data ~ garch(1, 1) [data = xt]

Conditional Distribution: std

Error Analysis:
          Estimate    Std. Error   t value    Pr(>|t|)
mu        -4.113e-04  2.254e-04    -1.824     0.06811 .
omega      1.922e-06  7.417e-07     2.592     0.00954 **
alpha1     6.448e-02  1.323e-02     4.874     1.09e-06 ***
beta1      9.286e-01  1.407e-02    65.993     < 2e-16 ***
shape      5.751e+00  6.080e-01     9.459     < 2e-16 ***
---
> predict(m2,3)
   meanForecast  meanError            standardDeviation
1 -0.0004112738 0.008100872          0.008100872
2 -0.0004112738 0.008191119          0.008191119
3 -0.0004112738 0.008279772          0.008279772
> m22=RMeasure(-.0004113,.0081009,cond.dist="std",df=5.751)

Risk Measures for selected probabilities:
       prob        VaR            ES
[1,] 0.950 0.01545311 0.02184843
[2,] 0.990 0.02542061 0.03294803
[3,] 0.999 0.04289786 0.05332908
```

7.4.1 Multiple Periods

Suppose that at the time index h we compute the k-horizon risk measures for a financial position with loss variable x_t. In this case, the variable of interest is the k-period cumulative loss at the forecast origin h, that is, $x_h[k] = x_{h+1} + \cdots + x_{h+k}$. If the loss variable x_t follows the time series model in Equations (7.12) and (7.13), then the conditional mean and variance of $x_h[k]$ given the information set F_h can be obtained by the forecasting methods discussed in Chapters 2 and 4.

Expected Loss and Forecast Error. The conditional mean $E(x_h[k]|F_h)$ can be obtained by the forecasting method of ARMA models in Chapter 2. Specifically, we have

$$\hat{x}_h[k] = x_h(1) + \cdots + x_h(k),$$

where $x_h(\ell)$ is the ℓ-step ahead forecast of the loss at the forecast origin h. These forecasts can be computed recursively as discussed in Chapter 2. Using the MA representation

$$x_t = \mu + a_t + \psi_1 a_{t-1} + \psi_2 a_{t-2} + \cdots$$

of the ARMA model in Equation (7.12), we can write the ℓ-step ahead forecast error at the forecast origin h as

$$e_h(\ell) = x_{h+\ell} - x_h(\ell) = a_{h+\ell} + \psi_1 a_{h+\ell-1} + \cdots + \psi_{\ell-1} a_{h+1};$$

see Equation (2.34) and the associated forecast error. The forecast error of the expected k-period return $\hat{x}_h[k]$ is the sum of 1-step to k-step forecast errors of x_t at the forecast origin h and can be written as

$$e_h[k] = e_h(1) + e_h(2) + \cdots + e_h(k)$$

$$= a_{h+1} + (a_{h+2} + \psi_1 a_{h+1}) + \cdots + \sum_{i=0}^{k-1} \psi_i a_{h+k-i}$$

$$= a_{h+k} + (1 + \psi_1) a_{h+k-1} + \cdots + \left(\sum_{i=0}^{k-1} \psi_i \right) a_{h+1}, \qquad (7.14)$$

where $\psi_0 = 1$.

Expected Volatility. The volatility forecast of the k-period loss variable at the forecast origin h is the conditional variance of $e_h[k]$ given F_h. Using the independent assumption of ϵ_{t+i} for $i = 1, \ldots, k$, where $a_{t+i} = \sigma_{t+i}\epsilon_{t+i}$, we have

$$V_h(e_h[k]) = V_h(a_{h+k}) + (1 + \psi_1)^2 V_h(a_{h+k-1}) + \cdots + \left(\sum_{i=0}^{k-1} \psi_i \right)^2 V_h(a_{h+1})$$

$$= \sigma_h^2(k) + (1 + \psi_1)^2 \sigma_h^2(k-1) + \cdots + \left(\sum_{i=0}^{k-1} \psi_i \right)^2 \sigma_h^2(1), \qquad (7.15)$$

where $V_h(z)$ denotes the conditional variance of z given F_h and $\sigma_h^2(\ell)$ is the ℓ-step ahead volatility forecast at the forecast origin h. If the volatility model is the GARCH model in Equation (7.13), then these volatility forecasts can be obtained recursively by the methods discussed in Chapter 4.

As an illustration, consider the special time series model

$$x_t = \mu + a_t, \quad a_t = \sigma_t \epsilon_t,$$

$$\sigma_t^2 = \alpha_0 + \alpha_1 a_{t-1}^2 + \beta_1 \sigma_{t-1}^2.$$

Then, we have $\psi_i = 0$ for all $i > 0$. The point forecast of the cumulative k-period loss at the forecast origin h is $\hat{x}_h[k] = k\mu$ and the associated forecast error is

$$e_h[k] = a_{h+k} + a_{h+k-1} + \cdots + a_{h+1}.$$

Consequently, the volatility forecast for the k-period loss at the forecast origin h is

$$\text{Var}(e_h[k]|F_h) = \sum_{\ell=1}^{k} \sigma_h^2(\ell).$$

Using the forecasting method of GARCH(1,1) models in Chapter 4, we have

$$\sigma_h^2(1) = \alpha_0 + \alpha_1 a_h^2 + \beta_1 \sigma_h^2,$$
$$\sigma_h^2(\ell) = \alpha_0 + (\alpha_1 + \beta_1)\sigma_h^2(\ell-1), \quad \ell = 2,\ldots,k. \tag{7.16}$$

Using Equation (7.16), we obtain that for the case of $\psi_i = 0$ for $i > 0$,

$$\text{Var}(e_h[k]|F_h) = \frac{\alpha_0}{1-\phi}\left[k - \frac{1-\phi^k}{1-\phi}\right] + \frac{1-\phi^k}{1-\phi}\sigma_h^2(1), \tag{7.17}$$

where $\phi = \alpha_1 + \beta_1 < 1$. If $\psi_i \neq 0$ for some $i > 0$, then one should use the general formula of $\text{Var}(e_h[k]|F_h)$ in Equation (7.15). If ϵ_t is Gaussian, then the conditional distribution of $x_h[k]$ given F_h is normal with mean $k\mu$ and variance $\text{Var}(e_h[k]|F_h)$. The quantiles needed in risk-measure calculation are readily available. If the conditional distribution of a_t is not Gaussian (e.g., a Student-t or generalized error distribution), simulation can be used to obtain the multiperiod VaR.

Example 7.5 (continued). Consider the Gaussian GARCH(1,1) model of Example 7.5 for the daily loss variable of IBM stock. That is, consider Model 1 of the example. Suppose that we are interested in the risk measure of a 15-day holding period starting at the forecast origin 2515 (i.e., December 31, 2010). We can use the fitted model to compute the conditional mean and variance for the 15-day cumulative loss via $x_{2515}[15] = \sum_{i=1}^{15} x_{2515+i}$ given F_{2515}. The conditional mean is -0.009015 and the conditional variance is 0.001261, which is obtained by the recursion in Equation (7.16). Thus, we have $x_{2515}[15] \sim N(-0.009015, 0.001261)$. From which, we can easily calculate the risk measures. For instance,

$$\text{VaR}_{0.95}(15) = 0.04939, \quad \text{ES}_{0.95}(15) = 0.06423.$$

For the financial position, these results give

$$\text{VaR}_{0.95}(15) = \$49,390, \quad \text{ES}_{0.95}(15) = \$64,230.$$

On the other hand, applying the square root of time rule to the 1-holding period VaR, we have $\sqrt{15} \times \text{VaR}_{0.95} = \$47,366$. Thus, this example demonstrates that the square root of time rule does not apply in general to risk measures based on econometric models.

Recall that in Example 7.2 we calculated the risk measures for the position using the RiskMetrics methodology and obtained $\text{VaR}_{0.95}(15) = \$45,430$ and $\text{ES}_{0.95}(15) = \$56,972$. These quantities are smaller than those obtained by the econometric approach even under the Gaussian innovations. Thus, different methods for calculating risk measures often lead to different results. Care must be exercised in selecting a proper method to assess risk in risk management.   □

R Demonstration. Multi-period risk measures

```
> M1=predict(m1,15) % Model m1 is defined in the output of Example 7.5.
> names(M1)
[1] "meanForecast"        "meanError"           "standardDeviation"
> mf=M1$meanForecast
> merr=M1$meanError
> pmean=sum(mf)
> pvar=sum(merr^2)
> pstd=sqrt(pvar)
> pmean
[1] -0.009014501
> pvar
[1] 0.001260837
> pstd
[1] 0.03550827
> M11=RMeasure(pmean,pstd)
 Risk Measures for selected probabilities:
        prob       VaR            ES
[1,] 0.950 0.04939141 0.06422887
[2,] 0.990 0.07359009 0.08562265
[3,] 0.999 0.10071431 0.11054505
```

Example 7.5 (continued). Consider next the calculation of multiperiod VaR and ES for the position on IBM stock using a GARCH(1,1) model with standardized Student-t innovations. As a linear combination of k standardized Student-t distributions with v degrees of freedom is not a standardized Student-t distribution with v degrees of freedom, we use GARCH simulation to compute multiperiod VaR and ES. Here, the fitted model is Model 2 of Example 7.5 so that we have $\mu = -4.113 \times 10^{-4}$, $(\alpha_0, \alpha_1) = (1.922 \times 10^{-6}, 0.06448)$, and $\beta = 0.9286$. In addition, from the fitted model, we have $\sigma_{2515} = 0.008282$, $x_{2515} = -6.138 \times 10^{-4}$, and $v = 5.751$. These initial values enable us to simulate $x_{2516} + \cdots + x_{2530}$ many times so that we have an empirical distribution of the loss variable for the next 15 trading days. Specifically, our simulation starts with

$$\sigma_{2516}^2 = 1.922 \times 10^{-6} + 0.06448(x_{2515} - \mu)^2 + \sigma_{2515}^2$$

$$a_{2516} = \sigma_{2516} \times \epsilon_{2516}, \quad \epsilon_{2516} \sim t^*_{5.751}$$
$$x_{2516} = -4.113 \times 10^{-3} + a_{2516},$$

where $\epsilon^*_{5.751}$ is a random draw from the standardized Student-t distribution with 5.751 degrees of freedom, which is obtained by the command `rstd` in the package `fGarch`. We then advance the time index by 1 repeatedly to simulate $x_{2517}, \ldots, x_{2530}$. The sum of these 15 realizations provides an estimate of the loss for the next 15 trading days. This procedure is repeated for many times, say 30,000 iterations, so that we have 30,000 estimates of the loss for the next 15 trading days. These estimates are, in turn, used to obtain an empirical distribution of the loss variable. From the empirical distribution, we can obtain VaR and ES. □

In practice, we use an R script `SimGarcht`, available on the book web, to perform the simulation. See the attached R demonstration. With 30,000 iterations, we obtain

$$\text{VaR}_{0.95}(15) = \$47,977, \quad \text{ES}_{0.95}(15) = \$67,136.$$

For this particular instance, the $\text{VaR}_{0.95}(15)$ is slightly lower than that of Gaussian GARCH(1,1) model, but the ES is higher, showing the effect of heavy tail.

R Demonstration for Simulating Student-t GARCH(1,1) model.

```
> source("SimGarcht.R")
> vol=volatility(m2)
> a1=c(1.922*10^(-6),0.06448); b1=0.9286; mu=-4.113*10^(-4)
> ini=c(ibm[2515],vol[2515])
> mm=SimGarcht(h=15,mu=mu,alpha=a1,b1=b1,df=5.751,ini=ini,nter=30000)
> rr=mm$rtn
> mean(rr)
[1] -0.006051321
> quantile(rr,c(0.95,0.99))   % Obtain VaR
       95%         99%
0.04797729 0.07839338
> idx=c(1:30000)[rr > 0.04797729] % Compute ES for p = 0.05
> mean(rr[idx])
[1] 0.06713603
> idx=c(1:30000)[rr > 0.07839338] % Compute ES for p = 0.01
> mean(rr[idx])
[1] 0.09739639
```

7.5 QUANTILE ESTIMATION

Quantile estimation provides a nonparametric approach to VaR calculation. It makes no specific distributional assumption on the loss variable of a portfolio except that the distribution continues to hold within the prediction period. There are two types of quantile methods. The first method is to use empirical quantile directly and the second method uses quantile regression.

7.5.1 Quantile and Order Statistics

Assuming that the distribution of the loss variable x_t in the prediction period is the same as that in the sample period, one can use the empirical quantile of x_t to calculate VaR and ES. Let $x_1, \ldots, x_n$ be the losses of a portfolio in the sample period. The *order statistics* of the sample are these values arranged in increasing order. We use the notation

$$x_{(1)} \le x_{(2)} \le \cdots \le x_{(n)}$$

to denote the arrangement and refer to $x_{(i)}$ as the ith order statistic of the sample. In particular, $x_{(1)}$ is the sample minimum and $x_{(n)}$ the sample maximum.

Assume that the loss variables are independent and identically distributed and have a continuous distribution with pdf $f(x)$ and CDF $F(x)$. Then, we have the following asymptotic result from the statistical literature (e.g., Cox and Hinkley, 1974, Appendix 2), for the order statistic $x_{(\ell)}$, where $\ell = np$ with $0 < q < 1$.

Result. Let x_q be the qth quantile of $F(x)$, that is, $x_q = F^{-1}(q)$. Assume that the pdf $f(x)$ is not zero at x_q (i.e., $f(x_q) \neq 0$). Then, the order statistic $x_{(\ell)}$ is asymptotically normal with mean x_q and variance $q(1-q)/[nf^2(x_q)]$. That is,

$$x_{(\ell)} \sim N\left[x_q, \frac{q(1-q)}{n[f(x_q)]^2}\right], \quad \ell = np. \tag{7.18}$$

On the basis of the prior result, one can use $x_{(\ell)}$ to estimate the quantile x_q, where $\ell = nq$. In practice, the probability of interest q may not satisfy that nq is a positive integer. In this case, one can use simple interpolation to obtain quantile estimates. More specifically, for noninteger nq, let ℓ_1 and ℓ_2 be the two neighboring positive integers such that $\ell_1 < nq < \ell_2$. Define $q_i = \ell_i/n$. The previous result shows that $x_{(\ell_i)}$ is a consistent estimate of the quantile x_{q_i}. From the definition, $q_1 < q < q_2$. Therefore, the quantile x_q can be estimated by

$$\hat{x}_q = \frac{q_2 - q}{q_2 - q_1} x_{(\ell_1)} + \frac{q - q_1}{q_2 - q_1} x_{(\ell_2)}. \tag{7.19}$$

In practice, sample quantiles can easily be obtained from most statistical packages, including R. Different packages may use slightly different procedures to calculate quantiles when nq is not an integer, but these procedures should be consistent when sample size n increases. In our demonstration, we use the command `quantile` in R to obtain empirical quantiles of a loss variable.

For the ES, we simply use the sample average of data that is greater than or equal to the VaR. A simple way to calculate this sample average is as follows. Let $q = 1 - p$, where p is the small upper tail probability of interest and i_q be the largest integer satisfying $i_q < nq$. Then, the sample ES is

$$\widehat{ES}_{1-p} = \frac{\sum_{i=i_q+1}^{n} x_{(i)}}{n - i_q}, \quad q = 1 - p.$$

Example 7.6. Consider the daily log returns of IBM stock from January 2, 2001 to December 31, 2010. For a long position in the IBM stock, the loss variable is the negative log returns. As $2515 \times 0.95 = 2389.25$, we have $\ell_1 = 2389$, $\ell_2 = 2390$, $p_1 = 2389/2515$, and $p_2 = 2390/2515$. The empirical 95% quantile of the negative log returns can be obtained as

$$\hat{x}_{0.95} = 0.75x_{(2389)} + 0.25x_{(2390)} = 0.02654,$$

$x_{(i)}$ is the ith order statistic of the loss variable x_t. In this particular instance, $x_{(2389)} = 0.02652$ and $x_{(2390)} = 0.02657$. Finally, with $p = 0.05$, the sample ES is $\widehat{ES}_{0.95} = \$39{,}949$ for the next trading day.   □

R Demonstration

```
> da=read.table("d-ibm-0110.txt",header=T)
> ibm=-log(da[,2]+1)
> prob1=c(0.9,0.95,0.99,0.999) % probabilities of interest
> quantile(ibm,prob1)
        90%        95%        99%       99.9%
0.01736836 0.02653783 0.05013151 0.07198369
> sibm=sort(ibm) % Sorting into increasing order
> 0.95*2515
[1] 2389.25
> es=sum(sibm[2390:2515])/(2515-2389)
> es
[1] 0.03994857
```

Discussion. Advantages of using the empirical quantile method to VaR calculation include (a) simplicity and (b) using no specific distributional assumption. However, the approach has several drawbacks. First, it assumes that the distribution of the loss x_t remains unchanged from the sample period to the prediction period. Given that VaR is concerned mainly with tail probability, this assumption implies that the predicted loss cannot be greater than that of the historical loss. It is definitely not so in practice. Second, when the tail probability p is small, the empirical quantile is not an efficient estimate of the theoretical quantile. This can be seen from the theoretical result mentioned earlier. With small p, $f(x_q)$ will be close to zero so that the variance can be very large, indicating huge uncertainty in the quantile estimate. Third, the direct quantile estimation fails to take into account the effect of explanatory variables that are relevant to the portfolio under study.

7.5.2 Quantile Regression

In real application, one often has explanatory variables available that are important to the problem under study. For example, the action taken by Federal Reserve Banks on interest rates could have important impacts on the returns (and loss variables) of

U.S. stocks. It is then more appropriate to consider the distribution function $x_{t+1}|F_t$, where F_t includes the explanatory variables. In other words, we are interested in the quantiles of the distribution function of x_{t+1} given F_t. Such a quantile is referred to as a *regression quantile* in the literature; see Koenker and Bassett (1978).

To understand regression quantile, it is helpful to cast the empirical quantile of the previous section as an estimation problem. For a given probability $q = 1 - p$, the qth quantile of $\{x_t\}$ is obtained by

$$\hat{x}_q = \text{argmin}_\beta \sum_{i=1}^n w_q(x_i - \beta),$$

where $w_q(z)$ is defined by

$$w_q(z) = \begin{cases} qz & \text{if } z \geq 0, \\ (q-1)z & \text{if } z < 0. \end{cases}$$

Regression quantile is a generalization of such an estimate.

To see the generalization, suppose that we have the linear regression

$$x_t = \beta' z_t + a_t, \tag{7.20}$$

where β is a k-dimensional vector of parameters and z_t is a vector of predictors that are elements of F_{t-1}. The conditional distribution of x_t given F_{t-1} is a translation of the distribution of a_t because $\beta' z_t$ is known. Viewing the problem this way, Koenker and Bassett (1978) suggest estimating the conditional quantile $x_q|F_{t-1}$ of x_t given F_{t-1} as

$$\hat{x}_q|F_{t-1} \equiv \inf\{\beta'_o z | R_q(\beta_o) = \min\}, \tag{7.21}$$

where "$R_q(\beta_o) = \min$" means that β_o is obtained by

$$\beta_o = \text{argmin}_\beta \sum_{t=1}^n w_q(x_t - \beta' z_t),$$

where $w_q(.)$ is defined as before. A computer program to obtain such an estimated quantile can be found in Koenker and D'Orey (1987). The package `quantreg` of R performs quantile regression analysis.

Example 7.7. Again, consider a long position of one million dollars on the IBM stock and the sampling period is from January 2, 2001 to December 31, 2010. It is clear that quantiles of a stock return is related to the stock volatility. The quantiles might also be related to the volatility of the market. To explore these possibilities, we shall employ a quantile regression with two predictors. The first predictor is the lag-1 daily volatility of the IBM stock and the second predictor is the lag-1 VIX index of Chicago Board Options Exchange (CBOE). More specifically, we consider the quantile regression

$$Q(q|z_t) = \sum_{t=2}^{2515} w_q(x_t - \beta_0 - \beta_1 s_{t-1} - \beta_2 v_{t-1}), \qquad (7.22)$$

where $x_t = -r_t$ with r_t being the daily log return of IBM stock, s_{t-1} is the lag-1 daily IBM stock volatility obtained from fitting a Gaussian GARCH(1,1) model to x_t, and v_{t-1} is the lag-1 VIX index obtained from CBOE. Here, we use the VIX index, not the percentage VIX.

Applying the quantile regression in Equation (7.22) with $q = 0.95$, we obtain

$$\hat{\beta}_0 = -0.001(0.003), \quad \hat{\beta}_1 = 1.17724(0.22268), \quad \hat{\beta}_2 = 0.02809(0.01615),$$

where the number in parentheses denotes standard error. The constant term is statistically insignificant, and the p-values for the two remaining coefficients are 0.0 and 0.08, respectively. Thus, as expected the 95th quantile of the IBM negative daily log returns depends critically on the lag-1 IBM daily volatility and marginally on the lag-1 VIX index. As $s_{2515} = 0.00802$ and $v_{2515} = 0.1775$, we have $\widehat{Q}(0.95|z_{2515}) = 0.013385$. This implies that VaR$_{0.95}$ = \$13,385 for the financial position. Figure 7.7 shows the

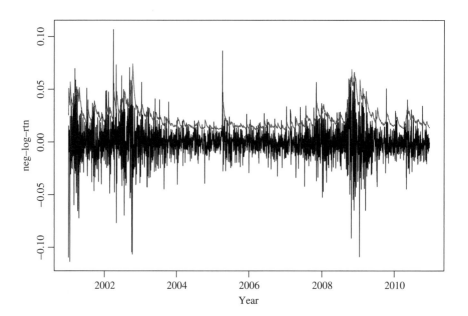

Figure 7.7. Time plot of the negative daily log returns of IBM stock from January 3, 2001 to December 31, 2010. The upper line shows the 95th quantiles obtained by the quantile regression of Equation (7.22).

time plot of the loss variable $x_t = -r_t$ and the fitted values of the quantile regression with probability $q = 0.95$. The fitted quantiles behave in a similar manner as the volatility. This is understandable because the coefficient of lag-1 volatility is highly significant. The plot also shows that VaR is time varying and highlights the fact that the actual loss may vary when the loss exceeds VaR.   □

Finally, the quantile regression becomes harder to estimate when the probability q is close to one. This is due to data limitation because for a large q, the number of relevant observations becomes small. For the IBM daily log returns considered, the quantile regression with probability 0.99 becomes noninformative. See the attached output. All coefficient estimates become statistically insignificant at the usual 5% level.

R Demonstration

```
> dd=read.table("d-ibm-rq.txt",header=T) % Load data
> head(dd)
          nibm         vol      vix
1 -0.109478400 0.01700121 29.99
2  0.015308580 0.01614694 26.60
 .....
6 -0.009408600 0.03211091 27.99
> dim(dd)
[1] 2514      3
> dd[,3]=dd[,3]/100
> library(quantreg)
> mm=rq(nibm~vol+vix,tau=0.95,data=dd) % Quantile regression
> summary(mm)
Call: rq(formula = nibm ~ vol + vix, tau = 0.95, data = dd)
tau: [1] 0.95   % probability
Coefficients:
            Value    Std. Error t value  Pr(>|t|)
(Intercept) -0.00104  0.00257   -0.40317  0.68686
vol          1.17724  0.22268    5.28660  0.00000
vix          0.02809  0.01615    1.73977  0.08202
> names(mm)
 [1] "coefficients"   "x"               "y"         "residuals"
 [5] "dual"           "fitted.values" "formula"   "terms"
 [9] "xlevels"        "call"           "tau"       "rho"
[13] "method"         "model"
> fit=mm$fitted.values
> tdx=c(2:2515)/252+2001
> plot(tdx,dd$nibm,type='l',xlab='year',ylab='neg-log-rtn')
> lines(tdx,fit,col='red')
> v1[2515]
[1] 0.008018202
> vix[2515]
[1] 17.75
> vfit=-.00104+1.17724*v1[2515]+0.02809*vix[2515]/100
> vfit
```

```
[1] 0.01338532
> mm=rq(xt~vol+vix,tau=0.99,data=dd) % 99th quantile
> summary(mm)
Call: rq(formula = xt ~ vol + vix, tau = 0.99, data = dd)
tau: [1] 0.99
Coefficients:
             Value    Std. Error  t value  Pr(>|t|)
(Intercept) 0.01182 0.00831       1.42190  0.15518
vol         1.03129 0.73125       1.41031  0.15857
vix         0.04409 0.05335       0.82641  0.40865
```

7.6 EXTREME VALUE THEORY

Risk measures are used primarily to safeguard a financial position against big losses. To better understand the rare, but important, events of big loss, extreme value theory (EVT) becomes highly relevant. In this section, we review some EVT developed in the statistical literature. Denote the loss variable of a financial position, measured in a fixed time interval such as daily, by x_t. For stock returns, x_t is either log returns or negative log returns; see Equation (7.9). Consider the collection of n measurements of loss, $\{x_1, \ldots, x_n\}$. The maximum loss is $x_{(n)}$, the maximum order statistic. Specifically, $x_{(n)} = \max_{1 \le j \le n}\{x_j\}$. The statistical theory that governs properties of a properly normalized $x_{(n)}$ as n increases is called the *EVT* (see Beirlant et al., 2004; Longin, 1996, 1999a, b).

7.6.1 Review of Extreme Value Theory

Assume that the losses x_t are serially independent with a common cumulative distribution function $F(x)$ and that the range of the return x_t is $[l, u]$. For log returns of a stock, we have $l = -\infty$ and $u = \infty$. Then, the CDF of $x_{(n)}$, denoted by $F_{n,n}(x)$, is given by

$$F_{n,n}(x) = \Pr[x_{(n)} \le x]$$

$$= \Pr(x_1 \le x, x_2 \le x, \ldots, x_n \le x) \quad \text{(by definition of maximum)}$$

$$= \prod_{j=1}^{n} \Pr(x_j \le x) \quad \text{(by independence)}$$

$$= \prod_{j=1}^{n} F(x) = [F(x)]^n. \tag{7.23}$$

In practice, the CDF $F(x)$ of x_t is unknown and, hence, $F_{n,n}(x)$ of $x_{(n)}$ is unknown. However, as n increases to infinity, $F_{n,n}(x)$ becomes degenerated – namely, $F_{n,n}(x) \to 0$ if $x < u$ and $F_{n,n}(x) \to 1$ if $x \ge u$ as n goes to infinity. This degenerated CDF has no practical value. Therefore, the EVT is concerned with finding two sequences $\{\mu_n\}$ and $\{\sigma_n\}$, where $\sigma_n > 0$, such that the distribution of $x^*_{(n)} \equiv (x_{(n)} - \mu_n)/\sigma_n$ converges to a nondegenerate distribution as n goes to infinity. The sequence $\{\mu_n\}$ is a location series and $\{\sigma_n\}$ is a series of scaling factors. Under the independent assumption, the limiting distribution of the normalized maximum $r^*_{(n)}$ is given by

$$F_*(x) = \begin{cases} \exp[-(1 + \xi x)^{-1/\xi}] & \text{if } \xi \neq 0 \\ \exp[-\exp(-x)] & \text{if } \xi = 0, \end{cases} \tag{7.24}$$

for $x < -1/\xi$ if $\xi < 0$ and for $x > -1/\xi$ if $\xi > 0$, where the subscript $*$ signifies the normalized maximum. The case of $\xi = 0$ is taken as the limit when $\xi \to 0$. The parameter ξ is referred to as the *shape parameter* that governs the tail behavior of the limiting distribution. The parameter $\alpha = 1/\xi$ is called the *tail index* of the distribution.

The limiting distribution in Equation (7.24) is the *generalized extreme value* (GEV) *distribution* of Jenkinson (1955) for the maximum. It encompasses the three types of limiting distribution of Gnedenko (1943):

- Type I: $\xi = 0$, the Gumbel family. The CDF is

$$F_*(x) = \exp[-\exp(-x)], \quad -\infty < x < \infty. \tag{7.25}$$

- Type II: $\xi > 0$, the Fréchet family. The CDF is

$$F_*(x) = \begin{cases} \exp[-(1 + \xi x)^{-1/\xi}] & \text{if } x > -1/\xi, \\ 0 & \text{otherwise.} \end{cases} \tag{7.26}$$

- Type III: $\xi < 0$, the Weibull family. The CDF here is

$$F_*(x) = \begin{cases} \exp[-(1 + \xi x)^{-1/\xi}] & \text{if } x < -1/\xi, \\ 1 & \text{otherwise.} \end{cases}$$

Gnedenko (1943) gave necessary and sufficient conditions for the CDF $F(x)$ of x_t to be associated with one of the three types of limiting distribution. Briefly speaking, the tail behavior of $F(x)$ determines the limiting distribution $F_*(x)$ of the maximum. The right tail of the distribution declines exponentially for the Gumbel family, by a power function for the Fréchet family, and is finite for the Weibull family (Fig. 7.8). Readers are referred to Embrechts et al. (1997) for a comprehensive treatment of the EVT. See also Gumbel (1958). For risk management, we are mainly interested in the Fréchet family that includes stable and Student-t distributions. The Gumbel family consists of thin-tailed distributions such as normal and lognormal distributions. The pdf of the generalized limiting distribution in Equation (7.24) can be obtained easily by differentiation:

$$f_*(x) = \begin{cases} (1 + \xi x)^{-1/\xi - 1} \exp[-(1 + \xi x)^{-1/\xi}] & \text{if } \xi \neq 0, \\ \exp[-x - \exp(-x)] & \text{if } \xi = 0, \end{cases} \tag{7.27}$$

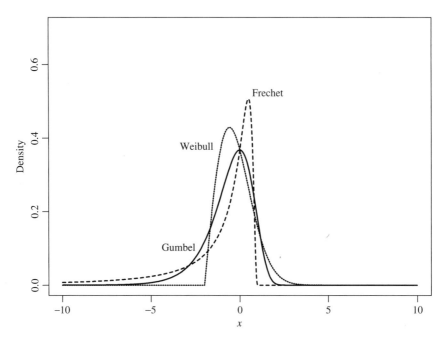

Figure 7.8. Probability density functions of extreme value distributions for normalized maximum. The solid line is for a Gumbel distribution, the dotted line is for the Weibull distribution
with $\xi = -0.5$, and the dashed line is for the Fréchet distribution with $\xi = 0.9$.

where $-\infty < x < \infty$ for $\xi = 0$, and $x < -1/\xi$ for $\xi < 0$, and $x > -1/\xi$
for $\xi > 0$.

The aforementioned EVT has two important implications. First, the tail behavior
of the CDF $F(x)$ of x_t, not the specific distribution, determines the limiting distribution
$F_*(x)$ of the normalized maximum. Thus, the theory is generally applicable to a wide
range of distributions for the loss variable x_t. The sequences $\{\mu_n\}$ and $\{\sigma_n\}$, however,
depend on the CDF $F(x)$. McNeil et al. (2005, Chapter 7) provide explicit values
of μ_n and σ_n for exponential and Pareto distributions. Second, Feller (1971, p. 279)
shows that the tail index ξ does not depend on the time interval of x_t. That is, the tail
index (or equivalently the shape parameter) is invariant under time aggregation. This
second feature of the limiting distribution becomes handy in the VaR calculation.

The EVT has been extended to serially dependent observations $\{x_t\}_{t=1}^{n}$ provided
that the dependence is weak. Berman (1964) shows that the same form of the limiting
extreme value distribution holds for stationary normal sequences provided that the
autocorrelation function of x_t is squared summable (i.e., $\sum_{i=1}^{\infty} \rho_i^2 < \infty$), where ρ_i
is the lag-i autocorrelation function of x_t. For further results concerning the effect
of serial dependence on the EVT, readers are referred to Leadbetter et al. (1983,
Chapter 3). Tsay (2010, Section 7.8) discusses the *extremal index* for a strictly stationary time series. Here, we provide a brief introduction of the concept of extremal
index (Leadbetter, 1974, 1983; Hsing et al., 1988).

Let $\{x_t\}$ be a strictly stationary time series, where the distribution of x_t is $F(x)$. Let $\{\tilde{x}_t\}$ be an associated series of independent and identically distributed random variables $\tilde{x}_t$, which follows the same distribution $F(x)$. That is, $\{\tilde{x}_t\}$ is a strict white noise series with the same distribution $F(x)$. Let $x_{(n)} = \max(x_1, \ldots, x_n)$ and $\tilde{x}_{(n)} = \max(\tilde{x}_1, \ldots, \tilde{x}_n)$. For many processes $\{x_t\}$, one can show that there exists a real number $\theta \in (0, 1]$ such that

$$\lim_{n \to \infty} P[(\tilde{x}_{(n)} - \mu_n)/\sigma_n \leq x] = F_*(x), \tag{7.28}$$

for a nondegenerated limit $F_*(x)$ if and only if

$$\lim_{n \to \infty} P[(x_{(n)} - \mu_n)/\sigma_n \leq x] = F_*^{\theta}(x). \tag{7.29}$$

This value θ is called the *extremal index* of $\{x_t\}$.

When the results (Eqs. 7.28 and 7.29) hold, we see that, from Equation (7.23), for a sufficiently large n we have

$$P(x_{(n)} \leq y) \approx P^{\theta}(\tilde{x}_{(n)} \leq y) = F^{n\theta}(y),$$

where $y = \sigma_n x + \mu_n$. Therefore, for large n, the probability distribution of the maximum of n observations from the strictly stationary time series with extremal index θ can be approximately by the distribution of the maximum of $n\theta$ observations from the associated *iid* series. Notice that $n\theta \leq n$. This approximation says that $n\theta$ can be thought of as the number of independent clusters of observations in n observations of x_t. In this sense, $n\theta$ can be treated as the equivalent number of observations of a dependent series with respect to its *iid* counterpart and $1/\theta$ can be interpreted as the reciprocal of the mean cluster size. It should be noted, however, that not every strictly stationary time series has an extremal index.

7.6.2 Empirical Estimation

Before applying EVT to assessing risk, we consider estimation of the unknown parameters of the generalized extreme value distribution. The distribution contains three parameters. They are the *shape* parameter ξ, the *location* parameter μ, and the *scale* σ. These parameters can be estimated by using either parametric or nonparametric methods. We review some of the estimation methods.

The Block Maxima Method. For a given sample, there is only a single maximum, and we cannot estimate the three parameters with an extreme observation. Alternative ideas must be used. One of the ideas used in the literature is to divide the sample into subsamples and apply the EVT to the subsamples. Assume that there are T observations available, say $\{x_j\}_{j=1}^{T}$. We divide the sample into g nonoverlapping subsamples each with n observations, assuming for simplicity that $T = ng$. In other words, we divide the data as

$$\{x_1, \ldots, x_n | x_{n+1}, \ldots, x_{2n} | x_{2n+1}, \ldots, x_{3n} | \cdots | x_{(g-1)n+1}, \ldots, x_{ng}\},$$

and write the observed returns as x_{in+j}, where $1 \le j \le n$ and $i = 0, \ldots, g-1$. Note that each subsample corresponds to a subperiod of the data span. When n is sufficiently large, we hope that the EVT applies to each subsample. In application, the choice of n can be guided by practical considerations. For example, for daily returns, $n = 21$ corresponds approximately to the number of trading days in a month and $n = 63$ denotes the number of trading days in a quarter.

Let $x_{n,i}$ be the maximum of the ith subsample (i.e., $x_{n,i}$ is the largest loss of the ith subsample), where the subscript n is used to denote the size of the subsample. When n is sufficiently large, $y_{n,i} = (x_{n,i} - \mu)/\sigma$ should follow an extreme value distribution, and the collection of subsample maxima $\{x_{n,i} | i = 1, \ldots, g\}$ can then be regarded as a random sample of g observations from that extreme value distribution. Specifically, we define

$$x_{n,i} = \max_{1 \le j \le n} \{x_{(i-1)n+j}\}, \quad i = 1, \ldots, g. \tag{7.30}$$

The collection of subsample maxima $\{x_{n,i}\}$ is the data we use to estimate the unknown parameters of the extreme value distribution. Clearly, the estimates obtained may depend on the choice of subperiod length n.

Remark. When T is not a multiple of the subsample size n, several methods have been used to deal with this issue. First, one can allow the last subsample to have a smaller size. Second, one can ignore the first few observations so that each subsample has size n. The evir package of R uses the first method.   □

Maximum Likelihood Method. Assuming that the subperiod maxima $\{x_{n,i}\}$ follow a generalized extreme value distribution such that the pdf of $y_i = (x_{n,i} - \mu)/\sigma$ is given in Equation (7.27), we can obtain the pdf of $x_{n,i}$ by a simple transformation as

$$f(x_{n,i}) =$$

$$\begin{cases} \dfrac{1}{\sigma} \left(1 + \dfrac{\xi(x_{n,i} - \mu)}{\sigma} \right)^{-(1+\xi)/\xi} \exp\left[-\left(1 + \dfrac{\xi(x_{n,i} - \mu)}{\sigma} \right)^{-1/\xi} \right] & \text{if } \xi \ne 0, \\[4mm] \dfrac{1}{\sigma} \exp\left[-\dfrac{x_{n,i} - \mu}{\sigma} - \exp\left(-\dfrac{x_{n,i} - \mu}{\sigma} \right) \right] & \text{if } \xi = 0, \end{cases}$$

where it is understood that $1 + \xi(x_{n,i} - \mu)/\sigma > 0$ if $\xi \ne 0$. Under the independence assumption, the likelihood function of the subperiod maxima is

$$\ell(x_{n,1}, \ldots, x_{n,g} | \xi, \sigma, \mu) = \prod_{i=1}^{g} f(x_{n,i}).$$

Nonlinear estimation procedures can then be used to obtain maximum likelihood estimates of ξ, μ, and σ. These estimates are unbiased, asymptotically normal, and of minimum variance under proper assumptions. See Embrechts et al. (1997) and Coles (2001) for details. We apply this approach to some stock return series later.

The Nonparametric Approach. The shape parameter ξ can be estimated using some nonparametric methods. Two such methods are mentioned here. These two methods are proposed by Hill (1975) and Pickands (1975) and are referred to as the *Hill estimator* and *Pickands estimator*, respectively. Both estimators apply directly to the returns $\{x_t\}_{t=1}^{T}$. Thus, there is no need to consider subsamples. Denote the order statistics of the sample as

$$x_{(1)} \leq x_{(2)} \leq \cdots \leq x_{(T)}.$$

Let q be a positive integer. The two estimators of ξ are defined as

$$\xi_p(q) = \frac{1}{\ln(2)} \ln \left(\frac{x_{(T-q+1)} - x_{(T-2q+1)}}{x_{(T-2q+1)} - x_{(T-4q+1)}} \right), \quad q \leq T/4, \tag{7.31}$$

$$\xi_h(q) = \frac{1}{q} \sum_{i=1}^{q} \left[\ln(x_{(T-i+1)}) - \ln(x_{(T-q)}) \right], \tag{7.32}$$

where the argument (q) is used to emphasize that the estimators depend on q and the subscripts p and h denote Pickands and Hill estimators, respectively. The choice of q differs between Hill and Pickands estimators. It has been investigated by several researchers, but there is no general consensus on the best choice available. Dekkers and De Haan (1989) show that $\xi_p(q)$ is consistent if q increases at a properly chosen pace with the sample size T. In addition, $\sqrt{q}[\xi_p(q) - \xi]$ is asymptotically normal with mean zero and variance $\xi^2(2^{2\xi+1} + 1)/[2(2^{\xi} - 1)\ln(2)]^2$. The Hill estimator is applicable to the Fréchet distribution only, but it is more efficient than the Pickands estimator when applicable. Goldie and Smith (1987) show that $\sqrt{q}[\xi_h(q) - \xi]$ is asymptotically normal with mean zero and variance ξ^2. In practice, one may plot the Hill estimator $\xi_h(q)$ against q and find a proper q such that the estimate appears to be stable. The estimated tail index $\alpha = 1/\xi_h(q)$ can then be used to obtain extreme quantiles of the return series; see Zivot and Wang (2003).

7.6.3 Application to Stock Returns

We apply the EVT to the daily log returns of IBM stock from January 2, 2001 to December 31, 2010. The returns are measured in percentages, and the sample size is 2515 (i.e., $T = 2515$). Figure 7.9 shows the time plots of block maximum of the daily log returns when the length of the subperiod is 21 days, which corresponds approximately to a month. The upper plot is for the positive returns while the lower plot the negative returns. Empirical analysis of this section is carried out in R using the command `gev` of the package `evir`, which stands for extreme value in R, and a simple R script that calculates the Hill estimator.

Table 7.1 summarizes some estimation results of the shape parameter ξ via the Hill estimator. Two choices of q are reported in Table 7.1, and the results are stable. To provide an overall picture of the performance of the Hill estimator, Figure 7.10

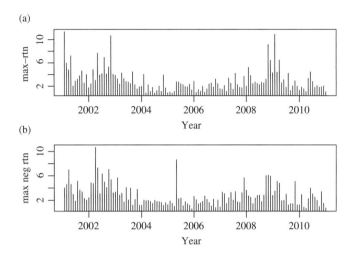

Figure 7.9. Block maximum of daily log returns of IBM stock, in percentages, when the subperiod is 21 trading days. The data span is from January 2, 2001 to December 31, 2010 so that there are 120 blocks. (a) positive returns and (b) negative returns.

TABLE 7.1. Results of the Hill Estimator for Daily Log Returns of IBM Stock from July 3, 1962 to December 31, 1998

q	110	130	150
r_t	0.380(0.036)	0.399(0.035)	0.398(0.032)
$-r_t$	0.356(0.034)	0.383(0.034)	0.405(0.033)

Standard errors are in parentheses.

shows the scatterplots of the Hill estimator $\xi_h(q)$ and its pointwise 95% confidence interval against q. For the extreme positive daily log returns, the estimator is stable ranging between 0.30 and 0.4 for a wide range of q. For the extreme negative returns, the estimate of ξ is approximately 0.3 for q in [40,126]. However, the estimate drops significantly for smaller q. Overall, the estimated shape parameters are significantly different from zero at the asymptotic 5% level for a wide range of q, indicating that the distribution of daily log returns of IBM stock belongs to the Fréchet family. The analysis thus rejects the normality assumption commonly used in practice. This is consistent with our experience in fitting volatility models to the IBM daily stock returns.

Next, we apply the maximum likelihood method to estimate parameters of the generalized extreme value distribution for IBM daily log returns. Table 7.2 summarizes the estimation results for two choices of the length of subperiods, namely, 1 month ($n = 21$) and 2 months ($n = 42$). We do not use longer subperiods because of the sample size constraint. From the table, we make the following observations:

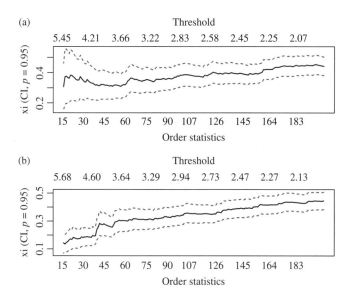

Figure 7.10. Scatterplots of the Hill estimator for the daily log returns of IBM stock. The sample period is from January 2, 2001 to December 31, 2010: (a) positive returns and (b) negative returns.

TABLE 7.2. Maximum Likelihood Estimates of the Extreme Value Distribution for Daily Log Returns of IBM Stock, in Percentages, from January 2, 2001 to December 31, 2010

Length of Subperiod	Shape Parameter ξ	Scale σ	Location μ
	Maximal Positive Returns		
1 mo ($n = 21$, $g = 120$)	0.278(0.087)	1.046(0.092)	2.046(0.111)
2 mo ($n = 42$, $g = 60$)	0.315(0.109)	1.168(0.145)	2.622(0.170)
	Maximal Negative Returns		
1 mo ($n = 21$, $g = 120$)	0.251(0.088)	1.029(0.090)	1.966(0.109)
2 mo ($n = 42$, $g = 60$)	0.287(0.142)	1.100(0.143)	2.489(0.170)

[a] Standard errors are in parentheses.

- Estimates of the location and scale parameters $\hat{\mu}$ and $\hat{\sigma}$ increase in modulus as n increases. This is expected as magnitudes of the subperiod minimum and maximum are nondecreasing functions of n.
- Estimates of the shape parameter (or equivalently the tail index) are relatively stable, but they become less accurate when the number of subperiods g is small.

These results are similar to those of Tsay (2010, Chapter 7) that uses daily log returns of IBM stock from 1962 to 1998.

R Demonstration for Extreme Value Analysis. The series is IBM daily log
returns from 2001 to 2010. Output edited.

```
> da=read.table("d-ibm-0110.txt",header=T)
> ibm=log(da$return+1)*100
> xt=-ibm
> source("Hill.R") % compile R script
> Hill
function(x,q){
# Compute the Hill estimate of the shape parameter.
sx=sort(x); T=length(x); ist=T-q
y=log(sx[ist:T])
hill=sum(y[2:length(y)])/q
hill=hill-y[1]
sd=sqrt(hill^2/q)
cat("Hill estimate & std-err:",c(hill,sd),"\ n")
Hill <- list(est=hill,std=sd)
}
> Hill(ibm,110)
Hill estimate & std-err: 0.3800632 0.0362376
> Hill(xt,110)
Hill estimate & std-err: 0.3555175 0.03389727

> library(evir) % Load package
> par(mfcol=c(2,1))
> hill(ibm,option=c("xi"),end=200)
> hill(xt,option=c("xi"),end=200)
> help(hill)

> m1=gev(xt,block=21)
> m1
$n.all
[1] 2515
$n
[1] 120
$data
  [1]   4.0335654   4.6038703   6.9818569   ......
$block
[1] 21
$par.ests
      xi      sigma          mu
0.251353 1.028910 1.965850
$par.ses
        xi        sigma          mu
0.08847742 0.09013351 0.10932034
$varcov
              [,1]            [,2]            [,3]
[1,]   0.007828254 -0.001080741 -0.003453668
[2,]  -0.001080741  0.008124049  0.006145413
```

```
[3,] -0.003453668  0.006145413   0.011950936
$converged
[1] 0

> plot(m1)
Make a plot selection (or 0 to exit):
1: plot: Scatterplot of Residuals
2: plot: QQ-plot of Residuals
Selection: 1
```

Define the residuals of a GEV distribution fit as

$$w_i = \left(1 + \hat{\xi}\frac{x_{n,i} - \hat{\mu}}{\hat{\sigma}}\right)^{-1/\hat{\xi}}.$$

Using the pdf of the GEV distribution and transformation of variables, one can easily show that $\{w_i\}$ should form an iid random sample of exponentially distributed random variables if the fitted model is correctly specified. Figure 7.11 shows the residual plots of the GEV distribution fit to the negative IBM daily log returns with subperiods of 21 days. Figure 7.11a gives the residuals and Figure 7.11b shows a quantile-to-quantile (QQ) plot against an exponential distribution. The plots indicate that the fit is reasonable.

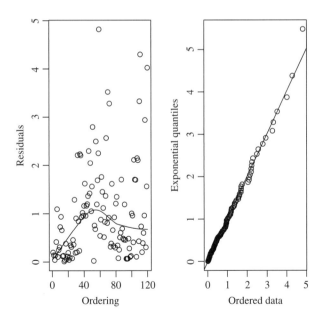

Figure 7.11. Residual analysis of fitting a GEV distribution to the negative IBM daily log returns, in percentages, from January 2, 2001 to December 31, 2010. The subperiod length used is 21 days.

Remark. Besides `evir`, several other packages are also available in R to perform extreme value analysis. They are `evd`, `POT`, and `extRemes`.   □

7.7   AN EXTREME VALUE APPROACH TO VAR

In this section, we discuss an approach to VaR calculation using the extreme value theory. We divide the discussion into two parts. The first part is concerned with parameter estimation using the method discussed in the previous section. The second part focuses on VaR calculation by relating the probabilities of interest to the choice of the length of subperiods.

Part I. Assume that there are T observations of a loss variable available in the sample period. We partition the sample period into g nonoverlapping subperiods of length n such that $T = ng$. If $T = ng + m$ with $1 \leq m < n$, then we can simply delete the first m observations from the sample. The EVT discussed in the previous section enables us to obtain estimates of the location, scale, and shape parameters μ_n, σ_n, and ξ_n for the subperiod maxima $\{x_{n,i}\}$. Here, we use the subscript n to signify that the parameters are estimated using subperiods of length n. Plugging the maximum likelihood estimates into the CDF in Equation (7.24) with $y = (x - \mu_n)/\sigma_n$, we can obtain the quantile of a given probability of the generalized extreme value distribution. Let p^* be a small upper tail probability and y_n^* be the $(1 - p^*)$th quantile of the subperiod maxima under the limiting generalized extreme value distribution. Then, we have

$$
1 - p^* = \begin{cases} \exp\left[-\left(1 + \dfrac{\xi_n(y_n^* - \mu_n)}{\sigma_n}\right)^{-1/\xi_n}\right] & \text{if } \xi_n \neq 0, \\[3mm] \exp\left[-\exp\left(-\dfrac{y_n^* - \mu_n}{\sigma_n}\right)\right] & \text{if } \xi_n = 0, \end{cases}
$$

where it is understood that $1 + \xi_n(y_n^* - \mu_n)/\sigma_n > 0$ for $\xi_n \neq 0$. Rewriting this equation as

$$
\ln(1 - p^*) = \begin{cases} -\left[1 + \dfrac{\xi_n(y_n^* - \mu_n)}{\sigma_n}\right]^{-1/\xi_n} & \text{if } \xi_n \neq 0, \\[3mm] -\exp\left[-\dfrac{y_n^* - \mu_n}{\sigma_n}\right] & \text{if } \xi_n = 0, \end{cases}
$$

we obtain the quantile as

$$
y_n^* = \begin{cases} \mu_n - \dfrac{\sigma_n}{\xi_n}\left\{1 - \left[-\ln(1 - p^*)\right]^{-\xi_n}\right\} & \text{if } \xi_n \neq 0, \\[3mm] \mu_n - \sigma_n \ln[-\ln(1 - p^*)] & \text{if } \xi_n = 0. \end{cases} \tag{7.33}
$$

In financial applications, the case of $\xi_n \neq 0$ is of major interest.

Part II. For a given upper tail probability p^*, the quantile y_n^* of Equation (7.33) is the VaR based on the EVT for the subperiod maximum. The next step is to study the explicit relationship between quantiles of the subperiod maximum and the observed loss x_t series.

Because most asset losses are either serially uncorrelated or have weak serial correlations, we may use the relationship in Equation (7.23) and obtain

$$1 - p^* = P(x_{n,i} \le y_n^*) = [P(x_t \le y_n^*)]^n. \tag{7.34}$$

This relationship between probabilities allows us to obtain VaR for the original loss variable x_t. More precisely, for a specified small upper probability p, the $(1 - p)$th quantile of x_t is y_n^* if the upper tail probability p^* of the subperiod maximum is chosen based on Equation (7.34), where $P(x_t \le y_n^*) = 1 - p$. Consequently, for a given small upper tail probability p, the VaR of a financial position with loss variable x_t is

$$\text{VaR} = \begin{cases} \mu_n - \dfrac{\sigma_n}{\xi_n} \left\{ 1 - \left[-n \ln(1 - p) \right]^{-\xi_n} \right\} & \text{if } \xi_n \neq 0 \\[2mm] \mu_n - \sigma_n \ln[-n \ln(1 - p)] & \text{if } \xi_n = 0, \end{cases} \tag{7.35}$$

where n is the length of subperiods.

Summary. We summarize the approach of applying the traditional EVT to VaR calculation as follows:

1. Select the length of the subperiod n and obtain subperiod maxima $\{x_{n,i}\}$, $i = 1, \ldots, g$, where $g = [T/n]$.
2. Obtain the maximum likelihood estimates of μ_n, σ_n, and ξ_n.
3. Check the adequacy of the fitted extreme value model; see the next section for some methods of model checking.
4. If the extreme value model is adequate, apply Equation (7.35) to calculate VaR.

Example 7.8. Consider the daily log return, in percentage, of IBM stock from January 2, 2001 to December 31, 2010. Again, assume that we hold a long position of one million dollars on the stock. In this case, the loss variable is $x_t = -r_t$. From Table 7.2, we have $\hat{\alpha}_n = 1.029$, $\hat{\beta}_n = 1.966$, and $\hat{\xi}_n = 0.251$ for $n = 21$. Therefore, for the left-tail probability $p = 0.05$, the corresponding VaR is

$$\text{VaR} = 1.966 - \frac{1.029}{0.251} \left\{ 1 - [-21 \ln(1 - 0.05)]^{-0.251} \right\}$$

$$= 1.8902.$$

Thus, for negative daily log returns of the stock, the upper 1% quantile is 1.8902%. Consequently, we have $\text{VaR}_{0.95} = \$1,000,000 \times 0.018902 = \$18,902$. If the probability is 0.01, then the corresponding VaR is \$39,242.

If we chose $n = 42$ (i.e., 2 months), then $\hat{\alpha}_n = 1.1$, $\hat{\beta}_n = 2.489$, and $\hat{\xi}_n = 0.287$. The upper 1% quantile of the loss variable based on the extreme value distribution is

$$\text{VaR} = 2.489 - \frac{1.1}{0.287}\{1 - [-42\ln(1 - 0.01)]^{-0.287}\} = 3.5655.$$

Therefore, for a long position of \$1,000,000, the corresponding 1-day horizon VaR is \$35,655 at the 1% risk level. If the probability is 0.05, then the corresponding VaR is \$17,313. In this particular case, the choice of $n = 21$ gives higher VaR values.   □

Remark. As shown by the results of Example 7.6, the VaR calculation based on the traditional EVT depends on the choice of n, which is the length of subperiods. For the limiting extreme value distribution to hold, one would prefer a large n. But a larger n means a smaller g when the sample size T is fixed, where g is the effective sample size used in estimating the three parameters σ_n, μ_n, and ξ_n. Therefore, some compromise between the choices of n and g is needed. A proper choice may depend on the returns of the asset under study. We recommend that one should check the stability of the resulting VaR in applying the traditional EVT.   □

7.7.1 Discussion

We have applied various methods of VaR calculation to the daily log returns of IBM stock for a long position of \$1 million. Consider the VaR of the position for the next trading day. If the probability is 5%, which means that with probability 0.95 the loss will be less than or equal to the VaR for the next trading day, then the results obtained are

1. \$11,730 for the RiskMetrics,
2. \$12,270 for a Gaussian GARCH(1,1) model,
3. \$15,450 for a GARCH(1,1) model with a standardized Student-t distribution with 5.75 degrees of freedom,
4. \$26,540 for using the empirical quantile,
5. \$13,385 for using quantile regression, and
6. \$18,901 for applying the traditional EVT using $n = 21$ for the length of subperiods.

If the tail probability is 1%, then the VaR is

1. \$16,590 for the RiskMetrics,
2. \$15,540 for a Gaussian GARCH(1,1) model,

3. $25,420 for a GARCH(1,1) model with a standardized Student-t distribution with 5.75 degrees of freedom,
4. $50,132 for using the empirical quantile, and
5. $39,242 for applying the traditional EVT using $n = 21$.

There are substantial differences among different approaches. This is not surprising because there exists substantial uncertainty in estimating tail behavior of a statistical distribution. As there is no truc VaR available to compare the accuracy of different approaches, we recommend that one applies several methods to gain insight into the range of VaR.

7.7.2 Multiperiod VaR

The square root of time rule of the RiskMetrics methodology becomes a special case under the EVT. The proper relationship between ℓ-day and 1-day horizons is

$$\text{VaR}(\ell) = \ell^{1/\alpha}\text{VaR} = \ell^{\xi}\text{VaR},$$

where α is the tail index and ξ is the shape parameter of the extreme value distribution; see Danielsson and de Vries (1997a). This relationship is referred to as the α-*root* of time rule.

For illustration, consider the daily log returns of IBM stock in Example 7.8. If we use $p = 0.01$ and the results of $n = 21$, then for a 15-day horizon we have

$$\text{VaR}(15) = (15)^{0.251}\text{VaR} = 1.973 \times \$39,242 = \$77,437.$$

Because $\ell^{0.251} < \ell^{0.5}$, the α-root of time rule produces lower ℓ-day horizon VaR than the square root of time rule does.

7.7.3 Return Level

Another risk measure based on the extreme values of subperiods is the *return level*. The g n-subperiod return level, $L_{n,g}$, is defined as the *level* that is exceeded in one out of every g subperiods of length n. That is,

$$P(x_{n,i} > L_{n,g}) = \frac{1}{g},$$

where $x_{n,i}$ denotes subperiod maximum. The subperiod in which the return level is exceeded is called a *stress period*. If the subperiod length n is sufficiently large so that normalized $x_{n,i}$ follows the GEV distribution, then the return level is

$$L_{n,g} = \mu_n - \frac{\sigma_n}{\xi_n}\{1 - [-\ln(1 - 1/g)]^{-\xi_n}\},$$

provided that $\xi_n \neq 0$. Note that this is precisely the quantile of extreme value distribution given in Equation (7.33) with tail probability $p^* = 1/g$, even though we write it in a slightly different way. Thus, return level applies to the subperiod maximum, not to the underlying returns. This marks the difference between VaR and return level.

For the negative daily IBM log returns with subperiod length of 21 days, we can use the fitted model to obtain the return level for 12 such subperiods (i.e., $g = 12$). The return level is 5.434%.

R Demonstration for Obtaining Return Level

```
> da=read.table("d-ibm-0110.txt",header=T)
> xt=-log(da[,2]+1)*100
> library(evir)
> m1=gev(xt,block=21)     % GEV estimation with sub-period length 21.
> rl.21.12=rlevel.gev(m1,k.block=12)
> rl.21.12                       % Output plot is not shown.
[1]  4.653307 5.434319 6.756033
      % return-level = 5.434 with 95% C.I. (4.653, 6.756)
```

7.8   PEAKS OVER THRESHOLDS

The traditional EVT approach to risk calculation encounters some difficulties. First, the choice of subperiod length n is not clearly defined. Second, by using only subperiod maxima, the approach does not make efficient use of the loss data. Third, the approach is unconditional and, hence, does not take into consideration effects of other explanatory variables. To overcome these difficulties, an alternative approach to using EVT has been proposed in the statistical literature; see Davison and Smith (1990) and Smith (1989, 1999). Instead of focusing on the extremes (maximum or minimum), the new approach focuses on exceedances of the loss over some high threshold and the times at which the exceedances occur. Thus, this new approach is also referred to as *peaks over thresholds* (POT). For illustration, consider the daily returns of IBM stock used in this chapter and a long position on the stock. Denote the negative daily log return by x_t. Let η be a prespecified high threshold. We may, for instance, choose $\eta = 2.5\%$. Suppose that the ith exceedance occurs at day t_i (i.e., $x_{t_i} \geq \eta$). Then, the new approach focuses on the data $(t_i, x_{t_i} - \eta)$. Here, $x_{t_i} - \eta$ is the exceedance over the threshold η and t_i is the time at which the ith exceedance occurs.

In practice, the occurrence times $\{t_i\}$ provide useful information about the intensity of the occurrence of important "rare events" (e.g., greater than the threshold η for a loss variable). A cluster of t_i indicates a period of large market declines. The exceeding amount (or exceedance) $x_{t_i} - \eta$ is also of importance as it provides the actual quantity of interest.

On the basis of the prior introduction, the POT approach does not require the choice of a subperiod length n, but it requires the specification of threshold η. Different choices of the threshold η lead to different estimates of the shape parameter ξ (and hence the tail index $1/\xi$). In the literature, some researchers believe that the choice of

η is a statistical problem as well as a financial one, and it cannot be determined based purely on statistical theory. For example, different financial institutions (or investors) have different risk tolerances. As such, they may select different thresholds even for an identical financial position. For the daily log returns of IBM stock considered in this chapter, the calculated VaR is not sensitive to the choice of η.

The choice of threshold η also depends on the observed log returns. For a stable return series, $\eta = 2.5\%$ may fare well for a long position. For a volatile return series (e.g., daily returns of a dot-com stock), η may be as high as 10%. Limited experience shows that η can be chosen so that the number of exceedances is sufficiently large (e.g., about 5% of the sample). For a more formal study on the choice of η, see Danielsson and de Vries (1997b).

7.8.1 Statistical Theory

Again consider the loss variable x_t of an asset. Suppose that the ith exceedance occurs at t_i. Focusing on the exceedance $y = x_t - \eta$ and exceeding time t_i results in a fundamental change in statistical thinking. Instead of using the marginal distribution (e.g., the limiting distribution of the minimum or maximum), the POT approach employs a conditional distribution to handle the magnitude of exceedance given that the measurement exceeds a threshold. The chance of exceeding the threshold is governed by a probability law. In other words, the POT approach considers the conditional distribution of $y = x_t - \eta$ given $x_t \geq \eta$ for a long position. Occurrence of the event $\{x_t \geq \eta\}$ follows a point process (e.g., a Poisson process). In particular, if the intensity parameter λ of the process is time invariant, then the Poisson process is homogeneous. If λ is time variant, then the process is nonhomogeneous. The concept of Poisson process can be generalized to the multivariate case.

The basic theory of the POT approach is to consider the conditional distribution of $x = y + \eta$ given $x > \eta$ for the limiting distribution of the maximum given in Equation (7.24). Then, the conditional distribution of $x \leq y + \eta$ given $x > \eta$ is

$$\Pr(x \leq y + \eta | x > \eta) = \frac{\Pr(\eta \leq x \leq y + \eta)}{\Pr(x > \eta)} = \frac{\Pr(x \leq y + \eta) - \Pr(x \leq \eta)}{1 - \Pr(x \leq \eta)}. \quad (7.36)$$

Using the CDF $F_*(.)$ of Equation (7.24) and the approximation $e^{-z} \approx 1 - z$ and after some algebra, we obtain that

$$\Pr(x \leq y + \eta | x > \eta) = \frac{F_*(y + \eta) - F_*(\eta)}{1 - F_*(\eta)}$$

$$= \frac{\exp\left[-\left(1 + \frac{\xi(y+\eta-\mu)}{\sigma}\right)^{-1/\xi}\right] - \exp\left[-\left(1 + \frac{\xi(\eta-\mu)}{\sigma}\right)^{-1/\xi}\right]}{1 - \exp\left[-\left(1 + \frac{\xi(\eta-\mu)}{\sigma}\right)^{-1/\xi}\right]}$$

$$\approx 1 - \left(1 + \frac{\xi y}{\sigma + \xi(\eta - \mu)}\right)^{-1/\xi}, \quad (7.37)$$

where $y > 0$ and $1 + \xi(\eta - \mu)/\sigma > 0$. As is seen later, this approximation makes explicitly the connection of the POT approach to the traditional EVT. The case of $\xi = 0$ is taken as the limit of $\xi \to 0$ so that

$$Pr(x \le y + \eta | x > \eta) \approx 1 - \exp(-y/\sigma).$$

The probability distribution function with CDF

$$G_{\xi,\psi(\eta)}(y) = \begin{cases} 1 - \left[1 + \dfrac{\xi y}{\psi(\eta)} \right]^{-1/\xi} & \text{for } \xi \ne 0, \\ 1 - \exp[-y/\psi(\eta)] & \text{for } \xi = 0, \end{cases} \tag{7.38}$$

where $\psi(\eta) > 0$, $y \ge 0$ when $\xi \ge 0$, and $0 \le y \le -\psi(\eta)/\xi$ when $\xi < 0$, which is called the *generalized Pareto distribution* (GPD). Thus, the result of Equation (7.37) shows that the conditional distribution of x given $x > \eta$ is well approximated by a GPD with parameters ξ and $\psi(\eta) = \sigma + \xi(\eta - \mu)$. See Embrechts et al. (1997) for further information. An important property of the GPD is as follows. Suppose that the excess distribution of x given a threshold η_0 is a GPD with shape parameter ξ and scale parameter $\psi(\eta_0)$. Then, for an arbitrary threshold $\eta > \eta_0$, the excess distribution over the threshold η is also a GPD with shape parameter ξ and scale parameter $\psi(\eta) = \psi(\eta_0) + \xi(\eta - \eta_0)$.

When $\xi = 0$, the GPD in Equation (7.38) reduces to an exponential distribution. This result motivates the use of a QQ-plot of excess returns over a threshold against exponential distribution to infer the tail behavior of the returns. If $\xi = 0$, then the QQ-plot should be linear. Figure 7.12a shows the QQ-plot of daily negative IBM log returns used in this chapter with threshold 0.01. The plot shows a clear deviation from being a straight line, indicating that the left tail of the daily IBM log returns is heavier than that of a normal distribution, that is, $\xi \ne 0$.

R Commands Used to Produce Figure 7.12.

```
> da=read.table("d-ibm-0110.txt",header=T)
> ibm=log(da[,2]+1)
> library(evir)
> par(mfcol=c(2,1))
> xt=-ibm
> qplot(xt,threshold=0.01,pch='*',cex=0.8,
               main="Loss variable of daily IBM log returns")
> meplot(ibm)
> title(main="Daily IBM log returns")
```

7.8.2 Mean Excess Function

Given a high threshold η_0, suppose that the excess $y = x - \eta_0$ follows a GPD with parameter ξ and $\psi(\eta_0)$, where $0 < \xi < 1$. Then, the *mean excess* of x over the

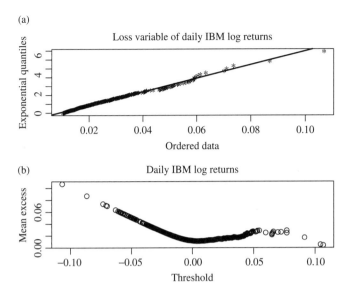

(a)

Loss variable of daily IBM log returns

(b)

Daily IBM log returns

Figure 7.12. Plots for IBM daily log returns from January 2, 2001 to December 31, 2010. (a) QQ-plot of excess returns over the threshold 1% for negative returns and (b) the mean excess plot.

threshold η_0 is the expectation of a random variable Y whose CDF is given in Equation (7.38). With some algebra, one can show that

$$E(x - \eta_0 | x > \eta_0) = \frac{\psi(\eta_0)}{1 - \xi}.$$

For any $\eta > \eta_0$, define the *mean excess function* $e(\eta)$ as

$$e(\eta) = E(x - \eta | x > \eta) = \frac{\psi(\eta)}{1 - \xi} = \frac{\psi(\eta_0) + \xi(\eta - \eta_0)}{1 - \xi}.$$

In other words, for any $y > 0$,

$$e(\eta_0 + y) = E[x - (\eta_0 + y) | x > \eta_0 + y] = \frac{\psi(\eta_0) + \xi y}{1 - \xi}.$$

Thus, for a fixed ξ, the mean excess function is a linear function of $y = \eta - \eta_0$. This result leads to a simple graphical method to infer the appropriate threshold value η_0 for the GPD. Define the *empirical mean excess function* as

$$e_T(\eta) = \frac{1}{N_\eta} \sum_{i=1}^{N_\eta} (x_{t_i} - \eta), \tag{7.39}$$

where N_η is the number of returns that exceed η and x_{t_i} are the values of the corresponding returns. See the next section for more information on the notation. The scatterplot of $e_T(\eta)$ against η is called the *mean excess plot*, which should be linear in η for $\eta > \eta_0$ under the GPD. The plot is also called *mean residual life plot*. Figure 7.12b shows the mean excess plot of the negative daily IBM log returns. It shows that, among others, a threshold of about 1% is reasonable for the negative return series. In the `evir` package of R, the command for mean excess plot is `meplot`.

7.8.3 Estimation

Consider a given high threshold η, we can obtain the probability density function of the generalized Pareto distribution by taking derivative of CDF in Equation (7.38) with respect to y. This pdf can, in turn, be used to estimate the parameters ξ, μ, and σ via the maximum likelihood method under the framework of point processes. See Smith (1989), Smith and Shively (1995), and Tsay (2010, Chapter 7) for more details. To demonstrate, consider again the loss variable $x_t = -r_t$, where r_t is the daily log returns of IBM stock from January 2, 2001 to December 31, 2010. Table 7.3 shows the estimation results for various thresholds. From the table, the estimate of ξ is only marginally significant at the 5% level when the threshold is 1.2%.

The results of Table 7.3 are obtained by using the comment `pot` of the `evir` package of R. The program also provides various plots for model checking. In this particular instance, the plots indicate that the assumption of independent observations is questionable. This is understandable in light of the volatility clustering commonly seen in asset returns. With the estimated parameters, one can calculate the risk measures. This is done by the command `riskmeasures` in R. For the long position of $1 million on IBM stock, we have

$$\mathrm{VaR}_{0.95} = \$25,855, \quad \mathrm{ES}_{0.95} = \$39,625,$$

for the first trading day of 2011 when the threshold of 1% is used. If the threshold is 1.2%, we have

$$\mathrm{VaR}_{0.95} = \$26,115, \quad \mathrm{ES}_{0.95} = \$39,603.$$

Finally, for threshold of 0.8%, we have

$$\mathrm{VaR}_{0.95} = \$25,866, \quad \mathrm{ES}_{0.95} = \$39,620.$$

TABLE 7.3. Maximum Likelihood Estimates of the Generalized Pareto Distribution For Negative Daily Log Returns of IBM Stock from January 2, 2001 to December 31, 2010

η	n.exceed	Shape ξ	Scale σ	Location μ	$\psi(\eta)$
1	504	0.107(0.042)	0.009(0.001)	−0.006(0.001)	0.011
1.2	410	0.075(0.044)	0.010(0.001)	−0.007(0.002)	0.011
0.8	610	0.106(0.039)	0.009(0.001)	−0.006(0.001)	0.010

From these results, we see that the risk measures are relatively stable for the three choices of threshold. Compared with the VaR of Example 7.8 that uses the traditional EVT , the POT approach provides a more stable VaR calculation.

R Demonstration Using POT Commands. Output edited. Also, beta = sigma +xi(eta-mu), which is $\psi(\eta)$ of the text.

```
> da=read.table("d-ibm-0110.txt",header=T)
> ibm=log(da[,2]+1)
> xt=-ibm
> m1=pot(xt,threshold=0.01)
> m1
$n
[1] 2515
$period
[1]    1 2515
$data
  [1] 0.01530858 0.01074553 0.01139063 .....
attr(,"times")
  [1]    3    6   10 .....
$span
[1] 2514
$threshold
[1] 0.01
$p.less.thresh
[1] 0.7996024
$n.exceed
[1] 504
$par.ests
        xi          sigma              mu             beta
 0.107268254   0.008914461  -0.005634968   0.010591597
$par.ses
        xi          sigma              mu
0.0415025597 0.0009052881 0.0012156539
$intensity
[1] 0.2004773
$converged
[1] 0
> plot(m1)
Make a plot selection (or 0 to exit):

1: plot: Point Process of Exceedances
2: plot: Scatterplot of Gaps
3: plot: Qplot of Gaps
4: plot: ACF of Gaps
5: plot: Scatterplot of Residuals
6: plot: Qplot of Residuals
7: plot: ACF of Residuals
```

```
8: plot: Go to GPD Plots
Selection: 0

> riskmeasures(m1,c(0.95,0.99))
          p   quantile        sfall
[1,] 0.950 0.02585540 0.03962479
[2,] 0.990 0.04744964 0.06381374
> riskmeasures(m2,c(0.95,0.99)) % Threshold=0.012
          p   quantile        sfall
[1,] 0.950 0.02611524 0.03960353
[2,] 0.990 0.04745886 0.06267327
> riskmeasures(m3,c(0.95,0.99)) % Threshold=0.008
          p   quantile        sfall
[1,] 0.950 0.02586561 0.03962012
[2,] 0.990 0.04744180 0.06376612
```

7.8.4 An Alternative Parameterization

As mentioned before, for a given threshold η, the GPD can also be parameterized by the shape parameter ξ and the scale parameter $\psi(\eta) = \alpha + \xi(\eta - \beta)$. This is the parameterization commonly used in GPD estimation. For instance, it is used by the command gpd in the `evir` package of R. Specifically, (xi,beta) of R corresponds to $(\xi, \psi(\eta))$ of this chapter. For illustration, consider the daily negative IBM log return series from 2001 to 2010. The results of R with command gpd are given below.

R Demonstration. Data are negative daily IBM log returns. Output edited.

```
> library(evir)
> da=read.table("d-ibm-0110.txt",header=T)
> ibm=log(da[,2]+1)
> xt=-ibm
> m1gpd=gpd(xt,threshold=0.01)
> m1gpd
$n
[1] 2515
$data
   [1] 0.01530858 0.01074553 0.01139063  .....
$threshold
[1] 0.01
$p.less.thresh
[1] 0.7996024
$n.exceed
[1] 504
$method
[1] "ml"
$par.ests
          xi          beta
```

```
0.10703752 0.01059601
$par.ses
          xi            beta
0.0544269528 0.0007255951
$converged
[1] 0
$nllh.final
[1] -1733.994

> names(m1gpd)
 [1] "n"              "data"           "thresh-
old"      "p.less.thresh"
 [5] "n.exceed"       "method"         "par.ests"       "par.ses"
 [9] "varcov"         "information"    "converged"      "nllh.final"
> par(mfcol=c(2,2))
> plot(m1gpd)
Make a plot selection (or 0 to exit):

1: plot: Excess Distribution
2: plot: Tail of Underlying Distribution
3: plot: Scatterplot of Residuals
4: plot: QQ-plot of Residuals

Selection: 0
> riskmeasures(m1gpd,c(0.95,0.99))
          p    quantile        sfall
[1,] 0.950 0.02585941 0.03962658
[2,] 0.990 0.04745161 0.06380699
```

As expected, the results are very close to those in Table 7.3. The estimates of ξ and $\psi(\eta)$ are 0.107(0.054) and 0.0106(0.0007), respectively, where the number in parentheses denotes standard error. The minor difference is likely due to different optimization methods used in estimation. Figure 7.13 shows the diagnostic plots for the GPD fit to the daily negative log returns of IBM stock. The QQ-plot (Fig. 7.13d) and the tail probability estimate (in log scale and in the Fig. 7.13b) show some minor deviation from a straight line, indicating that further improvement is possible.

From the conditional distributions in Equations (7.36) and (7.37) and the GPD in Equation (7.38), we have

$$\frac{F(x) - F(\eta)}{1 - F(\eta)} \approx G_{\eta,\psi(\eta)}(y),$$

where $x = y + \eta$ with $y > 0$. If we estimate the CDF $F(\eta)$ of the losses by the empirical CDF, then

$$\hat{F}(\eta) = \frac{T - N_\eta}{T},$$

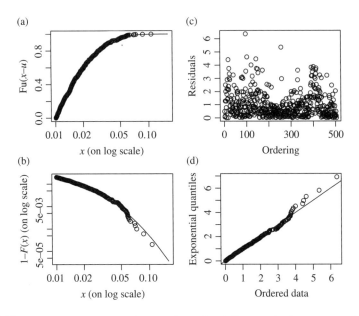

Figure 7.13. (a–d) Diagnostic plots for GPD fit to the daily negative log returns of IBM stock from January 2, 2001 to December 31, 2010.

where N_η is the number of exceedances of the threshold η and T is the sample size. Consequently, by Equation (7.38),

$$F(x) = F(\eta) + G(y)[1 - F(\eta)]$$

$$\approx 1 - \frac{N_\eta}{T}\left[1 + \frac{\xi(x - \eta)}{\psi(\eta)}\right]^{-1/\xi}.$$

This leads to an alternative estimate of the quantile of $F(x)$ for use in VaR calculation. Specifically, for a small upper tail probability p, let $q = 1 - p$. Then, by solving for x, we can estimate the qth quantile of $F(x)$, denoted by VaR_q, by

$$\text{VaR}_q = \eta - \frac{\psi(\eta)}{\xi}\left\{1 - \left[\frac{T}{N_\eta}(1 - q)\right]^{-\xi}\right\}, \tag{7.40}$$

where, as before, η is the threshold, T is the sample size, N_η is the number of exceedances, and $\psi(\eta)$ and ξ are the scale and shape parameters of the GPD distribution. This method to VaR calculation is used in R.

For the generalized Pareto distribution, ES assumes a simple form. Specifically, for a given tail probability p, let $q = 1 - p$ and denote the VaR by VaR_q. Then, the ES is defined by

$$\text{ES}_q = E(x|x > \text{VaR}_q) = \text{VaR}_q + E(x - \text{VaR}_q|x > \text{VaR}_q). \tag{7.41}$$

Using properties of the GPD shown in Section 7.8.2, we have

$$E(x - \text{VaR}_q | x > \text{VaR}_q) = \frac{\psi(\eta) + \xi(\text{VaR}_q - \eta)}{1 - \xi},$$

provided that $0 < \xi < 1$. Consequently, we have

$$\text{ES}_q = \frac{\text{VaR}_q}{1 - \xi} + \frac{\psi(\eta) - \xi\eta}{1 - \xi}.$$

To illustrate the VaR and ES calculations, we again use the daily negative log returns of IBM stock with threshold 1%. The results are given in the prior R demonstration. From the output, the VaR values for the financial position of $1 million are $25,859 and $47,452 for the tail probability of 0.05 and 0.01, respectively. These two values are essentially the same as those given in Example 7.8 that are based on the command `pot`. The ESs for the financial position are $39,627 and $63,807, respectively, for the tail probability of 0.05 and 0.01.

Remark. The POT and traditional EVT methods discussed assume that the loss variable x_t is homogeneous, indicating that the risk measures are evaluated by the loss data alone. In practice, the loss variable x_t could form a nonhomogeneous process that depends on some explanatory variables such as the VIX index or the volatility of the asset under study. The POT approach can be generalized to handle nonhomogeneous loss process x_t. See, for instance, Smith (1989) and Tsay (2010, Chapter 7). Interested readers are referred to Tsay (2010) for demonstration of such an application. □

7.9 THE STATIONARY LOSS PROCESSES

Finally, we briefly mention the adjustment needed in calculating risk measures when the loss process $\{x_t\}$ is a strictly stationary time series. See O'Brien (1987). The relationship between $F_*(x)$ of the maximum of a stationary time series and $\tilde{F}_*(x)$ of its iid counterpart given in Equations (7.29) and (7.28) can be used to calculate the VaR of a financial position when the associated log returns form a stationary time series. Specifically, from $P(x_{(n)} \leq u_n) \approx [F(x)]^{n\theta}$, the $(1 - p)$th quantile of $F(x)$ is the $(1 - p)^{n\theta}$th quantile of the limiting extreme value distribution of $x_{(n)}$. Consequently, the VaR of Equation (7.35) based on the EVT becomes

$$\text{VaR} = \begin{cases} \mu_n - \dfrac{\sigma_n}{\xi_n} \left\{ 1 - \left[-n\theta \ln(1 - p) \right]^{-\xi_n} \right\} & \text{if } \xi_n \neq 0 \\ \mu_n - \sigma_n \ln[-n\theta \ln(1 - p)] & \text{if } \xi_n = 0, \end{cases} \tag{7.42}$$

where n is the length of subperiod. From the formula, we may underestimate the VaR if the extremal index is overlooked.

As an illustration, again consider the negative daily log returns of IBM stock from January 2, 2001 to December 31, 2010. Figure 7.14 shows the estimates of extremal

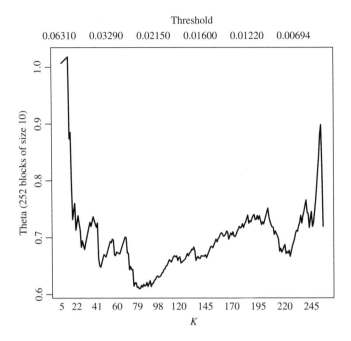

Figure 7.14. Estimates of extremal index for the daily negative log returns of IBM stock from January 2, 2001 to December 31, 2010. Block size is 10 and K of the lower horizontal axis of the plot denotes number of blocks whose maximum exceeds the threshold.

index obtained by the command `exindex` of the package `evir` in R. Because the serial correlations of daily IBM log returns are small, we chose a block size 10 in the estimation. Note that this block size is different from that used in the traditional EVT approach to estimate the generalized extreme value distribution. From the plot, we see that with threshold of 1%, the extremal index is about $\hat{\theta} = 0.72$. With this estimate, the 1% VaR for the long position of $1 million on the stock for the next trading day becomes $44,449 for the case of choosing $n = 21$ days in parameter estimation. As expected, this is higher than the value $39,242 of Example 7.8 when the extremal index is neglected.

R Demonstration

```
> library(evir)
> help(exindex)
> m1=exindex(xt,10) % Estimate the extremal index of Figure 7.10.
>          % VaR calculation.
> v1=1.966-(1.029/.251)*(1-(-21*.72*log(.99))^(-.251))
> v1
[1] 4.444898
```

EXERCISES

1. Consider a long position of $1 million on the Apple stock. To assess the risk of the position, we employ daily returns of the stock from January 2, 2001 to September 30, 2011 for 2704 observations. The daily simple returns are obtained from CRSP and in the file d-aapl-0111.txt. Let the tail probability be $p = 0.01$. Compute the VaR and ES of the position for the next trading day and the next 10 trading days using the following methods:

   (a) The RiskMetrics method. Write down the fitted special IGARCH(1,1) model.
   (b) A Gaussian GARCH model. Write down the fitted model.
   (c) A GARCH model with a standardized Student-t innovations. Write down the fitted model. [You should use simulation to compute VaR and ES for the next 10 trading days based on the fitted model.]

2. Again, consider the position and data of Exercise 1. Compute the VaR for the next trading day using the following methods:

   (a) Empirical quantile with $p = 0.05$ and 0.01.
   (b) Quantile regression with $1 - p = 0.95$ and 0.99 using lag-1 volatility of a Gaussian GARCH(1,1) model and lag-1 absolute value of the log return as predictors. Write down the fitted quantile regression. Are the estimates statistically significant at the usual 5% level?

3. Again, consider the position and data of Exercise 1. Answer the following questions:

   (a) Obtain estimates of μ, σ, and ξ using the traditional EVT with subperiod length $n = 21$ and 42, respectively.
   (b) Use the estimates to compute $\text{VaR}_{0.95}$ and $\text{VaR}_{0.99}$ for the next trading day and the next 10 trading days.

4. Again, consider the position and data of Exercise 1. Apply the POT method to calculate risk measures for the position:

   (a) Use threshold 2.5%. Write down the estimates, including standard error, and compute the corresponding $\text{VaR}_{0.99}$ and $\text{ES}_{0.99}$.
   (b) Use threshold 2%. Write down the estimates and their standard errors, obtain the QQ-plot of the data versus exponential quantiles, and compute the corresponding $\text{VaR}_{0.99}$ and $\text{ES}_{0.99}$.
   (c) Are the risk measures sensitive to the choice of threshold? Why?

5. Again, consider the position and data of Exercise 1. In this exercise, apply the GPD with threshold 2.5% to obtain the parameter estimates and the corresponding diagnostic plots. Compute $\text{VaR}_{0.99}(10)$ of the position, that is, the VaR for the next 10 trading days.

6. Consider the daily log returns of Apple stock and the log returns of Bank of America Merrill Lynch U.S. Corp AAA Index Value from January 3, 2001 to September 30, 2011. The bond index has more daily observations, but we

match the dates before computing the daily log returns of bond index. The data are in `d-aaplbnd-0111.txt`. Consider a long position of $1 million consisting of 50-50 on stock and AAA bond. Compute VaR and ES of the position for the next trading day using the following methods:

(a) The RiskMetrics method.

(b) The econometric approach using Gaussian GARCH(1,1) model.

REFERENCES

Artzner P, Delbaen F, Eber J, Health D. Thinking coherently. Risk 1997; 10(11): 68–71.

Beirlant J, Goegebeur Y, Segers J, Teugels J. Statistics of Extremes: Theory and Applications. New Jersey: John Wiley & Sons; 2004.

Berman SM. Limiting theorems for the maximum term in stationary sequences. Ann Math Stat 1964; 35: 502–516.

Coles S. An Introduction to Statistical Modeling of Extreme Values. New York: Springer-Verlag; 2001.

Cox DR, Hinkley DV. Theoretical Statistics. London: Chapman and Hall; 1974.

Danielsson J, de Vries CG. Value at risk and extreme returns. Working paper, London School of Economics, London, UK; 1997a.

Danielsson J, de Vries CG. Tail index and quantile estimation with very high frequency data. J Empir Finance 1997b; 4: 241–257.

Davison AC, Smith RL. Models for exceedances over high thresholds (with discussion). J R Stat Soc Ser B 1990; 52: 393–442.

Dekkers ALM, De Haan L. On the estimation of extreme value index and large quantile estimation. Ann Stat 1989; 17: 1795–1832.

Duffie D, Pan J. An overview of value at risk. J Deriv 1997; 4: 7–48.

Embrechts P, Kuppelberg C, Mikosch T. Modelling Extremal Events. Berlin: Springer Verlag; 1997.

Feller W. An Introduction to Probability Theory and Its Applications. Volume 2. Hoboken (NJ): John Wiley & Sons; 1971.

Goldie CM, Smith RL. Slow variation with remainder: theory and applications. Q J Math 1987; 38: 45–71.

Gnedenko BV. Sur la distribution limite du terme maximum of d'une série Aléatorie. Ann Math 1943; 44: 423–453.

Gumbel EJ. Statistics of Extremes. New York: Columbia University Press; 1958.

Hill BM. A simple general approach to inference about the tail of a distribution. Ann Stat 1975; 3: 1163–1173.

Hsing T, Hüsler J, Leadbetter MR. On the exceedance point process for a stationary sequence. Probab Theory Relat Fields 1988; 78: 97–112.

Jenkinson AF. The frequency distribution of the annual maximum (or minimum) of meteorological elements. Q J R Meteorol Soc 1955; 81: 158–171.

Jorion P. Value at Risk: The New Benchmark for Managing Financial Risk. 3rd ed. Chicago: The McGraw-Hill; 2006.

Koenker RW, Bassett GW. Regression quantiles. Econometrica 1978; 46: 33–50.

Koenker RW, D'Orey V. Computing regression quantiles. Appl Stat 1987; 36: 383–393.

Klugman SA, Panjer HH, Willmot GE. Loss Models: From Data to Decisions. Hoboken (NJ): John Wiley & Son; 2008.

Leadbetter MR. On extreme values in stationary sequences. Zeitschrift für Wahrscheinlichkeitsthorie und Verwandte Gebiete 1974; 28: 289–303.

Leadbetter MR. Extremes and local dependence in stationary sequences. Zeitschrift für Wahrscheinlichkeitsthorie und Verwandte Gebiete 1983; 65: 291–306.

Leadbetter MR, Lindgren G, Rootzén H. Extremes and Related Properties of Random Sequences and Processes. New York: Springer Verlag; 1983.

Longerstaey J, More L. Introduction to RiskMetrics™. 4th ed. New York: Morgan Guaranty Trust Company; 1995.

Longin FM. The asymptotic distribution of extreme stock market returns. J Bus 1996; 69: 383–408.

Longin FM. Optimal margin level in futures markets: extreme price movements. J Futures Mark 1999a; 19: 127–152.

Longin FM. From value at risk to stress testing: the extreme value approach. Working paper, Centre for Economic Policy Research, London, UK; 1999b.

McNeil AJ, Frey R, Embrechts P. Quantitative Risk Management: Concepts, Techniques and Tools. Princeton (NJ): Princeton University Press; 2005.

O'Brien GL. Extreme values for stationary and Markov sequences. Ann Probab 1987; 15: 281–291.

Pickands J. Statistical inference using extreme order statistics. Ann Stat 1975; 3: 119–131.

Smith RL. Extreme value analysis of environmental time series: an application to trend detection in ground-level ozone (with discussion). Stat Sci 1989; 4: 367–393.

Smith RL. Measuring risk with extreme value theory. Working paper, Department of Statistics, University of North Carolina at Chapel Hill; 1999.

Smith RL, Shively TS. A point process approach to modeling trends in tropospheric ozone. Atmos Environ 1995; 29: 3489–3499.

Tsay RS. Analysis of Financial Time Series. 3rd ed. Hoboken (NJ): John Wiley & Sons; 2010.

Zivot E, Wang J. Modeling Financial Time Series with S-Plus. New York: Springer-Verlag; 2003.

INDEX

An Introduction to Analysis of Financial Data with R, First Edition. Ruey S. Tsay.
© 2013 John Wiley & Sons, Inc. Published 2013 by John Wiley & Sons, Inc.

WILEY SERIES IN PROBABILITY AND STATISTICS
ESTABLISHED BY WALTER A. SHEWHART AND SAMUEL S. WILKS

Editors: *David J. Balding, Noel A. C. Cressie, Garrett M. Fitzmaurice,*
Harvey Goldstein, Iain M. Johnstone, Geert Molenberghs, David W. Scott,
Adrian F. M. Smith, Ruey S. Tsay, Sanford Weisberg
Editors Emeriti: *Vic Barnett, J. Stuart Hunter, Joseph B. Kadane, Jozef L. Teugels*

The *Wiley Series in Probability and Statistics* is well established and authoritative. It covers many topics of current research interest in both pure and applied statistics and probability theory. Written by leading statisticians and institutions, the titles span both state-of-the-art developments in the field and classical methods.

Reflecting the wide range of current research in statistics, the series encompasses applied, methodological and theoretical statistics, ranging from applications and new techniques made possible by advances in computerized practice to rigorous treatment of theoretical approaches.

This series provides essential and invaluable reading for all statisticians, whether in academia, industry, government, or research.

† ABRAHAM and LEDOLTER · Statistical Methods for Forecasting
  AGRESTI · Analysis of Ordinal Categorical Data, *Second Edition*
  AGRESTI · An Introduction to Categorical Data Analysis, *Second Edition*
  AGRESTI · Categorical Data Analysis, *Second Edition*
  ALTMAN, GILL, and McDONALD · Numerical Issues in Statistical Computing for the
    Social Scientist
  AMARATUNGA and CABRERA · Exploration and Analysis of DNA Microarray and
    Protein Array Data
  ANDĚL · Mathematics of Chance
  ANDERSON · An Introduction to Multivariate Statistical Analysis, *Third Edition*
* ANDERSON · The Statistical Analysis of Time Series
  ANDERSON, AUQUIER, HAUCK, OAKES, VANDAELE, and WEISBERG ·
    Statistical Methods for Comparative Studies
  ANDERSON and LOYNES · The Teaching of Practical Statistics
  ARMITAGE and DAVID (editors) · Advances in Biometry
  ARNOLD, BALAKRISHNAN, and NAGARAJA · Records
* ARTHANARI and DODGE · Mathematical Programming in Statistics
* BAILEY · The Elements of Stochastic Processes with Applications to the Natural
    Sciences
  BAJORSKI · Statistics for Imaging, Optics, and Photonics
  BALAKRISHNAN and KOUTRAS · Runs and Scans with Applications
  BALAKRISHNAN and NG · Precedence-Type Tests and Applications
  BARNETT · Comparative Statistical Inference, *Third Edition*
  BARNETT · Environmental Statistics
  BARNETT and LEWIS · Outliers in Statistical Data, *Third Edition*
  BARTHOLOMEW, KNOTT, and MOUSTAKI · Latent Variable Models and Factor
    Analysis: A Unified Approach, *Third Edition*
  BARTOSZYNSKI and NIEWIADOMSKA-BUGAJ · Probability and Statistical
    Inference, *Second Edition*
  BASILEVSKY · Statistical Factor Analysis and Related Methods: Theory and
    Applications
  BATES and WATTS · Nonlinear Regression Analysis and Its Applications
  BECHHOFER, SANTNER, and GOLDSMAN · Design and Analysis of Experiments for
    Statistical Selection, Screening, and Multiple Comparisons

*Now available in a lower priced paperback edition in the Wiley Classics Library.
†Now available in a lower priced paperback edition in the Wiley–Interscience Paperback Series.

BEIRLANT, GOEGEBEUR, SEGERS, TEUGELS, and DE WAAL · Statistics of
Extremes: Theory and Applications

BELSLEY · Conditioning Diagnostics: Collinearity and Weak Data in Regression

† BELSLEY, KUH, and WELSCH · Regression Diagnostics: Identifying Influential
Data and Sources of Collinearity

BENDAT and PIERSOL · Random Data: Analysis and Measurement Procedures,
Fourth Edition

BERNARDO and SMITH · Bayesian Theory

BHAT and MILLER · Elements of Applied Stochastic Processes, *Third Edition*

BHATTACHARYA and WAYMIRE · Stochastic Processes with Applications

BIEMER, GROVES, LYBERG, MATHIOWETZ, and SUDMAN · Measurement Errors
in Surveys

BILLINGSLEY · Convergence of Probability Measures, *Second Edition*

BILLINGSLEY · Probability and Measure, *Anniversary Edition*

BIRKES and DODGE · Alternative Methods of Regression

BISGAARD and KULAHCI · Time Series Analysis and Forecasting by Example

BISWAS, DATTA, FINE, and SEGAL · Statistical Advances in the Biomedical Sciences:
Clinical Trials, Epidemiology, Survival Analysis, and Bioinformatics

BLISCHKE and MURTHY (editors) · Case Studies in Reliability and Maintenance

BLISCHKE and MURTHY · Reliability: Modeling, Prediction, and Optimization

BLOOMFIELD · Fourier Analysis of Time Series: An Introduction, *Second Edition*

BOLLEN · Structural Equations with Latent Variables

BOLLEN and CURRAN · Latent Curve Models: A Structural Equation Perspective

BOROVKOV · Ergodicity and Stability of Stochastic Processes

BOSQ and BLANKE · Inference and Prediction in Large Dimensions

BOULEAU · Numerical Methods for Stochastic Processes

* BOX and TIAO · Bayesian Inference in Statistical Analysis

BOX · Improving Almost Anything, *Revised Edition*

* BOX and DRAPER · Evolutionary Operation: A Statistical Method for Process
Improvement

BOX and DRAPER · Response Surfaces, Mixtures, and Ridge Analyses, *Second Edition*

BOX, HUNTER, and HUNTER · Statistics for Experimenters: Design, Innovation,
and Discovery, *Second Editon*

BOX, JENKINS, and REINSEL · Time Series Analysis: Forcasting and Control, *Fourth
Edition*

BOX, LUCEÑO, and PANIAGUA-QUIÑONES · Statistical Control by Monitoring
and Adjustment, *Second Edition*

* BROWN and HOLLANDER · Statistics: A Biomedical Introduction

CAIROLI and DALANG · Sequential Stochastic Optimization

CASTILLO, HADI, BALAKRISHNAN, and SARABIA · Extreme Value and Related
Models with Applications in Engineering and Science

CHAN · Time Series: Applications to Finance with R and S-Plus®, *Second Edition*

CHARALAMBIDES · Combinatorial Methods in Discrete Distributions

CHATTERJEE and HADI · Regression Analysis by Example, *Fourth Edition*

CHATTERJEE and HADI · Sensitivity Analysis in Linear Regression

CHERNICK · Bootstrap Methods: A Guide for Practitioners and Researchers,
Second Edition

CHERNICK and FRIIS · Introductory Biostatistics for the Health Sciences

CHILÈS and DELFINER · Geostatistics: Modeling Spatial Uncertainty, *Second Edition*

CHOW and LIU · Design and Analysis of Clinical Trials: Concepts and Methodologies,
Second Edition

CLARKE · Linear Models: The Theory and Application of Analysis of Variance

CLARKE and DISNEY · Probability and Random Processes: A First Course with
Applications, *Second Edition*

*Now available in a lower priced paperback edition in the Wiley Classics Library.

†Now available in a lower priced paperback edition in the Wiley–Interscience Paperback Series.

*Now available in a lower priced paperback edition in the Wiley Classics Library.

†Now available in a lower priced paperback edition in the Wiley–Interscience Paperback Series.

---

*Now available in a lower priced paperback edition in the Wiley Classics Library.

†Now available in a lower priced paperback edition in the Wiley–Interscience Paperback Series.

HOCHBERG and TAMHANE · Multiple Comparison Procedures

HOCKING · Methods and Applications of Linear Models: Regression and the Analysis of Variance, *Second Edition*

HOEL · Introduction to Mathematical Statistics, *Fifth Edition*

HOGG and KLUGMAN · Loss Distributions

HOLLANDER and WOLFE · Nonparametric Statistical Methods, *Second Edition*

HOSMER and LEMESHOW · Applied Logistic Regression, *Second Edition*

HOSMER, LEMESHOW, and MAY · Applied Survival Analysis: Regression Modeling of Time-to-Event Data, *Second Edition*

HUBER · Data Analysis: What Can Be Learned From the Past 50 Years

HUBER · Robust Statistics

† HUBER and RONCHETTI · Robust Statistics, *Second Edition*

HUBERTY · Applied Discriminant Analysis, *Second Edition*

HUBERTY and OLEJNIK · Applied MANOVA and Discriminant Analysis, *Second Edition*

HUITEMA · The Analysis of Covariance and Alternatives: Statistical Methods for Experiments, Quasi-Experiments, and Single-Case Studies, *Second Edition*

HUNT and KENNEDY · Financial Derivatives in Theory and Practice, *Revised Edition*

HURD and MIAMEE · Periodically Correlated Random Sequences: Spectral Theory and Practice

HUSKOVA, BERAN, and DUPAC · Collected Works of Jaroslav Hajek— with Commentary

HUZURBAZAR · Flowgraph Models for Multistate Time-to-Event Data

JACKMAN · Bayesian Analysis for the Social Sciences

† JACKSON · A User's Guide to Principle Components

JOHN · Statistical Methods in Engineering and Quality Assurance

JOHNSON · Multivariate Statistical Simulation

JOHNSON and BALAKRISHNAN · Advances in the Theory and Practice of Statistics: A Volume in Honor of Samuel Kotz

JOHNSON, KEMP, and KOTZ · Univariate Discrete Distributions, *Third Edition*

JOHNSON and KOTZ (editors) · Leading Personalities in Statistical Sciences: From the Seventeenth Century to the Present

JOHNSON, KOTZ, and BALAKRISHNAN · Continuous Univariate Distributions, Volume 1, *Second Edition*

JOHNSON, KOTZ, and BALAKRISHNAN · Continuous Univariate Distributions, Volume 2, *Second Edition*

JOHNSON, KOTZ, and BALAKRISHNAN · Discrete Multivariate Distributions

JUDGE, GRIFFITHS, HILL, LÜTKEPOHL, and LEE · The Theory and Practice of Econometrics, *Second Edition*

JUREK and MASON · Operator-Limit Distributions in Probability Theory

KADANE · Bayesian Methods and Ethics in a Clinical Trial Design

KADANE AND SCHUM · A Probabilistic Analysis of the Sacco and Vanzetti Evidence

KALBFLEISCH and PRENTICE · The Statistical Analysis of Failure Time Data, *Second Edition*

KARIYA and KURATA · Generalized Least Squares

KASS and VOS · Geometrical Foundations of Asymptotic Inference

† KAUFMAN and ROUSSEEUW · Finding Groups in Data: An Introduction to Cluster Analysis

KEDEM and FOKIANOS · Regression Models for Time Series Analysis

KENDALL, BARDEN, CARNE, and LE · Shape and Shape Theory

KHURI · Advanced Calculus with Applications in Statistics, *Second Edition*

KHURI, MATHEW, and SINHA · Statistical Tests for Mixed Linear Models

* KISH · Statistical Design for Research

---

*Now available in a lower priced paperback edition in the Wiley Classics Library.
†Now available in a lower priced paperback edition in the Wiley–Interscience Paperback Series.

MARONNA, MARTIN and YOHAI · Robust Statistics: Theory and Methods

MASON, GUNST, and HESS · Statistical Design and Analysis of Experiments with Applications to Engineering and Science, *Second Edition*

McCULLOCH, SEARLE, and NEUHAUS · Generalized, Linear, and Mixed Models, *Second Edition*

McFADDEN · Management of Data in Clinical Trials, *Second Edition*

* McLACHLAN · Discriminant Analysis and Statistical Pattern Recognition

McLACHLAN, DO, and AMBROISE · Analyzing Microarray Gene Expression Data

McLACHLAN and KRISHNAN · The EM Algorithm and Extensions, *Second Edition*

McLACHLAN and PEEL · Finite Mixture Models

McNEIL · Epidemiological Research Methods

MEEKER and ESCOBAR · Statistical Methods for Reliability Data

MEERSCHAERT and SCHEFFLER · Limit Distributions for Sums of Independent Random Vectors: Heavy Tails in Theory and Practice

MENGERSEN, ROBERT, and TITTERINGTON · Mixtures: Estimation and Applications

MICKEY, DUNN, and CLARK · Applied Statistics: Analysis of Variance and Regression, *Third Edition*

* MILLER · Survival Analysis, *Second Edition*

MONTGOMERY, JENNINGS, and KULAHCI · Introduction to Time Series Analysis and Forecasting

MONTGOMERY, PECK, and VINING · Introduction to Linear Regression Analysis, *Fifth Edition*

MORGENTHALER and TUKEY · Configural Polysampling: A Route to Practical Robustness

MUIRHEAD · Aspects of Multivariate Statistical Theory

MULLER and STOYAN · Comparison Methods for Stochastic Models and Risks

MURTHY, XIE, and JIANG · Weibull Models

MYERS, MONTGOMERY, and ANDERSON-COOK · Response Surface Methodology: Process and Product Optimization Using Designed Experiments, *Third Edition*

MYERS, MONTGOMERY, VINING, and ROBINSON · Generalized Linear Models. With Applications in Engineering and the Sciences, *Second Edition*

NATVIG · Multistate Systems Reliability Theory With Applications

† NELSON · Accelerated Testing, Statistical Models, Test Plans, and Data Analyses

† NELSON · Applied Life Data Analysis

NEWMAN · Biostatistical Methods in Epidemiology

NG, TAIN, and TANG · Dirichlet Theory: Theory, Methods and Applications

OKABE, BOOTS, SUGIHARA, and CHIU · Spatial Tesselations: Concepts and Applications of Voronoi Diagrams, *Second Edition*

OLIVER and SMITH · Influence Diagrams, Belief Nets and Decision Analysis

PALTA · Quantitative Methods in Population Health: Extensions of Ordinary Regressions

PANJER · Operational Risk: Modeling and Analytics

PANKRATZ · Forecasting with Dynamic Regression Models

PANKRATZ · Forecasting with Univariate Box-Jenkins Models: Concepts and Cases

PARDOUX · Markov Processes and Applications: Algorithms, Networks, Genome and Finance

PARMIGIANI and INOUE · Decision Theory: Principles and Approaches

* PARZEN · Modern Probability Theory and Its Applications

PEÑA, TIAO, and TSAY · A Course in Time Series Analysis

PESARIN and SALMASO · Permutation Tests for Complex Data: Applications and Software

PIANTADOSI · Clinical Trials: A Methodologic Perspective, *Second Edition*

POURAHMADI · Foundations of Time Series Analysis and Prediction Theory

---

POWELL · Approximate Dynamic Programming: Solving the Curses of Dimensionality, *Second Edition*
POWELL and RYZHOV · Optimal Learning
PRESS · Subjective and Objective Bayesian Statistics, *Second Edition*
PRESS and TANUR · The Subjectivity of Scientists and the Bayesian Approach
PURI, VILAPLANA, and WERTZ · New Perspectives in Theoretical and Applied Statistics
† PUTERMAN · Markov Decision Processes: Discrete Stochastic Dynamic Programming
QIU · Image Processing and Jump Regression Analysis
* RAO · Linear Statistical Inference and Its Applications, *Second Edition*
RAO · Statistical Inference for Fractional Diffusion Processes
RAUSAND and HØYLAND · System Reliability Theory: Models, Statistical Methods, and Applications, *Second Edition*
RAYNER, THAS, and BEST · Smooth Tests of Goodnes of Fit: Using R, *Second Edition*
RENCHER and SCHAALJE · Linear Models in Statistics, *Second Edition*
RENCHER and CHRISTENSEN · Methods of Multivariate Analysis, *Third Edition*
RENCHER · Multivariate Statistical Inference with Applications
RIGDON and BASU · Statistical Methods for the Reliability of Repairable Systems
* RIPLEY · Spatial Statistics
* RIPLEY · Stochastic Simulation
ROHATGI and SALEH · An Introduction to Probability and Statistics, *Second Edition*
ROLSKI, SCHMIDLI, SCHMIDT, and TEUGELS · Stochastic Processes for Insurance and Finance
ROSENBERGER and LACHIN · Randomization in Clinical Trials: Theory and Practice
ROSSI, ALLENBY, and McCULLOCH · Bayesian Statistics and Marketing
† ROUSSEEUW and LEROY · Robust Regression and Outlier Detection
ROYSTON and SAUERBREI · Multivariate Model Building: A Pragmatic Approach to Regression Analysis Based on Fractional Polynomials for Modeling Continuous Variables
* RUBIN · Multiple Imputation for Nonresponse in Surveys
RUBINSTEIN and KROESE · Simulation and the Monte Carlo Method, *Second Edition*
RUBINSTEIN and MELAMED · Modern Simulation and Modeling
RYAN · Modern Engineering Statistics
RYAN · Modern Experimental Design
RYAN · Modern Regression Methods, *Second Edition*
RYAN · Statistical Methods for Quality Improvement, *Third Edition*
SALEH · Theory of Preliminary Test and Stein-Type Estimation with Applications
SALTELLI, CHAN, and SCOTT (editors) · Sensitivity Analysis
SCHERER · Batch Effects and Noise in Microarray Experiments: Sources and Solutions
* SCHEFFE · The Analysis of Variance
SCHIMEK · Smoothing and Regression: Approaches, Computation, and Application
SCHOTT · Matrix Analysis for Statistics, *Second Edition*
SCHOUTENS · Levy Processes in Finance: Pricing Financial Derivatives
SCOTT · Multivariate Density Estimation: Theory, Practice, and Visualization
* SEARLE · Linear Models
† SEARLE · Linear Models for Unbalanced Data
† SEARLE · Matrix Algebra Useful for Statistics
† SEARLE, CASELLA, and McCULLOCH · Variance Components
SEARLE and WILLETT · Matrix Algebra for Applied Economics
SEBER · A Matrix Handbook For Statisticians
† SEBER · Multivariate Observations
SEBER and LEE · Linear Regression Analysis, *Second Edition*
† SEBER and WILD · Nonlinear Regression
SENNOTT · Stochastic Dynamic Programming and the Control of Queueing Systems

*Now available in a lower priced paperback edition in the Wiley Classics Library.
†Now available in a lower priced paperback edition in the Wiley–Interscience Paperback Series.

* SERFLING · Approximation Theorems of Mathematical Statistics

SHAFER and VOVK · Probability and Finance: It's Only a Game!

SHERMAN · Spatial Statistics and Spatio-Temporal Data: Covariance Functions and Directional Properties

SILVAPULLE and SEN · Constrained Statistical Inference: Inequality, Order, and Shape Restrictions

SINGPURWALLA · Reliability and Risk: A Bayesian Perspective

SMALL and McLEISH · Hilbert Space Methods in Probability and Statistical Inference

SRIVASTAVA · Methods of Multivariate Statistics

STAPLETON · Linear Statistical Models, *Second Edition*

STAPLETON · Models for Probability and Statistical Inference: Theory and Applications

STAUDTE and SHEATHER · Robust Estimation and Testing

STOYAN · Counterexamples in Probability, *Second Edition*

STOYAN, KENDALL, and MECKE · Stochastic Geometry and Its Applications, *Second Edition*

STOYAN and STOYAN · Fractals, Random Shapes and Point Fields: Methods of Geometrical Statistics

STREET and BURGESS · The Construction of Optimal Stated Choice Experiments: Theory and Methods

STYAN · The Collected Papers of T. W. Anderson: 1943–1985

SUTTON, ABRAMS, JONES, SHELDON, and SONG · Methods for Meta-Analysis in Medical Research

TAKEZAWA · Introduction to Nonparametric Regression

TAMHANE · Statistical Analysis of Designed Experiments: Theory and Applications

TANAKA · Time Series Analysis: Nonstationary and Noninvertible Distribution Theory

THOMPSON · Empirical Model Building: Data, Models, and Reality, *Second Edition*

THOMPSON · Sampling, *Third Edition*

THOMPSON · Simulation: A Modeler's Approach

THOMPSON and SEBER · Adaptive Sampling

THOMPSON, WILLIAMS, and FINDLAY · Models for Investors in Real World Markets

TIERNEY · LISP-STAT: An Object-Oriented Environment for Statistical Computing and Dynamic Graphics

TSAY · Analysis of Financial Time Series, *Third Edition*

TSAY · An Introduction to Analysis of Financial Data with R

UPTON and FINGLETON · Spatial Data Analysis by Example, Volume II: Categorical and Directional Data

† VAN BELLE · Statistical Rules of Thumb, *Second Edition*

VAN BELLE, FISHER, HEAGERTY, and LUMLEY · Biostatistics: A Methodology for the Health Sciences, *Second Edition*

VESTRUP · The Theory of Measures and Integration

VIDAKOVIC · Statistical Modeling by Wavelets

VIERTL · Statistical Methods for Fuzzy Data

VINOD and REAGLE · Preparing for the Worst: Incorporating Downside Risk in Stock Market Investments

WALLER and GOTWAY · Applied Spatial Statistics for Public Health Data

WEISBERG · Applied Linear Regression, *Third Edition*

WEISBERG · Bias and Causation: Models and Judgment for Valid Comparisons

WELSH · Aspects of Statistical Inference

WESTFALL and YOUNG · Resampling-Based Multiple Testing: Examples and Methods for p-Value Adjustment

* WHITTAKER · Graphical Models in Applied Multivariate Statistics

WINKER · Optimization Heuristics in Economics: Applications of Threshold Accepting

WOODWORTH · Biostatistics: A Bayesian Introduction

*Now available in a lower priced paperback edition in the Wiley Classics Library.
†Now available in a lower priced paperback edition in the Wiley–Interscience Paperback Series.

WOOLSON and CLARKE · Statistical Methods for the Analysis of Biomedical Data, *Second Edition*

WU and HAMADA · Experiments: Planning, Analysis, and Parameter Design Optimization, *Second Edition*

WU and ZHANG · Nonparametric Regression Methods for Longitudinal Data Analysis

YIN · Clinical Trial Design: Bayesian and Frequentist Adaptive Methods

YOUNG, VALERO-MORA, and FRIENDLY · Visual Statistics: Seeing Data with Dynamic Interactive Graphics

ZACKS · Stage-Wise Adaptive Designs

* ZELLNER · An Introduction to Bayesian Inference in Econometrics

ZELTERMAN · Discrete Distributions—Applications in the Health Sciences

ZHOU, OBUCHOWSKI, and McCLISH · Statistical Methods in Diagnostic Medicine, *Second Edition*